Chirality at the Nanoscale

Edited by
David B. Amabilino

Further Reading

Carreira, E. M., Kvaerno, L.

Classics in Stereoselective Synthesis

2009
ISBN: 978-3-527-32452-1

Amouri, H., Gruselle, M.

Chirality in Transition Metal Chemistry

Molecules, Supramolecular Assemblies and Materials

2009
ISBN: 978-0-470-06053-7

Ding, K. / Uozumi, Y. (eds.)

Handbook of Asymmetric Heterogeneous Catalysis

2008
ISBN-13: 978-3-527-31913-8

Köhler, M., Fritzsche, W.

Nanotechnology

An Introduction to Nanostructuring Techniques

2007
ISBN: 978-3-527-31871-1

Wagnière, G. H.

On Chirality and the Universal Asymmetry

Reflections on Image and Mirror Image

2007
ISBN: 978-3-906390-38-3

Samori, P. (ed.)

Scanning Probe Microscopies Beyond Imaging

Manipulation of Molecules and Nanostructures

2006
ISBN: 978-3-527-31269-6

Chirality at the Nanoscale

Nanoparticles, Surfaces, Materials and more

Edited by
David B. Amabilino

WILEY-VCH Verlag GmbH & Co. KGaA

The Editor

Dr. David B. Amabilino
Institut de Ciència de Materials
de Barcelona (CSIC)
Campus Universitari
08193 Bellaterra
Spain

Graphic designer: Adam

All books published by Wiley-VCH are carefully produced. Nevertheless, authors, editors, and publisher do not warrant the information contained in these books, including this book, to be free of errors. Readers are advised to keep in mind that statements, data, illustrations, procedural details or other items may inadvertently be inaccurate.

Library of Congress Card No.: applied for

British Library Cataloguing-in-Publication Data
A catalogue record for this book is available from the British Library.

Bibliographic information published by the Deutsche Nationalbibliothek
The Deutsche Nationalbibliothek lists this publication in the Deutsche Nationalbibliografie; detailed bibliographic data are available on the Internet at http://dnb.d-nb.de.

© 2009 WILEY-VCH Verlag GmbH & Co. KGaA, Weinheim

All rights reserved (including those of translation into other languages). No part of this book may be reproduced in any form – by photoprinting, microfilm, or any other means – nor transmitted or translated into a machine language without written permission from the publishers. Registered names, trademarks, etc. used in this book, even when not specifically marked as such, are not to be considered unprotected by law.

Typesetting Thomson Digital, Noida, India
Printing betz-druck GmbH, Darmstadt
Binding Litges & Dopf GmbH, Heppenheim

Printed in the Federal Republic of Germany
Printed on acid-free paper

ISBN: 978-3-527-32013-4

Contents

Preface *XIII*
List of Contributors *XVII*
List of Abbreviations *XXI*

1   **An Introduction to Chirality at the Nanoscale**   *1*
       Laurence D. Barron
1.1   Historical Introduction to Optical Activity and Chirality   *1*
1.2   Chirality and Life   *4*
1.2.1   Homochirality   *4*
1.2.2   Pasteur's Conjecture   *7*
1.3   Symmetry and Chirality   *8*
1.3.1   Spatial Symmetry   *8*
1.3.2   Inversion Symmetry: Parity, Time Reversal and Charge Conjugation   *9*
1.3.3   True and False Chirality   *10*
1.3.4   Symmetry Violation   *14*
1.3.5   Symmetry Violation *versus* Symmetry Breaking   *16*
1.3.6   Chirality in Two Dimensions   *17*
1.4   Absolute Enantioselection   *18*
1.4.1   Truly Chiral Influences   *18*
1.4.2   Falsely Chiral Influences   *20*
1.5   Spectroscopic Probes of Chirality in Nanosystems   *21*
1.5.1   Electronic Optical Activity   *22*
1.5.2   Vibrational Optical Activity   *23*
1.6   Conclusion   *24*
       References   *24*

2   **Optically Active Supramolecules**   *29*
       Alessandro Scarso and Giuseppe Borsato
2.1   Introduction to Supramolecular Stereochemistry   *29*
2.1.1   Survey of Weak Intermolecular Attractive Forces   *31*
2.1.2   Timescale of Supramolecular Interactions and Racemization Processes   *33*

Chirality at the Nanoscale: Nanoparticles, Surfaces, Materials and more. Edited by David B. Amabilino
Copyright © 2009 WILEY-VCH Verlag GmbH & Co. KGaA, Weinheim
ISBN: 978-3-527-32013-4

2.2	Self-Assembly of Intrinsically Chiral Molecular Capsules 37
2.2.1	Hydrogen-Bonded Assemblies 37
2.2.1.1	Double Rosettes 37
2.2.1.2	Hydrogen-Bonded Capsules 39
2.2.2	Metal–ligand Assemblies 43
2.3	Chiral Induction in the Formation of Supramolecular Systems 46
2.3.1	Chiral Memory Effect in Hydrogen-Bonded Assemblies 46
2.3.2	Chiral Memory Effect in Metal–Ligand Assemblies 49
2.4	Chiral Spaces for Chiral Recognition 51
2.4.1	Enantioselective Recognition within Chiral Racemic Self-Assembled Hosts 52
2.4.1.1	Hydrogen-Bonded Hosts 52
2.4.1.2	Metal–Ligand Hosts 53
2.4.2	Interguests Chiral Sensing within Achiral Self-Assembled Hosts 56
2.4.2.1	Hydrogen-Bonded Hosts 57
2.4.2.2	Metal–Ligand Hosts 60
2.5	Conclusion and Outlook 61
	References 62
3	**Chiral Nanoparticles** 67
	Cyrille Gautier and Thomas Bürgi
3.1	Introduction 67
3.2	Nanoparticle Properties and Synthesis 68
3.2.1	Nanoparticle Properties 68
3.2.2	Preparation, Purification and Size Separation 71
3.2.2.1	Preparation 71
3.2.3	Purification and Separation of Nanoparticles 74
3.3	Chiroptical Properties of Inorganic Nanoparticles 74
3.3.1	Vibrational Circular Dichroism 74
3.3.2	Circular Dichroism 75
3.3.3	Origin of Optical Activity in Metal-Based Transitions 78
3.4	Optically Active Coordination Clusters 80
3.5	Nanoparticles of Chiral Organic Compounds 81
3.6	Applications 83
3.6.1	Asymmetric Catalysis 83
3.6.2	Nanoparticles in Liquid-Crystal Media 85
3.6.3	Chiral Discrimination 87
3.7	Outlook 87
	References 87
4	**Gels as a Media for Functional Chiral Nanofibers** 93
	Sudip Malik, Norifumi Fujita, and Seiji Shinkai
4.1	A Brief Introduction to Gels 93
4.1.1	Introduction 93
4.1.2	Definition of Gels 94

4.1.3	Classification of Gels	94
4.1.4	Chirality in Gels	95
4.2	Chiral Organogels	96
4.2.1	Steroid-Based Chiral Gelators	96
4.2.2	Pyrene-Based Chiral Gelators	103
4.2.3	Diaminoyclohexane-Based Chiral Gelators	103
4.2.4	OPV-Based Chiral Gelators	105
4.3	Chiral Hydrogels	108
4.3.1	Chiral Fatty Acids	108
4.3.2	Chiral Sugar-Based Gelators	109
4.3.3	Miscellaneous Chiral Hydrogelators	110
4.3.3.1	The Future of Chiral Gels in Nanoscience and Nanotechnology	111
	References	111
5	**Expression of Chirality in Polymers**	**115**
	Teresa Sierra	
5.1	Historical Perspective on Chiral Polymers	115
5.2	Chiral Architecture Control in Polymer Synthesis	117
5.2.1	Polymerization of Chiral Assemblies	117
5.2.1.1	Chiral Organization Through H-Bonding Interactions	118
5.2.1.2	Chiral Organization Through π-Stacking Interactions	120
5.2.1.3	Chiral Organization Through Mesogenic Driving Forces	121
5.2.2	Control of Chiral Architecture During Polymerization	123
5.2.2.1	Polymerization in Chiral Solvents	123
5.2.2.2	Polymerization with Chiral Templates	127
5.2.2.3	Polymerization of Chiral Assemblies by Circularly Polarized Radiation	128
5.2.3	Chiral Architecture Control upon Polymerization: Noncovalent Interactions	129
5.2.3.1	Control of the Chiral Architecture by H-Bonding Interactions	129
5.2.3.2	Control of the Chiral Architecture by π-Stacking and Steric Factors	133
5.2.3.3	Chiral Superstructures by π-Interactions: Chiral Aggregates	134
5.3	Asymmetry Induction in Nonchiral Polymers	137
5.3.1	Induction Through Noncovalent Interaction with Chiral Molecules	137
5.3.1.1	Chiral Induction by Acid–Base Interactions	137
5.3.1.2	Chiral Induction by Host–Cation Interactions	143
5.3.1.3	Chiral Induction by Metal Coordination	143
5.3.2	Induction Through Noncovalent Interaction with Chiral Polymers	146
5.3.3	Induction Through the Formation of Inclusion Complexes	147
5.3.4	Induction by a Chiral External Stimulus	150
5.3.4.1	Solvent-Induced Chirality	150
5.3.4.2	Light-Induced Chirality	151
5.4	Chiral Memory Effects. Tuning Helicity	154
5.4.1	Memory Effects from Chiral Polymers	154
5.4.1.1	Temperature- and/or Solvent-Driven Memory Effects	154

5.4.1.2	Light-Driven Memory Effects	157
5.4.2	Memory Effects from Achiral Polymers	158
5.5	Chiral Block-Copolymers and Nanoscale Segregation	161
5.5.1	Chiral Block-Copolymers: Nanoscale Segregation in the Bulk	162
5.5.2	Chiral Block-Copolymers: Nanoscale Segregation in the Mesophase	162
5.5.3	Chiral Block-Copolymers: Nanoscale Segregation in Solvents. Amphiphilic Block-Copolymers	165
5.6	Templates for Chiral Objects	169
5.6.1	Templates for Chiral Supramolecular Aggregates	169
5.6.1.1	Templating with Natural Helical Polymers	169
5.6.1.2	Templating with Synthetic Helical Polymers	172
5.6.2	Molecular Imprinting with Helical Polymers	174
5.6.3	Templating by Wrapping with Helical Polymers	175
5.6.4	Alignment of Functional Groups	176
5.6.4.1	Polyisocyanides	176
5.6.4.2	Polypeptides	178
5.6.4.3	Polyacetylenes	178
5.6.4.4	Foldamers	179
5.7	Outlook	180
	References	181

6 Nanoscale Exploration of Molecular and Supramolecular Chirality at Metal Surfaces under Ultrahigh-Vacuum Conditions 191
Rasmita Raval

6.1	Introduction	191
6.2	The Creation of Surface Chirality in 1D Superstructures	192
6.3	The Creation of 2D Surface Chirality	196
6.3.1	2D Supramolecular Chiral Clusters at Surfaces	196
6.3.2	2D Covalent Chiral Clusters at Surfaces	199
6.3.3	Large Macroscopic 2-D Chiral Arrays	200
6.3.4	Chiral Nanocavity Arrays	204
6.4	Chiral Recognition Mapped at the Single-Molecule Level	205
6.4.1	Homochiral Self-Recognition	205
6.4.2	Diastereomeric Chiral Recognition	207
6.4.2.1	Diastereomeric Chiral Recognition by Homochiral Structures	207
6.4.2.2	Diastereomeric Chiral Recognition by Heterochiral Structures	209
6.5	Summary	211
	References	212

7 Expression of Chirality in Physisorbed Monolayers Observed by Scanning Tunneling Microscopy 215
Steven De Feyter, Patrizia Iavicoli, and Hong Xu

7.1	Introduction	215
7.2	How to Recognize Chirality at the Liquid/Solid Interface	217
7.2.1	Chirality at the Level of the Monolayer Symmetry	217

7.2.2	Chirality at the Level of the Monolayer – Substrate Orientation 219
7.2.3	Determination Absolute Configuration 220
7.3	Chirality in Monolayers Composed of Enantiopure Molecules 221
7.4	Polymorphism 228
7.5	Is Chirality Always Expressed? 230
7.6	Racemic Mixtures: Spontaneous Resolution? 231
7.6.1	Chiral Molecules 231
7.6.2	Achiral Molecules 234
7.7	Multicomponent Structures 237
7.8	Physical Fields 240
7.9	Outlook 240
	References 243

8	**Structure and Function of Chiral Architectures of Amphiphilic Molecules at the Air/Water Interface** 247
	Isabelle Weissbuch, Leslie Leiserowitz, and Meir Lahav
8.1	An introduction to Chiral Monolayers on Water Surface 247
8.2	Two-Dimensional Crystalline Self-Assembly of Enantiopure and Racemates of Amphiphiles at the Air/Water Interface; Spontaneous Segregation of Racemates into Enantiomorphous 2D Domains 248
8.3	Langmuir Monolayers of Amphiphilic α-Amino Acids 249
8.3.1	Domain Morphology and Energy Calculations in Monolayers of N-acyl-α-Amino Acids 253
8.4	Stochastic Asymmetric Transformations in Two Dimensions at the Water Surface 254
8.5	Self-Assembly of Diastereoisomeric Films at the Air/Water Interface 255
8.6	Interactions of the Polar Head Groups with the Molecules of the Aqueous Environment 256
8.7	Interdigitated Bi- or Multilayer Films on the Water Surface 261
8.8	Structural Transfer from 2D Monolayers to 3D Crystals 263
8.9	Homochiral Peptides from Racemic Amphiphilic Monomers at the Air/Water Interface 265
8.10	Conclusions 268
	References 268

9	**Nanoscale Stereochemistry in Liquid Crystals** 271
	Carsten Tschierske
9.1	The Liquid-Crystalline State 271
9.2	Chirality in Liquid Crystals Based on Fixed Molecular Chirality 273
9.2.1	Chiral Nematic Phases and Blue Phases 274
9.2.2	Chirality in Smectic Phases 276
9.2.3	Polar Order and Switching in Chiral LC Phases 276
9.2.3.1	Ferroelectric and Antiferroelectric Switching 276
9.2.3.2	Electroclinic Effect 279

9.2.3.3	Electric-Field-Driven Deracemization *279*
9.2.4	Chirality Transfer via Guest–Host Interactions *279*
9.2.5	Induction of Phase Chirality by External Chiral Stimuli *281*
9.2.6	Chirality in Columnar LC Phases *282*
9.3	Chirality Due to Molecular Self-Assembly of Achiral Molecules *284*
9.3.1	Helix Formation in Columnar Phases *284*
9.3.2	Helical Filaments in Lamellar Mesophases *287*
9.4	Polar Order and Chirality in LC Phases Formed by Achiral Bent-Core Molecules *288*
9.4.1	Phase Structures and Polar Order *288*
9.4.2	Superstructural Chirality and Diastereomerism *290*
9.4.3	Switching of Superstructural Chirality *291*
9.4.4	Macroscopic Chirality and Spontaneous Reflection Symmetry Breaking in "Banana Phases" *292*
9.4.4.1	Layer Chirality *292*
9.4.4.2	Dark Conglomerate Phases *292*
9.5	Spontaneous Reflection-Symmetry Breaking in Other LC Phases *295*
9.5.1	Chirality in Nematic Phases of Achiral Bent-Core Molecules *295*
9.5.2	Spontaneous Resolution of Racemates in LC Phases of Rod-Like Mesogens *295*
9.5.3	Deracemization of Fluxional Conformers via Diastereomeric Interactions *296*
9.5.4	Chirality in Nematic, Smectic and Cubic Phases of Achiral Rod-Like Molecules *296*
9.5.5	Segregation of Chiral Conformers in Fluids, Fact or Fiction? *296*
9.6	Liquid Crystals as Chiral Templates *298*
9.7	Perspective *299*
	References *299*

10	**The Nanoscale Aspects of Chirality in Crystal Growth: Structure and Heterogeneous Equilibria** *305*
	Gérard Coquerel and David B. Amabilino
10.1	An introduction to Crystal Symmetry and Growth for Chiral Systems. Messages for Nanoscience *305*
10.2	Supramolecular Interactions in Crystals *308*
10.2.1	Hydrogen Bonds *309*
10.2.2	Interaromatic Interactions *310*
10.2.3	Electrostatic Interactions *311*
10.2.4	Modulation of Noncovalent Interactions with Solvent *312*
10.2.5	Polymorphism *312*
10.3	Symmetry Breaking in Crystal Formation *312*
10.3.1	Spontaneous Resolution of Chiral Compounds *313*
10.3.2	Spontaneous Resolution of Achiral Compounds *315*
10.4	Resolutions of Organic Compounds *317*

10.5	Resolutions of Coordination Compounds with Chiral Counterions *320*	
10.6	Thermodynamic Considerations in the Formation of Chiral Crystals *322*	
10.6.1	What is the Order of a System Composed of Two Enantiomers? *322*	
10.6.2	Resolution by Diastereomeric Associations *331*	
10.7	Influencing the Crystallization of Enantiomers *335*	
10.7.1	Solvent *335*	
10.7.2	Preferential Nucleation and Inhibition *336*	
10.8	Chiral Host–Guest Complexes *338*	
10.9	Perspectives *341*	
	References *341*	
11	**Switching at the Nanoscale: Chiroptical Molecular Switches and Motors** *349*	
	Wesley R. Browne, Dirk Pijper, Michael M. Pollard, and Ben L. Feringa	
11.1	Introduction *349*	
11.2	Switching of Molecular State *351*	
11.3	Azobenzene-Based Chiroptical Photoswitching *354*	
11.4	Diarylethene-Based Chiroptical Switches *359*	
11.5	Electrochiroptical Switching *364*	
11.6	Molecular Switching with Circularly Polarized Light *366*	
11.7	Diastereomeric Photochromic Switches *368*	
11.8	Chiroptical Switching of Luminescence *370*	
11.9	Switching of Supramolecular Organization and Assemblies *372*	
11.10	Molecular Motors *373*	
11.11	Chiral Molecular Machines *374*	
11.12	Making Nanoscale Machines Work *380*	
11.13	Challenges and Prospects *386*	
	References *387*	
12	**Chiral Nanoporous Materials** *391*	
	Wenbin Lin and Suk Joong Lee	
12.1	Classes of Chiral Nanoporous Materials *391*	
12.2	Porous Chiral Metal-Organic Frameworks *392*	
12.3	Porous Oxide Materials *397*	
12.4	Chiral Immobilization of Porous Silica Materials *400*	
12.5	Outlook *406*	
	References *407*	

Index *411*

Preface

The left- or right-handedness of things – *chirality* to the scientist – surrounds us on Earth. The importance of the phenomenon is clear when one considers that, at the submicroscopic scale, it can have either dramatic and triumphal or tragic consequences in and around us. Preparation of chiral systems and the effects they produce are vital for certain chemical processes, such as catalysis, and physical phenomena, such as the switching in displays. Understanding and influencing these processes at the atomic and molecular level – the nanometer scale – is essential for their development. This book sets out to explain the foundations of the formation and characterization of asymmetric structures as well as the effects they produce, and reveals the tremendous insight the tenets and tools of nanoscience provide to help in understanding them. The chapters trace the development of the preparative methods used for the creation of chiral nanostructures, in addition to the experimental techniques used to characterize them, and the surprising physical effects that can arise from these minuscule materials. Every category of material is covered, from organic, to coordination compounds, metals and composites, in zero, one, two and three dimensions. The structural, chemical, optical, and other physical properties are reviewed, and the future for chiral nanosystems is considered. In this interdisciplinary area of science, the book aims to combine physical, chemical and material science views in a synergistic way, and thereby to stimulate further this rapidly growing area of science.

The first chapter is an overview of chirality and all the phenomena related with it, written by one of the most eminent present-day authorities on stereochemistry, Laurence Barron from the University of Glasgow. With the scene set, the views of chirality in different systems of increasing dimensionality are covered. In "zero dimensions", well-defined supramolecular clusters formed by purely organic and metallo-organic complexes are elegantly presented by Alessandro Scarso and Giuseppe Borsato (Università Cá Foscari di Venezia) and the preparation and properties of chiral nanoparticles of all types, and the many exciting challenges associated with them, are reviewed comprehensively by Cyrille Gautier and Thomas Bürgi (Université de Neuchâtel).

The expression of chirality in essentially one-dimensional objects of a supramolecular or covalent kind has been observed widely in gels and polymers. For the gel systems Sudip Malik, Norifumi Fujita and Seiji Shinkai from Kyushu University (Japan) provide an enlightening vision of when, where and how chirality is seen. My close colleague Teresa Sierra from the Materials Science Institute in Saragossa (CSIC) provides an authoritative and comprehensive view of the many aspects of chiral induction in polymeric systems, one of the most prolific areas of research in terms of chiral induction phenomena, and one that affords many opportunities that remain to be exploited in terms of nanoscale materials.

Two-dimensional systems are extremely interesting for exploring the transmission of chirality, both because of their symmetry requirements, which limit packing possibilities, as well as for the range of techniques that exist for probing them. This situation is made patently clear in the chapters by Rasmita Raval (University of Liverpool) who describes research done on metal surface–adsorbate systems, Steven De Feyter, Patrizia Iavicoli, and Hong Xu (Katholieke Universiteit Leuven and ICMAB CSIC), who summarize chirality in physisorbed monolayers in solution, and Isabelle Weissbuch, Leslie Leiserowitz and Meir Lahav (Weizmann Institute of Science, Rehovot) who provide an overview of the tremendous contributions they and others have made to the exploration of chirality in Langmuir-type monolayer systems. These complementary chapters show just how much the tools of nanoscience can reveal about the transfer and expression of chirality in low-dimensional systems, an area that is truly blossoming at the present time.

The creation and manifestations of handedness in bulk fluids and solids are then reviewed, with special emphasis on the mechanisms of induction of chirality with a view at the scale of nanometers. Carsten Tschierske (Martin-Luther-University Halle-Wittenberg, Germany) provides an instructive overview of the occurrence of chirality in liquid-crystal systems, in which many remarkable effects are witnessed, and perhaps where nanoscientists can draw inspiration. The supramolecular and thermodynamic aspects of chiral bulk crystals, where a wealth of valuable information can be gleaned in terms of structure and phenomenology, are the subject of an extensive review by Gérard Coquerel (Université de Rouen) and myself. In particular, the construction of phase diagrams is shown to be a crucial part of understanding chiral selection in crystalline systems. This part concludes the path through the structures of different chiral systems.

In the remaining chapters, particular properties of chiral nanoscale systems are divulged. Wesley R. Browne, Dirk Pijper, Michael M. Pollard and Ben Feringa (University of Groningen) provide an accessible expert view of chiral molecular machines and switches, perhaps one of the most attractive areas in contemporary stereochemistry. Finally, Wenbin Lin and Suk Joong Lee (University of North Carolina, USA) review another fascinating family of materials, that of chiral nanoporous solids, in which spaces available for molecular recognition and catalysis are available. Thus, the exceptional contributions and their combination in this volume make a unique and useful resource for those entering or established in research concerning stereochemical aspects of nanoscale systems.

This book came about largely because of the Marie Curie Research Training Network CHEXTAN (Chiral Expression and Transfer at the Nanoscale) funded by the European Commission. The Network, coming to its end as these lines are written, brought together eight groups – some of which contribute to this book – with the aim of training young scientists in this interdisciplinary area of science. I thank wholeheartedly all those who participated in the Network – the senior scientists and excellent group of young researchers – for helping to give an impetus to the area. As a consequence of the Network, the International Conference Chirality at the Nanoscale was held (in Sitges, Spain in September 2007) and proved to be a significant stimulus to thinking for many of the groups working on nanosystems and chirality. I thank everyone who helped make that meeting a success, the lecturers and all the participants, and for such a special moment.

I have to thank the Spanish Research Council (the CSIC) who employs me, the staff of the Barcelona Materials Science Institute (ICMAB) for providing such a pleasant environment to work in, and everyone in the Molecular Nanoscience and Organic Materials Department for the healthy environment in which to carry out research. Finally, and most importantly, I am indebted to all the authors for the great effort they have put into producing these excellent summaries that make up the book. With the many pressures we have to write nowadays it is difficult to dedicate time to this kind of enterprise, but they collaborated magnificently and the combined effort is one that I hope you, the readers will appreciate.

Institut de Ciència de Materials de Barcelona (CSIC)
September 2008
David Amabilino

List of Contributors

David B. Amabilino
Institut de Ciència de Materials de
Barcelona (CSIC)
Campus Universitari de Bellaterra
08193 Cerdanyola del Vallès
Catalonia
Spain

Laurence D. Barron
Department of Chemistry
University of Glasgow
Glasgow G12 8QQ
UK

Giuseppe Borsato
Università Ca' Foscari di Venezia
Dipartimento Chimica
Dorsoduro 2137
30123 Venezia
Italy

Wesley R. Browne
Stratingh Institute for Chemistry &
Zernike Institute for Advanced
Materials
Faculty of Mathematics and Natural
Sciences
University of Groningen, Nijenborgh 4
9747 AG
Groningen
The Netherlands

Thomas Bürgi
Institute for Physical Chemistry
Rupert-Karls University Heidelberg
Im Neuenheimer Feld 253
69120 Heidelberg
Germany

Gérard Coquerel
UC2M2, UPRES EA 3233
Université de Rouen-IRCOF
76821 Mont Saint Aignan Cedex
France

Steven De Feyter
Laboratory of Photochemistry and
Spectroscopy
Molecular and Nano Materials
Department of Chemistry, and INPAC -
Institute for Nanoscale Physics and
Chemistry
Katholieke Universiteit Leuven
Celestijnenlaan 200-F
3001 Leuven
Belgium

Ben L. Feringa
Stratingh Institute for Chemistry &
Zernike Institute for Advanced
Materials
Faculty of Mathematics and Natural
Sciences
University of Groningen
Nijenborgh 4
9747 AG
Groningen
The Netherlands

Norifumi Fujita
Department of Chemistry and
Biochemistry
Graduate School of Engineering
Kyushu University
Moto-oka 744, Nishi-ku
Fukuoka 819-0395
Japan

Cyrille Gautier
Université de Neuchâtel
Institut de Microtechnique
Rue Emile-Argand 11
2009 Neuchâtel
Switzerland

Patrizia Iavicoli
Institut de Ciència de Materials de
Barcelona (CSIC)
Campus Universitari
08193 Bellaterra
Catalonia
Spain

Meir Lahav
Department of Materials and Interfaces
Weizmann Institute of Science
76100-Rehovot
Israel

Suk Joong Lee
Department of Chemistry
CB#3290
University of North Carolina at Chapel
Hill
NC 27599
USA

Leslie Leiserowitz
Department of Materials and Interfaces
Weizmann Institute of Science
76100-Rehovot
Israel

Sudip Malik
Department of Chemistry and
Biochemistry
Graduate School of Engineering
Kyushu University
Moto-oka 744, Nishi-ku
Fukuoka 819-0395
Japan

Dirk Pijper
Stratingh Institute for Chemistry &
Zernike Institute for Advanced
Materials
Faculty of Mathematics and Natural
Sciences
University of Groningen
Nijenborgh 4
9747 AG
Groningen
The Netherlands

Michael M. Pollard
Stratingh Institute for Chemistry &
Zernike Institute for Advanced
Materials
Faculty of Mathematics and Natural
Sciences
University of Groningen
Nijenborgh 4
9747 AG
Groningen
The Netherlands

Rasmita Raval
The Surface Science Research Centre
and Department of Chemistry
University of Liverpool
Liverpool, L69 3BX
UK

Alessandro Scarso
Università Ca' Foscari di Venezia
Dipartimento di Chimica
Dorsoduro 2137
30123 Venezia
Italy

Seiji Shinkai
Department of Chemistry and
Biochemistry
Graduate School of Engineering
Kyushu University
Moto-oka 744, Nishi-ku
Fukuoka 819-0395
Japan

Teresa Sierra
Instituto de Ciencia de Materiales de
Aragón
Facultad de Ciencias
Universidad de Zaragoza-CSIC
Zaragoza-50009
Spain

Wenbin Lin
Department of Chemistry
CB#3290
University of North Carolina at Chapel
Hill
NC 27599
USA

Carsten Tschierske
Institute of Chemistry
Martin-Luther University Halle
Kurt-Mothes Str. 2
06120 Halle
Germany

Isabelle Weissbuch
Department of Materials and Interfaces
Weizmann Institute of Science
76100-Rehovot
Israel

Hong Xu
Laboratory of Photochemistry and
Spectroscopy
Molecular and Nano Materials
Department of Chemistry, and INPAC -
Institute for Nanoscale Physics and
Chemistry
Katholieke Universiteit Leuven
Celestijnenlaan 200-F
3001 Leuven
Belgium

List of Abbreviations

AFM	atomic force microscopy
AIEE	aggregate-induced enhanced emission
APS	aminopropyltrimethoxysilane
BAR	barbiturates
CCW	counterclockwise
CD	circular dichroism
CLG	cholesteryl-S-glutamate
CN	cinchonine
CPL	circularly polarized light
CPL	circular polarization of luminescence
CW	clockwise
CYA	cyanurate
1D	one dimensional
2D	two-dimensional
3D	three-dimensional
DFT	Density functional theory
DSC	differential scanning calorimetry
2DSD	two-dimensional structural database
ECD	electronic circular dichroism
ee	enantiomeric excess
EM	electron microscopy
EPJ	European Physical Journal
EPL	elliptically polarized light
FE-SEM	field emission scanning electron microscopy
FLC	ferroelectric liquid crystals
GIXD	grazing-incidence X-ray diffraction
HBC	hexabenzocoronenes
HTP	helical twisting power
IUPAC	International Union of Pure and Applied Chemistry
LB	Langmuir-Blodgett
LC	liquid crystal(line)
LC	liquid crystalline
LDH	layered double hydroxides

Chirality at the Nanoscale: Nanoparticles, Surfaces, Materials and more. Edited by David B. Amabilino
Copyright © 2009 WILEY-VCH Verlag GmbH & Co. KGaA, Weinheim
ISBN: 978-3-527-32013-4

LEED	low-energy electron diffraction
LMW	Low molecular weight
LMWG	low molecular weight gelators
MALDI-TOF MS	matrix-assisted laser desorption-ionization time-of-flight mass spectrometry
MBETs	metal-based electronic transitions
MD	Marks decahedron
ML	monolayers
MOFs	metalorganic frameworks
N-LC	nematic LC
NIC	N-isobutyryl-cysteine
NIR	near-infrared
NPs	nanoparticles
NRDs	nanorods
ONPs	organic nanoparticles
ORD	optical rotatory dispersion
PAGE	polyacrylamide gel electrophoresis
PS	polystyrene
PVA	poly(vinyl alcohol)
PVED	parity-violating energy difference
QSEs	quantum size effects
RA	resolving agent
RAIRS	reflection absorption infrared spectroscopy
ROA	Raman optical activity
RW	re-writable
SEC	size exclusion chromatography
SP	surface plasmon
STM	scanning tunnelling microscopy
T_m	melting temperature
TA	tartaric acid
TEOS	tetraethoxysilane
THF	tetrahydrofuran
TOAB	tetraoctylammonium bromide
TPP	triphenylphosphine
TTF	tetrathiafulvalene
UHV	ultra-high vacuum
VCD	vibrational circular dichroism
VDSA	vapor-driven self-assembly
WORM	write-once read many
XPD	X-ray photoelectron diffraction
XPS	X-ray photoelectron spectroscopy

1
An Introduction to Chirality at the Nanoscale
Laurence D. Barron

1.1
Historical Introduction to Optical Activity and Chirality

Scientists have been fascinated by chirality, meaning right- or left-handedness, in the structure of matter ever since the concept first arose as a result of the discovery, in the early years of the nineteenth century, of natural optical activity in refracting media. The concept of chirality has inspired major advances in physics, chemistry and the life sciences [1, 2]. Even today, chirality continues to catalyze scientific and technological progress in many different areas, nanoscience being a prime example [3–5].

The subject of optical activity and chirality started with the observation by Arago in 1811 of colors in sunlight that had passed along the optic axis of a quartz crystal placed between crossed polarizers. Subsequent experiments by Biot established that the colors originated in the rotation of the plane of polarization of linearly polarized light (optical rotation), the rotation being different for light of different wavelengths (optical rotatory dispersion). The discovery of optical rotation in organic liquids such as turpentine indicated that optical activity could reside in individual molecules and could be observed even when the molecules were oriented randomly, unlike quartz where the optical activity is a property of the crystal structure, because molten quartz is not optically active. After his discovery of circularly polarized light in 1824, Fresnel was able to understand optical rotation in terms of different refractive indices for the coherent right- and left-circularly polarized components of equal amplitude into which a linearly polarized light beam can be resolved. This led him to suggest that optical activity may result from "a helicoidal arrangement of the molecules of the medium, which would present inverse properties according to whether these helices were dextrogyrate or laevogyrate." This early work culminated in Pasteur's epoch-making separation in 1848 of crystals of sodium ammonium paratartrate, an optically inactive form of sodium ammonium tartrate, into two sets that, when dissolved in water, gave optical rotations of equal magnitude but opposite sign. This demonstrated that paratartaric acid was a mixture, now known as a *racemic* mixture, of equal numbers of mirror-image molecules. Pasteur was lucky in that his racemic solution crystallized into equal amounts of crystals containing exclusively one or other of the mirror-image molecules, a process known as *spontaneous resolution*. (Such mixtures of crystals are called conglomerates, as distinct from racemic compounds where each crystal contains equal amounts of the mirror-image molecules.)

Although a system is called "optically active" if it has the power to rotate the plane of polarization of a linearly polarized light beam, optical rotation is in fact just one of a number of optical activity phenomena that can all be reduced to the common origin of a different response to right- and left-circularly polarized light. Substances that are optically active in the absence of external influences are said to exhibit *natural* optical activity.

In 1846, Faraday discovered that optical activity could be induced in an otherwise inactive sample by a magnetic field. He observed optical rotation in a rod of lead borate glass placed between the poles of an electromagnet with holes bored through the pole pieces to enable a linearly polarized light beam to pass through. This effect is quite general: a Faraday rotation is found when linearly polarized light is transmitted through any crystal or fluid in the direction of a magnetic field, the sense of rotation being reversed on reversing the direction of either the light beam or the magnetic field. At the time, the main significance of this discovery was to demonstrate conclusively the intimate connection between electromagnetism and light; but it also became a source of confusion to some scientists (including Pasteur) who failed to appreciate that there is a fundamental distinction between magnetic optical rotation and the natural optical rotation that is associated with handedness in the microstructure. That the two phenomena have fundamentally different symmetry characteristics is intimated by the fact that the magnetic rotation is additive when the light is reflected back though the medium, whereas the natural rotation cancels.

Although he does not provide a formal definition, it can be inferred [6] from his original article that described in detail his experiments with salts of tartaric acid that Pasteur in 1848 introduced the word *dissymmetric* to describe hemihedral crystals of a tartrate "which differ only as an image in a mirror differs in its symmetry of position from the object which produces it" and used this word to describe handed figures and handed molecules generally. The two distinguishable mirror-image crystal forms were subsequently called *enantiomorphs* by Naumann in 1856. Current usage reserves *enantiomorph* for macroscopic objects and *enantiomer* for molecules [7], but because of the ambiguity of scale in general physical systems, these two terms are often used as synonyms [8]. This is especially pertinent in nanoscience that embraces such a large range of scales, from individual small molecules to crystals, polymers and supramolecular assemblies.

The word dissymmetry was eventually replaced by *chirality* (from the Greek *cheir*, meaning hand) in the literature of stereochemistry. This word was first introduced into science by Lord Kelvin [9], Professor of Natural Philosophy in the University of Glasgow, to describe a figure "if its image in a plane mirror, ideally realized, cannot be brought to coincide with itself." The two mirror-image enantiomers of the small archetypal molecule bromochlorofluoromethane are illustrated in Figure 1.1a, together with the two enantiomers of hexahelicene in Figure 1.1b. The modern system for specifying the absolute configurations of most chiral molecules is based on the *R*

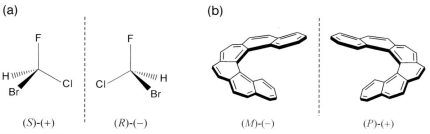

Figure 1.1 The two mirror-image enantiomers of bromochlorofluoromethane (a) and hexahelicene (b).

(for *rectus*) and S (for *sinister*) system of Cahn, Ingold and Prelog, supplemented with the P (for plus) and M (for minus) designation for molecules that have a clear helical structure [7]. The older D,L designation, based on Fischer planar projections, is still used for amino acids and carbohydrates. The sense of optical rotation (usually measured at the sodium D-line wavelength of 598 nm) associated with a particular absolute configuration is given in brackets.

Although Lord Kelvin's definition of chiral is essentially the same as that used earlier by Pasteur for dissymmetric, the two words are not strictly synonymous in the broader context of modern chemistry and physics. Dissymmetry means the absence of certain symmetry elements, these being improper rotation axes in Pasteur's usage. Chirality has become a more positive concept in that it refers to the possession of the attribute of handedness, which has a physical content. In molecular physics this is the ability to support time-even pseudoscalar observables; in elementary particle physics chirality is defined as the eigenvalue of the Dirac matrix operator γ_5.

To facilitate a proper understanding of the structure and properties of chiral molecules and of the factors involved in their synthesis and transformations, this chapter uses some principles of modern physics, especially fundamental symmetry arguments, to provide a description of chirality deeper than that usually encountered in the literature of stereochemistry. A central result is that, although dissymmetry is sufficient to guarantee chirality in a stationary object such as a finite helix, dissymmetric systems are not necessarily chiral when motion is involved. The words "true" and "false" chirality, corresponding to time-invariant and time-noninvariant enantiomorphism, respectively, were introduced by this author to draw attention to this distinction [10], but it was not intended that this would become standard nomenclature. Rather, it was suggested that the word "chiral" be reserved in future for systems that are truly chiral. The terminology of true and false chirality has, however, been taken up by others, especially in the area of absolute enantioselection, so for consistency it will be used in this chapter. We shall see that the combination of linear motion with a rotation, for example, generates true chirality, but that a magnetic field alone does not (in fact it is not even falsely chiral). Examples of systems with false chirality include a stationary rotating cone, and collinear electric and magnetic fields. The term "false" should not be taken to be perjorative in any sense; indeed, false chirality can generate fascinating new phenomena that are even more subtle than those associated with true chirality.

The triumph of theoretical physics in unifying the weak and electromagnetic forces into a single "electroweak" force by Weinberg, Salam and Glashow in the 1960s provided a new perspective on chirality. Because the weak and electromagnetic forces turned out to be different aspects of the same, but more fundamental, unified force, the absolute parity violation associated with the weak force is now known to infiltrate to a tiny extent into all electromagnetic phenomena so that free atoms, for example, exhibit very small optical rotations, and a tiny energy difference exists between the enantiomers of a chiral molecule.

1.2
Chirality and Life

1.2.1
Homochirality

Since chirality is a *sine qua non* for the amazing structural and functional diversity of biological macromolecules, the chemistry of life provides a paradigm for the potential roles of chirality in supramolecular chemistry and nanoscience [3]. Accordingly, a brief survey is provided of current knowledge on the origin and role of chirality in the chemistry of life.

A hallmark of life's chemistry is its *homochirality* [1, 11–15], which is well illustrated by the central molecules of life, namely proteins and nucleic acids. Proteins consist of polypeptide chains made from combinations of 20 different amino acids (primary structure), all exclusively the L-enantiomers. This homochirality in the monomeric amino acid building blocks of proteins leads to homochirality in higher-order structures such as the right-handed α-helix (secondary structure), and the fold (tertiary structure) that is unique to each different protein in its native state (Figure 1.2). Nucleic acids consist of chains of deoxyribonucleosides (for DNA) or ribonucleosides (for RNA), connected by phosphodiester links, all based exclusively on the D-deoxyribose or D-ribose sugar ring, respectively (Figure 1.3). This homochirality in the monomeric sugar building blocks of nucleic acids leads to homochirality in their secondary structures such as the right-handed B-type DNA double helix, and tertiary structures such as those found in catalytic and ribosomal RNAs. DNA itself is finding many applications in nanotechnology [5].

Homochirality is essential for an efficient biochemistry, rather like the universal adoption of right-handed screws in engineering. One example is Fischer's "lock and key" principle [16], which provides a mechanism for stereochemical selection in nature, as in enzyme catalysis. Small amounts of "non-natural" enantiomers such as the D-forms of some amino acids are in fact found in living organisms where they have specific roles [17, 18], but they have not been found in functional proteins (their detection in metabolically inert proteins like those found in lens and bone tissue is attributed to racemization during ageing [17]). Since molecules sufficiently large and

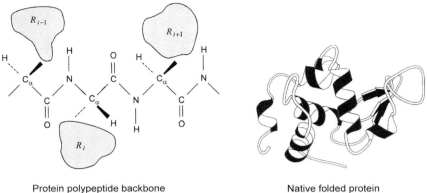

Protein polypeptide backbone

Native folded protein (hen lysozyme)

Figure 1.2 The polypeptide backbones of proteins are made exclusively from homochiral amino acids (all L). R_i represents side chains such as CH_3 for alanine. This generates homochiral secondary structures, such as the right-handed α-helix, within the tertiary structures of native folded proteins like hen lysozyme.

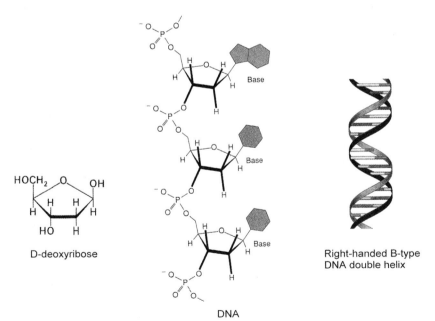

D-deoxyribose

DNA

Right-handed B-type DNA double helix

Figure 1.3 Nucleic acids are made exclusively from homochiral sugars (all D) such as D-deoxyribose for DNA. This generates homochiral secondary structures such as the right-handed B-type DNA double helix.

complex to support life are almost certain to exist in two mirror-image chiral forms, homochirality also appears to be essential for any molecule-based life on other worlds. Furthermore, since no element other than carbon forms such a huge variety of compounds, many of them chiral, the chemistry is expected to be organic. Last but not least, the liquid water that is essential for life on Earth is more than simply a medium: it acts as a "lubricant" of key biomolecular processes such as macromolecular folding, unfolding and interaction [19]. No other liquid solvent has the same balance of vital physicochemical properties. Hence homochirality associated with a complex organic chemistry in an aqueous environment would appear to be as essential for life on other worlds as it is on Earth. Nonetheless, the possibility of alternative scenarios based on elements other than carbon and solvents other than water should be kept in mind [20], and could be of interest in the context of synthetic homochiral supramolecular chemistry and nanoscience. Indeed, nanotechnology is already exploiting materials and devices that benefit explicitly from homochirality at the molecular level [5].

A central problem in the origin of life is which came first: homochirality in the prebiotic monomers or in the earliest prebiotic polymers [14, 21]. Homochiral nucleic acid polymers, for example, do not form efficiently in a racemic solution of the monomers [22]. Theoretical analysis suggested that addition of a nucleotide of the wrong handedness halts the polymerization [23], a process called enantiomeric cross-inhibition. However, homochirality in the chiral monomers is not essential for generating homochiral *synthetic* polymers [3]. Although polyisocyanates, for example, constructed from achiral monomers form helical polymers with equal numbers of right- and left-handed forms, the introduction of a chiral bias in the form of a small amount of a chiral version of a monomer can induce a high enantiomeric excess (ee), defined as the percentage excess of the enantiomer over the racemate [7], of one helical sense [24, 25]. This generation of an excess of the helical sense preferred by the small number of chiral units (the sergeants) is called the sergeants-and-soldiers effect. Furthermore, a polyisocyanate constructed from a random copolymerization of chiral monomers containing just a small percentage excess of one enantiomer over the other shows a large excess of the helical form generated from homopolymerization of the corresponding enantiopure monomer. This generation of an excess of the helical sense preferred by the excess enantiomer is called the majority rules effect.

Another example of the dramatic influence a small chiral bias may exert, this time in the generation of homochiral monomers, arises in solid–liquid phase equilibria of amino acids: a few per cent ee of one enantiomer in racemic compounds can lead to very high solution ees, including a virtually enantiopure solution for serine [26]. This is related to the well-known differences in relative solubilities of an enantiopure compound and the corresponding racemate, which forms the basis of enantioenrichment by crystallization [7]. An important feature of this system is that it is based on an equilibrium mechanism, as distinct from far-from-equilibrium mechanisms as previously invoked in kinetically induced amplification via autocatalytic reactions [27]. Also, sublimation of a near-racemic mixture of serine containing a small percentage ee of one enantiomer was recently found to generate a vapor with up to

98% ee that could be condensed into an almost enantiopure solid [28]. Apparently, clusters of the same enantiomer form preferentially over racemic clusters in the vapor, with those of the majority enantiomer forming faster and selectively plucking more of the majority enantiomer out of the subliming crystals. If all of the serine were allowed to sublime, it would segregate into homochiral clusters with the same overall slight initial ee as in the solid mixture. Similar results for a variety of other amino acids were reported shortly afterwards [29].

Crystal chemistry suggests further possibilities. In addition to spontaneous resolutions in crystallization, or the crystallization of achiral molecules in chiral spacegroups [30], the faces of chiral crystals such as quartz, or the chiral faces of nonchiral crystals such as calcite, could act as enantioselective templates [31]. Related to this is the demonstration that the chiral rims of racemic β-sheets can operate as templates for the generation of long homochiral oligopeptides from racemic monomers in aqueous solution [32].

1.2.2
Pasteur's Conjecture

From the above it is clear that small initial ees in chiral monomers can, in some circumstances, generate large ees in both chiral monomers and polymers. This small ee could be produced by some physical chiral influence. Since in Pasteur's time all substances found to be optically active in solution were natural products, Pasteur himself conjectured that molecular chirality in the living world is the product of some universal chiral force or influence in nature. Accordingly, he attempted to extend the concept of chirality (dissymmetry) to other aspects of the physical world [33]. For example, he thought that the combination of a translational with a rotational motion generated chirality; likewise a magnetic field. Curie [34] suggested that collinear electric and magnetic fields are chiral. However, as explained below, of these only a translating–rotating system exhibits "true chirality." Pasteur's incorrect belief that a static magnetic field alone is also a source of chirality has been shared by many other scientists. This misconception is based on the fact that a static magnetic field can induce optical rotation (the Faraday effect) in achiral materials (*vide supra*); but as Lord Kelvin [9] emphasized: "the magnetic rotation has no chirality." In a new twist to the story [35], a magnetic field was recently used in a more subtle fashion than that conceived by Pasteur by exploiting the novel phenomenon of magnetochiral dichroism (*vide infra*) to induce a small ee.

If it were ever proved that parity violation (*vide infra*) was linked in some way to the origin and role of homochirality in the living world, this would provide the ultimate source of a universal chiral force sensed by Pasteur. However, at the time of writing, there is no firm evidence to support the idea [36]. On a cosmic scale, enantioselective mechanisms depending on parity violation are the only ones that could predetermine a particular handedness, such as the L-amino acids and D-sugars found in terrestrial life; in all other mechanisms the ultimate choice would arise purely by chance.

1.3
Symmetry and Chirality

Chirality is an excellent subject for the application of symmetry principles [2, 37]. As well as conventional point group symmetry, the fundamental symmetries of space inversion, time reversal and even charge conjugation have something to say about chirality at all levels: the experiments that show up optical activity observables, the objects generating these observables and the nature of the quantum states that these objects must be able to support. Even the symmetry violations observed in elementary particle physics can infiltrate into the world of chiral molecules, with intriguing implications. These fundamental symmetry aspects, summarized briefly below, are highly relevant to considerations of molecular chirality and absolute enantioselection. They bring intrinsic physical properties of the universe to bear on the problem of the origin of homochirality and its role in the origin and special physicochemical characteristics of life, with important lessons for the generation and exploitation of chirality at the nanoscale.

1.3.1
Spatial Symmetry

A finite cylindrical helix is the archetype for all figures exhibiting chirality. Thus, a helix and its mirror image cannot be superposed since reflection reverses the screw sense. Chiral figures are not necessarily *asymmetric*, that is devoid of all symmetry elements, since they may possess one or more proper rotation axes: for example, a finite cylindrical helix has a twofold rotation axis C_2 through the midpoint of the coil, perpendicular to the long axis (Figure 1.4a). Hexahelicene (Figure 1.1b) provides a molecular example of this. However, chirality excludes improper symmetry elements, namely centers of inversion, reflection planes and rotation–reflection axes. Hence, chirality is supported by the point groups comprising only proper rotations, namely C_n, D_n, O, T and I.

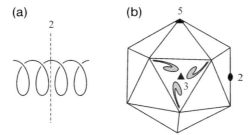

Figure 1.4 (a) A right-handed finite helix illustrating the twofold proper rotation axis. (b) A simple icosahedral virus capsid illustrating one each of the 6 fivefold, 10 threefold and 15 twofold proper rotation axes. Each triangular face contains three asymmetric protein subunits (only those in one face are shown for simplicity). The chirality of the protein subunits renders the entire capsid chiral but without destroying the proper rotation axes, thereby generating the point group I.

The chiral point group *I* has high rotational symmetry based on fivefold, threefold and twofold rotation axes (Figure 1.4b). The protein capsids of icosahedral viruses provide an interesting and pertinent example of this point group in the context of chiral supramolecular structures; indeed virus capsids are already widely used in nanotechnology. Although the folds of the many constituent coat protein subunits are intrinsically chiral, being completely asymmetric like the lysozyme fold shown in Figure 1.2, the protein subunits are identical in simple viruses and are tiled over the capsid surface in such a way as to preserve all the rotation axes of the icosahedron. However, their intrinsic chirality destroys any improper symmetry elements of the complete capsid, which renders it chiral overall. This example demonstrates how it is possible to construct a high-symmetry chiral supramolecular structure from the association of low-symmetry chiral macromolecules.

1.3.2
Inversion Symmetry: Parity, Time Reversal and Charge Conjugation

More fundamental than spatial (point group) symmetries are the symmetries in the laws of physics, and these in turn depend on certain uniformities that we perceive in the world around us. In quantum mechanics, the invariance of physical laws under an associated transformation often generates a conservation law or selection rule that follows from the invariance of the Hamiltonian *H* under the transformation. Three symmetry operations corresponding to distinct "inversions" are especially fundamental, namely parity, time reversal and charge conjugation [38].

Parity, represented by the operator *P* (not to be confused with the *P*-helicity specification of absolute configuration) inverts the coordinates of all the particles in a system through the coordinate origin. This is equivalent to a reflection of the physical system in any plane containing the coordinate origin followed by a rotation through 180° about an axis perpendicular to the reflection plane. If replacing the space coordinates (x,y,z) everywhere in equations describing physical laws (e.g., Newton's equations for mechanics or Maxwell's equations for electromagnetism) leaves the equations unchanged, all processes determined by such laws are said to conserve parity. Conservation of parity implies that *P* commutes with *H* so that, if ψ_k is an eigenfunction of *H*, then $P\psi_k$ is also an eigenfunction with the same energy.

Time reversal, represented by the operator *T*, reverses the motions of all the particles in a system. If replacing the time coordinate *t* by −*t* everywhere leaves equations describing physical laws unchanged, then all processes determined by such laws are said to conserve time-reversal invariance, or to have reversability. A process will have reversability as long as the process with all the motions reversed is in principle a possible process, however improbable (from the laws of statistics) it may be. Time reversal is therefore best thought of as motion reversal. It does not mean going backward in time! Conservation of time reversal implies that *T* and *H* commute so that, if *H* is time independent, the stationary state ψ_k and its time-reversed state $T\psi_k$ have the same energy.

Charge conjugation, represented by the operator *C*, interconverts particles and antiparticles. This operation from relativistic quantum field theory has conceptual value in studies of molecular chirality. It appears in the *CPT* theorem, which states that, even if one or more of *C*, *P*, or *T* are violated, invariance under the combined operation *CPT* will always hold. The *CPT* theorem has three important consequences: the rest mass of a particle and its antiparticle are equal; the particle and antiparticle lifetimes are the same; and the electromagnetic properties such as charge and magnetic moment of particles and antiparticles are equal in magnitude but opposite in sign.

A *scalar* physical quantity such as energy has magnitude but no directional properties; a *vector* quantity such as linear momentum **p** has magnitude and an associated direction; and a *tensor* quantity such as electric polarizability has magnitudes associated with two or more directions. Scalars, vectors and tensors are classified according to their behavior under *P* and *T*. A vector whose sign is reversed by *P* is called a *polar* or *true* vector; for example a position vector **r**. A vector whose sign is not changed by *P* is called an *axial* or *pseudo* vector; for example the angular momentum $\mathbf{L} = \mathbf{r} \times \mathbf{p}$ (since the polar vectors **r** and **p** change sign under *P*, their vector product **L** does not). A vector such as **r** whose sign is not changed by *T* is called *time even*; a vector such as **p** or **L** whose sign is reversed is called *time odd*.

Pseudoscalar quantities have magnitude with no directional properties, but they change sign under space inversion *P*. An example is the natural optical rotation angle.

1.3.3
True and False Chirality

There is no disagreement when the term "chiral" is applied to a static object displaying distinguishable enantiomers under space inversion *P* (or mirror reflection), like bromochlorofluoromethane or hexahelicene in Figure 1.1. But when the term is applied to less tangible enantiomorphous systems in which motion is an essential ingredient, time-reversal arguments are required to clarify the concept. The hallmark of a chiral system is that it can support time-even pseudoscalar observables, which are only supported by quantum states with mixed parity but that are invariant under time reversal. This leads to the following definition [2, 10].

True chirality is exhibited by systems existing in two distinct enantiomeric states that are interconverted by space inversion, but not by time reversal combined with any proper spatial rotation.

The spatial enantiomorphism shown by a truly chiral system is therefore time invariant. Spatial enantiomorpism that is time *non*invariant has different characteristics called "false chirality" to emphasize the distinction. Falsely chiral systems have quite different physical properties from truly chiral systems, which is due in part to their inability to support time-even pseudoscalar observables.

Consider an electron, which has a spin quantum number $s = 1/2$, with $m_s = \pm 1/2$ corresponding to the two opposite projections of the spin angular momentum onto a space-fixed axis. A stationary spinning electron is not a chiral object because space inversion *P* does not generate a distinguishable *P*-enantiomer (Figure 1.5a). However, an electron translating with its spin projection parallel or antiparallel to the direction of propagation has true chirality because *P* interconverts distinguishable

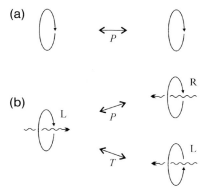

Figure 1.5 The effect of P and T on the motions of (a) a stationary spinning particle and (b) a translating spinning particle. Reprinted from Ref. [2] with permission.

left (L) and right (R) spin-polarized versions by reversing the propagation direction but not the spin sense, whereas time reversal T does not because it reverses both (Figure 1.5b). Similar considerations apply to a circularly polarized photon except that photons, being massless, are always chiral since they always move at the velocity of light in any reference frame.

Now consider a cone spinning about its symmetry axis. Because P generates a version that is not superposable on the original (Figure 1.6a), it might be thought that this is a chiral system. The chirality, however, is false because T followed by a rotation R_π through 180° about an axis perpendicular to the symmetry axis generates the same system as space inversion (Figure 1.6a). If, however, the spinning cone is also translating along the axis of spin, T followed by R_π now generates a system different from that generated by P alone (Figure 1.6b). Hence a *translating* spinning cone has true chirality. It has been argued that a nontranslating spinning cone belongs to the spatial point group C_∞ and so is chiral [39]. More generally, it was suggested that objects that exhibit enantiomorphism, whether T-invariant or not, belong to chiral point groups and hence that motion-dependent chirality is encompassed in the group-theoretical equivalent of Lord Kelvin's definition. However, a nontranslating spinning cone will have quite different physical properties from those of a finite helix. For example, the molecular realization of a spinning cone, namely a rotating symmetric top molecule such as CH_3Cl, does not support time-even pseudoscalar observables such as natural optical rotation (it supports magnetic optical rotation) [2]. To classify it as "chiral" the same as for a completely asymmetric molecule that does support natural optical rotation is therefore misleading as far as the physics is concerned, even though such a classification may be consistent within a particular mathematical description.

It is clear that neither a static uniform electric field **E** (a time-even polar vector) nor a static uniform magnetic field **B** (a time-odd axial vector) constitutes a chiral system. Likewise for time-dependent uniform electric and magnetic fields. Furthermore, no combination of a static uniform electric and a static uniform magnetic field can

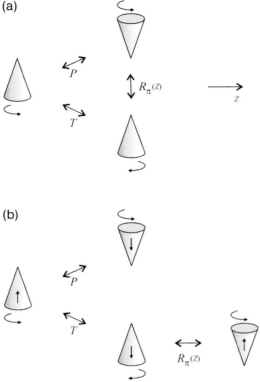

Figure 1.6 The effect of P, T and R_π on (a) a stationary spinning cone, which has false chirality, and on (b) a translating spinning cone, which has true chirality. The systems generated by P and T may be interconverted by a rotation $R_\pi(z)$ about an axis z perpendicular to the symmetry axis of the cone in (a) but not in (b). Adapted from Ref. [2] with permission.

constitute a chiral system. As Curie [34] pointed out, collinear electric and magnetic fields do indeed generate spatial enantiomorphism (dissymmetry). Thus, parallel and antiparallel arrangements are interconverted by space inversion and are not superposable. But they are also interconverted by time reversal combined with a rotation R_π through 180° about an axis perpendicular to the field directions and so the enantiomorphism corresponds to false chirality (Figure 1.7). Zocher and Török [40] also recognized that Curie's spatial enantiomorphism is not the same as that of a chiral molecule: they called the collinear arrangement of electric and magnetic fields a time-asymmetric enantiomorphism and said that it does not support time-symmetric optical activity. Tellegen [41] conceived of a medium with novel electromagnetic properties comprising microscopic electric and magnetic dipoles tied together with their moments either parallel or antiparallel. Such media clearly exhibit enantiomorphism corresponding to false chirality, and are potentially of great interest to nanotechnology. However, although much discussed [42, 43], the fabrica-

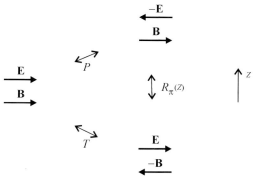

Figure 1.7 The effect of *P* and *T* on an arrangement of parallel electric and magnetic fields, which has false chirality. The opposite antiparallel arrangements generated by *P* and *T* may be interconverted by a rotation $R_\pi(z)$ about an axis *z* perpendicular to the field directions.

tion of Tellegen media proved elusive until very recently when the construction of particles with coupled electric and magnetic moments was reported for the first time [44]. These particles, made from white titanium oxide and black manganese ferrite suspended in polythene beads, were used to fabricate a switchable room-temperature magnetoelectric material that is isotropic in the absence of any field.

In fact, the basic requirement for two collinear vectorial influences to generate true chirality is that one transforms as a polar vector and the other as an axial vector, with both either time even- or time-odd. The second case is exemplified by magnetochiral phenomena [1, 2, 45] where a birefringence and a dichroism may be induced in an isotropic chiral sample by a uniform magnetic field **B** collinear with the propagation vector **k** of a light beam of arbitrary polarization, including unpolarized. The birefringence [46] and the dichroism [47] were first observed in the late 1990s. The magnetochiral dichroism experiment is illustrated in Figure 1.8. Here, the parallel

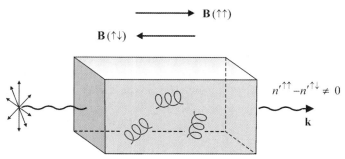

Figure 1.8 The magnetochiral dichroism experiment. The absorption index n' of a medium composed of chiral molecules is slightly different for *unpolarized* light when a static magnetic field is applied parallel (↑↑) and antiparallel (↑↓) to the direction of propagation of the beam. Reprinted from Ref. [2] with permission.

and antiparallel arrangements of **B** and **k**, which are interconverted by *P*, are true chiral enantiomers because they cannot be interconverted by *T* since **B** and **k** are both time odd. Magnetochiral phenomena are not confined to the realm of optics [1]. An important example for nanotechnology is an anisotropy in the electrical resistance through a chiral conductor in directions parallel and antiparallel to a static magnetic field, something that has been observed in both macroscopic chiral conductors in the form of helical bismuth wires [48], and microscopic helical conductors in the form of chiral single-walled nanotubes [49].

1.3.4
Symmetry Violation

Prior to the discovery of parity violation by Lee and Yang in 1956, it seemed self-evident that handedness is not built into the laws of nature. If two objects exist as nonsuperposable mirror images, such as the two enantiomers of a chiral molecule, it did not seem reasonable that nature should prefer one over the other. Any difference was thought to be confined to the sign of pseudoscalar observables: the mirror image of any complete experiment involving one enantiomer should be realizable, with any pseudoscalar observable (such as the natural optical rotation angle) changing sign but retaining exactly the same magnitude. Observations of asymmetries in phenomena such as radioactive β-decay demonstrated that this was not the case for processes involving the weak interactions. It was subsequently realized, however, that symmetry could be recovered by invoking invariance under the combined *CP* operation in which charge conjugation and space inversion are applied together [50].

The unification of the theory of the weak and electromagnetic interactions into a single electroweak interaction theory [50] revealed that the absolute parity violation associated with the weak interactions could infiltrate to a tiny extent into all electromagnetic phenomena and hence into the world of atoms and molecules. This is brought about by a "weak neutral current" that generates, *inter alia*, the following parity-violating electron–nucleus contact interaction term (in atomic units) in the Hamiltonian of the atom or molecule [36, 51]:

$$V_{eN}^{PV} = \frac{G\alpha}{4\sqrt{2}} Q_W \{\sigma_e \cdot \mathbf{p}_e, \rho_N(\mathbf{r}_e)\} \tag{1.1}$$

where {} denotes an anticommutator, G is the Fermi weak coupling constant, α is the fine structure constant, σ_e and $\mathbf{p}_e$ are the Pauli spin operator and linear momentum operator of the electron, $\rho_N(\mathbf{r}_e)$ is a normalized nuclear density function and Q_W is an effective weak charge. Since σ_e and $\mathbf{p}_e$ are axial and polar vectors, respectively, and both are time odd, their scalar product $\sigma_e \cdot \mathbf{p}_e$ and hence V_{eN}^{PV} are time-even pseudoscalars.

One manifestation of parity violation in atomic physics is a tiny natural optical rotation in vapors of free atoms [52]. *CP* invariance means that the equal and opposite sense of optical rotation would be shown by the corresponding atoms composed of antiparticles. Chiral molecules support a unique manifestation of parity violation in the form of a lifting of the exact degeneracy of the energy levels of mirror-image enantiomers, known as the parity-violating energy difference (PVED). Although not yet observed experimentally using, for example, ultrahigh resolution spectroscopy, this PVED may be calculated [14, 36, 53]. Since, on account of the PVED, the P-enantiomers of a truly chiral object are not exactly degenerate (isoenergetic), they are not strict enantiomers (because the concept of enantiomers implies the exact opposites). So where is the strict enantiomer of a chiral object to be found? In the antiworld, of course: strict enantiomers are interconverted by CP! In other words, the molecule with the opposite absolute configuration but composed of antiparticles should have exactly the same energy as the original [2, 37], which means that a chiral molecule is associated with two distinct pairs of strict enantiomers (Figure 1.9).

Violation of time reversal was first observed by Christenson *et al.* in 1964 in decay modes of the neutral *K*-meson, the K^0 [50]. The effects are very small; nothing like the parity-violating effects in weak processes, which can sometimes be absolute. In fact, *T* violation itself was not observed directly: rather, the observations showed *CP* violation from which *T* violation was implied from the *CPT* theorem. Direct *T* violation was observed in 1998 in the form of slightly different rates, and hence a breakdown in microscopic reversibility, for the particle to antiparticle process $K^0 \to K^{0*}$ and the inverse $K^{0*} \to K^0$. Since a particle and its antiparticle have the same rest mass if *CPT*

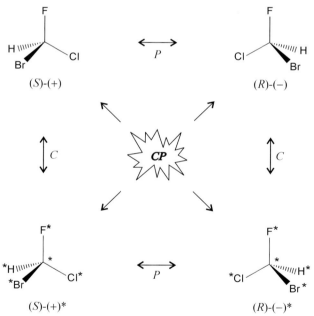

Figure 1.9 The two pairs of strict enantiomers (exactly degenerate) of a chiral molecule that are interconverted by CP. The structures with atoms marked by asterisks are antimolecules built from the antiparticle versions of the constituents of the original molecules. Adapted with corrections from Ref. [2] with permission.

invariance holds, *only the kinetics, but not the thermodynamics, are affected in CP- or T-violating process*. *CPT* invariance may also be used to show that the *CP*-enantiomers of a chiral molecule that appear in Figure 1.9 remain strictly degenerate even in the presence of *CP* violation [54]. Whether or not *CP* violation could have any direct manifestations in molecular physics is the subject of debate [54].

The concept that a spinning particle translating along the axis of spin possesses true chirality exposes a link between chirality and special relativity. Consider a particle with a right-handed chirality moving away from an observer. If the observer accelerates to a sufficiently high velocity that she starts to catch up with the particle, it will appear to be moving towards her and so takes on a left-handed chirality. The chirality of the particle vanishes in its rest frame. Only for massless particles such as photons and neutrinos is the chirality conserved since they always move at the velocity of light in any reference frame. This relativistic aspect of chirality is a central feature of elementary particle theory, especially in the weak interactions where the parity-violating aspects are velocity dependent [50].

1.3.5
Symmetry Violation *versus* Symmetry Breaking

The appearance of parity-violating phenomena is interpreted in quantum mechanics by saying that, contrary to what had been previously supposed, the Hamiltonian lacks inversion symmetry due to the presence of pseudoscalar terms such as the weak neutral current interaction. Such symmetry *violation*, sometimes called symmetry nonconservation, must be distinguished from symmetry *breaking* that applies when a system displays a lower symmetry than that of its Hamiltonian [2]. Natural optical activity, for example, is a phenomenon arising from parity (or mirror symmetry) breaking because a resolved chiral molecule displays a lower symmetry than its associated Hamiltonian: it lacks inversion symmetry (equivalent to mirror symmetry), whereas all the terms in the molecular Hamiltonian (ignoring tiny parity-violating terms) have inversion symmetry. It has been pointed out that the terms "chiral symmetry" and "chiral symmetry breaking," which are widely used to describe the appearance of chirality out of achiral precursors, are inappropriate because chirality is not a symmetry at all in molecular science [55, 56]. Rather, chirality is an attribute associated with special types of reduced spatial symmetry that enables an object to exist in two nonsuperposable mirror-image forms. "Mirror-symmetry breaking" is more correct. The term "chiral symmetry breaking" is, however, entirely appropriate in elementary particle physics, which requires relativistic quantum field theory within which chiral symmetry has a rigorous definition [57]. Chiral symmetry is an *internal* symmetry, rather than a geometrical symmetry, of massless particles, with mass associated with broken chiral symmetry.

In spontaneous resolutions such as that of sodium ammonium tartrate studied by Pasteur, mirror-symmetry breaking has not occurred at the bulk level because the sample remains optically inactive overall. However, bulk mirror-symmetry breaking can sometimes be induced to produce a large excess of one or other enantiomer. A famous example is the sodium chlorate ($NaClO_3$) system [58]. Solutions of this salt in water are optically inactive because the Na^+ and ClO_3^- ions into which it dissociates are achiral. $NaClO_3$ crystals, however, are chiral, but in the absence of perturbations a random distribution of the (+) and (−) enantiomeric crystals is obtained. Remarkably, when the evaporating $NaClO_3$ solution is stirred, mostly either (+) or (−) crystals are obtained; repeating the experiment many times gives equal numbers of (+) and (−) sets of crystals, as it must if parity is to be conserved. Chiral perturbations such as seeding with a small amount of the (+) or (−) crystals, or irradiation with energetic spin-polarized electrons (left-helical) or positrons (right-helical) from radioactive sources [59], can systematically induce bulk mirror-symmetry breaking in the form of a large excess of one or other of the chiral crystal forms. The formation of helical polymers with high ees via the sergeants-and-soldiers or majority rules phenomena (*vide supra*), and analogous phenomena in two dimensions in the context of the supramolecular assembly of molecules into homochiral domains on surfaces [60, 61], are further examples of mirror-symmetry breaking in the bulk induced by chiral perturbations.

1.3.6
Chirality in Two Dimensions

Since surfaces play an important role in nanoscience, a consideration of chirality in two dimensions is pertinent. This arises when there are two distinct enantiomers, confined to a plane or surface, that are interconverted by parity but not by any rotation within the plane about an axis perpendicular to the plane (symmetry operations out of the plane require an inaccessible third dimension). In two dimensions, however, the parity operation is no longer equivalent to an inversion through the coordinate origin as in three dimensions because this would not change the handedness of the two coordinate axes. Instead, an inversion of just one of the two axes is required [62]. For example, if the axes x, y are in the plane with z being perpendicular, then the parity operation could be taken as producing either $-x, y$ or $x, -y$, which are equivalent to mirror reflections across lines defined by the y- or x-axes, respectively. Hence, an object such as a scalene triangle (one with three sides of different length), which is achiral in three dimensions, becomes chiral in the two dimensions defined by the plane of the triangle because reflection across any line within the plane generates a triangle that cannot be superposed on the original by any rotation about the z-axis. Notice that a subsequent reflection across a second line, perpendicular to the first, generates a triangle superposable on the original, which demonstrates why an inversion of both axes, so that $x, y \rightarrow -x, -y$ is not acceptable as the parity operation in two dimensions.

Arnaut [63] has provided a generalization of the geometrical aspect of chirality to spaces of any dimensions. Essentially, an *N*-dimensional object is chiral in an *N*-dimensional space if it cannot be brought into congruence with its enantiomorph through a combination of translation and rotation within the *N*-dimensional space. As a consequence, an *N*-dimensional object with *N*-dimensional chirality loses its chirality in an *M*-dimensional space where $M > N$ because it can be rotated in the $(M - N)$-subspace onto its enantiomorph. Arnaut refers to chirality in one, two and three dimensions as axichirality, planochirality and chirality, respectively, and provides a detailed analysis of planochirality with examples such as a swastika, a logarithmic spiral and a jagged ring. He concludes that, for time-harmonic excitations, axichiral media have no significance, although the concept is significant for static and more general rectified fields. He also concludes that the notion of zero-dimensional chirality would be meaningless based on his view of chirality as a geometrical concept. However, these conclusions based on a strictly geometrical definition of chirality may need to be qualified if motion is an essential ingredient in the generation of the chirality.

One striking optical manifestation of planochirality is a large circular intensity difference in second-harmonic light scattering from chiral molecules on an isotropic surface [64]. Because the mechanism involves pure electric dipole interactions, the effect is three orders of magnitude larger than analogous phenomena observed in the bulk since the latter require interference between electric dipole and magnetic dipole interactions [2]. Other manifestations include rotation of the plane of polarization in light refracted from [65], and transmitted through [66], the surface of artificial chiral planar gratings based on swastika-like chiral surface nanostructures.

The concept of false chirality arises in two dimensions as well as in three. For example, the sense of a spinning electron on a surface with its axis of spin perpendicular to the surface is reversed under the two-dimensional parity operation (unlike in three dimensions). Because electrons with opposite spin sense are nonsuperposable in the plane, a spinning electron on a surface would seem to be chiral. However, the apparent chirality is false because the sense of spin is also reversed by time reversal. The enantiomorphism is therefore time-noninvariant, the system being invariant under the combined PT operation but not under P and T separately.

1.4
Absolute Enantioselection

The use of an external physical influence to produce an ee in what would otherwise be a racemic product in a chemical reaction is known as an absolute asymmetric synthesis. The production of an ee in more general situations is often referred to as absolute enantioselection or physical chiral induction. The subject still attracts much interest and controversy [12, 13, 15, 67]. The considerations of Section 1.3.3 above provide a sound foundation for the critical assessment of physical influences capable of inducing ees, however small.

1.4.1
Truly Chiral Influences

If an influence is classified as truly chiral it has the correct symmetry characteristics to induce absolute asymmetric synthesis, or some related process such as preferential asymmetric decomposition, in any conceivable situation, although of course the influence might be too weak to produce an observable effect. In this respect it is important to remember Jaeger's dictum [68]: "The necessary conditions will be that the externally applied forces are a *conditio sine qua non* for the initiation of the reaction which would be impossible without them."

The ability of a truly chiral influence to induce absolute asymmetric synthesis in a reaction process at equilibrium may be illustrated by a simple symmetry argument applied to the following unimolecular process

$$M \underset{k_b}{\overset{k_f}{\rightleftarrows}} R \underset{k_f^*}{\overset{k_b^*}{\rightleftarrows}} M^*$$

in which an achiral molecule R generates a chiral molecule M or its enantiomer M* and the ks are appropriate rate constants. In the absence of a chiral influence, M and M* have the same energy, so no ee can exist if the reaction reaches thermodynamic equilibrium. Consider a collection of single enantiomers M in the presence of a right-handed chiral influence $(Ch)_R$, say. Under parity P, the collection of enantiomers M becomes an equivalent collection of mirror-image enantiomers M* and the right-handed chiral influence $(Ch)_R$ becomes the equivalent left-handed chiral influence $(Ch)_L$. Assuming parity is conserved, this indicates that the energy of M in the presence of $(Ch)_R$ is equal to that of M* in the presence of $(Ch)_L$. But because parity (or any other symmetry operation) does not provide a relation between the energy of M and M* in the presence of the same influence, be it $(Ch)_R$ or $(Ch)_L$, they will in general have different energies. Hence, an ee can now exist at equilibrium (due to different Boltzmann populations of M and M*). There will also be kinetic effects because the enantiomeric transition states will also have different energies.

Circularly polarized photons, or longitudinal spin-polarized electrons associated with radioactive β-decay, are obvious examples of truly chiral influences, and their ability to induce absolute enantioselection has been demonstrated in a number of cases [12, 13, 15, 67]. Photochemistry with circularly polarized light is especially favorable because it conforms to Jaeger's dictum above. This photochemistry can occur by photoequilibration of a racemic mixture of molecules, or by selective destruction of one enantiomer over the other. A recent and impressive example, with important implications for astrobiology, was the use of intense circularly polarized synchrotron radiation in the vacuum ultraviolet to induce significant ees in racemic amino acids in the solid state via enantioselective photodecomposition, which models a realistic situation relevant to organic molecules in interstellar or circumstellar dust grains [69].

Vortex motion constitutes a truly chiral influence since it combines rotation with translation perpendicular to the rotation plane. There has been considerable interest in the possibility that vortex motion in a conical swirl might be exploited to induce absolute enantioselection, but until recently no convincing example has been demonstrated experimentally [67]. Then -several years ago reports appeared of mirror-symmetry breaking in homoassociation of achiral diprotonated porphyrins where helical conformations were generated by stirring in a rotary evaporator with the sense of chirality, detected by circular dichroism, being selected by the sense of stirring [70, 71]. In a later report, the same group claimed to have achieved similar results through magnetic stirring in a small tube, and went on to provide an explanation in terms of hydrodynamic effects of the vortex at the walls of the container [72]. Another recent result illustrates the complexity of such processes and the importance of the conditions. Thus it was found that vortexing an insulin solution at room temperature generated two distinct types of amyloid fibrils with opposite local chiral preferences, the dominance of one or other type of fibrils in a test tube being only stochastically determined; whereas vortexing at 60 °C always generated the same chiral form, presumably under the influence of the chiral bias of the exclusively L-amino acids in the protein [73]. Vortexing in the opposite sense made no difference to these results (W. Dzwolak, private communication). A further recent and highly relevant observation in this context concerns filamentous bacterial viruses: several types form cholesteric liquid crystals under the influence of their chiral protein and DNA constituents, while others form nematic liquid crystals that are apparently "oblivious" to the chirality of their molecular components [74].

Although a magnetic field alone has no chirality and so cannot induce absolute enantioselection, we have seen that a static magnetic field collinear with a light beam of arbitrary polarization (Figure 1.8) is a truly chiral system and hence can induce absolute enantioselection in all circumstances. This has been demonstrated experimentally in the form of small ees observed in an initially racemic solution of a chiral transition-metal complex in the presence of a static magnetic field collinear with an unpolarized light beam at photochemical equilibrium [75].

Being a time-even pseudoscalar, the weak neutral current interaction V_{eN}^{PV} responsible for the tiny PVED is the quintessential truly chiral influence in atomic and molecular physics. It lifts only the degeneracy of the space-inverted (P-) enantiomers of a truly chiral system; the P-enantiomers of a falsely chiral system such as a nontranslating rotating cone remain strictly degenerate. It has attracted considerable discussion as a possible source of biological homochirality [11, 14, 15, 36, 53, 76]. However, it is still not clear whether or not the PVED preferentially stabilizes the naturally occurring L-amino acids and D-sugars. Measurable differences reported in the physical properties of crystals of D- and L-amino acids and claimed to be due to parity violation have not been corroborated [77]: they have been shown instead to arise from traces of different impurities in the enantiomorphous crystals [78]. So far there is no convincing evidence that the PVED itself has any enantioselective influence on the crystallization of sodium chlorate (*vide supra*) or on that of any other system [30, 59].

1.4.2
Falsely Chiral Influences

It is important to appreciate that, unlike the case of a truly chiral influence, enantiomers M and M* remain strictly isoenergetic in the presence of a falsely chiral influence such as collinear electric and magnetic fields. Again this can be seen from a simple symmetry argument applied to the unimolecular reaction above. Under *P*, the collection of enantiomers M becomes the collection M* and the parallel arrangement, say, of **E** and **B** becomes antiparallel. The antiparallel arrangement of **E** and **B**, however, becomes parallel again under T; but these last two operations will have no affect on an isotropic collection of chiral molecules, even if paramagnetic. Hence, the energy of the collection M is the same as that of the collection M* in parallel (or antiparallel) electric and magnetic fields.

When considering the possibility or otherwise of absolute asymmetric synthesis being induced by a falsely chiral influence, a distinction must be made between reactions that have been left to reach thermodynamic equilibrium (*thermodynamic control*) and reactions that have not attained equilibrium (*kinetic control*). The case of thermodynamic control is quite clear: because M and M* remain strictly isoenergetic in the presence of a falsely chiral influence, such an influence cannot induce absolute asymmetric synthesis in a reaction that has been allowed to reach thermodynamic equilibrium. The case of kinetic control is more subtle. It has been suggested that processes involving chiral molecules in the presence of a falsely chiral influence such as collinear **E** and **B** may exhibit a breakdown of conventional microscopic reversibility, but preserve a new and deeper principal of *enantiomeric* microscopic reversibility [79]. Since only the kinetics, but not the thermodynamics, of the process are affected, this suggests an analogy with the breakdown in microscopic reversibility associated with CP- and T-violation in particle–antiparticle processes [37, 54, 79]. The force responsible for CP violation may be conceptualized as the quintessential falsely chiral influence in particle physics, being characterized by lack of CP and T invariance separately but possessing CPT invariance overall. This is analogous to a falsely chiral influence in the molecular case, which is characterized by a lack of P and T invariance separately but possessing PT invariance overall.

Since one effect of **E** in a falsely chiral influence such as collinear **E** and **B** is to partially align dipolar molecules [79], it is not required if the molecules are already aligned. Hence, a magnetic field alone might induce absolute enantioselection if the molecules are prealigned, as in a crystal or on a surface, and the process is far from equilibrium [80]. However, to date there has been no unequivocal demonstration of absolute enantioselection induced by this or any other falsely chiral influence [67].

1.5
Spectroscopic Probes of Chirality in Nanosystems

In order to detect chirality in molecular systems, a spectroscopic probe must be sensitive to absolute handedness. This usually means that it must exploit in some way the intrinsic chirality of circularly polarized light. The power of chiroptical spectroscopic techniques for applications to chiral macromolecules and supramolecular structures in general derives in part from their ability to cut through the complexity of conventional spectra (which are "blind" to chirality) to reveal three-dimensional information about the most rigid, twisted chiral parts of the structure, within the backbone in polymers, for example, since these often generate the largest chiroptical signals. Although chiroptical methods do not provide structures at atomic resolution like X-ray crystal and fiber diffraction, and multidimensional NMR, they are usually easier to apply and a much wider range of samples are accessible. Furthermore, the level of analysis is improving rapidly thanks to current progress in computational chemistry, particularly for the newer vibrational optical activity techniques.

1.5.1
Electronic Optical Activity

To date the most widely used chiroptical spectroscopies to study chiral nanosystems in solution are optical rotatory dispersion (ORD) and circular dichroism (CD), which originate in differential refraction and absorption, respectively, of right- and left-circularly polarized light. Closely related to CD are circular polarization of luminescence, and fluorescence-detected circular dichroism. The principles and applications of CD and other chiroptical spectroscopies, including some applications to supramolecular systems, are reviewed in Refs. [81, 82]. The observable in CD spectroscopy is the following *rotational strength* of the $j \leftarrow n$ electronic transition, which may be related to the area and sign of a corresponding CD spectral band [2, 81, 82]:

$$R(j \leftarrow n) = \text{Im}(\langle n|\mu|j\rangle \cdot \langle j|\mathbf{m}|n\rangle) \tag{1.2}$$

where μ and **m** are the electric and magnetic dipole moment operators, respectively. Since μ and **m** are time-even and time-odd polar and axial vectors, respectively, the imaginary part of their scalar product is a time-even pseudoscalar [2], as befits a chiral observable. Attempts to theoretically simulate observed UV-visible CD spectra focus on quantum-chemical calculations of the rotational strength (1.2) for the corresponding electronic transitions [81, 82]. At present, their use remains significantly limited by the molecular size, conformational flexibility, and difficulties in obtaining sufficiently accurate and reliable descriptions of the corresponding excited electronic states [81–83].

CD is immensely useful in nanoscience, but mainly in its qualitative aspects for monitoring conformation changes, especially inversion of chirality via reversal of CD signals. Theoretical simulations of the rotational strength (1.2) that are sufficiently accurate to provide information about absolute helical sense and conformational parameters from the experimental CD spectra of systems such as helical polyisocyanates [25] and β-peptides [84] have so far proved elusive.

One particularly valuable application of CD in supramolecular chemistry involves the observation of spectral signals arising from different types of intermolecular interactions. Four typical situations are encountered [83]. A chiral "guest" and an achiral chromophoric "host" compound such as a calixarene or a bisporphyrin can form a complex that exhibits an induced CD within the absorption bands of the host. Conversely, a small achiral chromophoric guest compound bound to a chiral host such as a cyclodextrin or an oligonucleotide may show a CD induced by the chiral host. Also, coupling between several achiral guest molecules such as carotenoids bound to different sites of a chiral macromolecular host such as serum albumin may show a diagnostic CD spectrum due to exciton coupling if the guests are held in a chiral orientation relative to each other. Finally, a chiral nonchromophoric ligand may bind to a metal ion with observable *d*- or *f*-type transitions making them CD-active.

1.5.2
Vibrational Optical Activity

ORD and CD at visible and ultraviolet wavelengths measure natural optical activity in the *electronic* spectrum. It had long been appreciated that extending natural optical activity into the *vibrational* spectrum could provide more detailed and reliable stereochemical information because a vibrational spectrum contains many more bands sensitive to the details of the molecular structure ($3N-6$ fundamentals, where N is the number of atoms) [2]. This was finally achieved in the early 1970s when vibrational optical activity was first observed in small chiral molecules in fluid media using two complementary techniques: a circular polarization dependence of vibrational Raman scattering of visible laser light [85], and circular dichroism of infrared radiation [86]. These are now known as Raman optical activity (ROA) and vibrational circular dichroism (VCD), respectively [2, 81, 82].

Vibrational optical activity is especially powerful for determining the absolute configuration together with conformational details, including relative populations, of smaller chiral molecules by means of *ab initio* quantum-chemical simulations of VCD and ROA spectra [87–89]. These are generally more reliable than the corresponding electronic CD calculations, one reason being that the calculations involve molecules in their ground electronic states that are usually well defined. Attempts to simulate VCD spectra focus on calculations of the rotational strength (1.2), with $j \leftarrow n$ now a fundamental vibrational transition, for all $3N-6$ normal modes of vibration. ROA simulations focus on calculations of products such as $\langle n|\alpha_{\alpha\beta}|j\rangle\langle j|G'_{\alpha\beta}|n\rangle$ and $\langle n|\alpha_{\alpha\beta}|j\rangle\varepsilon_{\alpha\gamma\delta}\langle j|A_{\gamma\delta\beta}|n\rangle$, which determine the intensity of an ROA band for the $j \leftarrow n$ fundamental vibrational transition, where $\alpha_{\alpha\beta}$ is an operator corresponding to the electric dipole–electric dipole polarizability tensor and $G'_{\alpha\beta}$ and $A_{\alpha\beta\gamma}$ are electric dipole–magnetic dipole and electric dipole–electric quadrupole optical activity tensor operators, respectively [2]. Like the rotational strength, these ROA intensity terms are again time-even pseudoscalars and so are only supported by chiral molecules.

The recent determination of the solution structure of a supramolecular tetramer of a chiral dimethyl-biphenyl-dicarboxylic acid [90] and of aromatic foldamers [91] from *ab initio* simulations of the observed VCD spectra provide examples of what can currently be achieved. Oligo- and polypeptides, including synthetic β-peptides, in model conformations are also becoming accessible to *ab initio* VCD and ROA simulations [92–94]. All this suggests that, although applications of ROA and VCD in nanoscience are still in their infancy, they appear to have significant potential for characterizing the absolute handedness and conformational details of synthetic chiral macromolecules and supramolecular structures in solution by means of theoretical simulations of observed spectra. ROA is especially promising in this respect because there appears to be no upper size limit to the structures that may be studied: even intact viruses are accessible to ROA measurements from which, *inter alia*, information such as coat protein folds and nucleic acid structure may be deduced [95].

1.6
Conclusion

Chirality is already a burgeoning topic in nanoscience and is expected to stimulate further new and fruitful developments. The homochirality of biological macromolecules and their many diverse structural and functional roles provide important lessons for chiral nanoscience, which is also developing its own themes with regard to synthetic supramolecular systems and nonaqueous solvents not encountered in the chemistry of life. Applications of chiroptical spectroscopies in nanoscience will continue to grow in importance, especially the vibrational optical activity techniques of ROA and VCD. It is hoped that the more fundamental aspects elaborated in this chapter will prove useful for understanding the generation, characterization and functional role of chirality in molecular systems and thereby facilitate the exploitation of its potential in nanoscience.

References

1 Wagniére, G.H. (2007) *On Chirality and the Universal Asymmetry*, Verlag Helvetica Chimica Acta, Zürich, and Wiley-VCH, Weinheim.
2 Barron, L.D. (2004) *Molecular Light Scattering and Optical Activity*, 2nd edn, Cambridge University Press, Cambridge.
3 Cornelissen, J.J.L.M., Rowan, A.E., Nolte, R.J.M. and Sommerdijk, N.A.J.M. (2001) *Chemical Reviews*, **101**, 4039–4070.
4 Green, M.M., Nolte, R.J.M. and Meijer, E.W. (eds) (2003) *Topics in Stereochemistry, Vol. 24, Materials-Chirality*, Wiley-VCH, New York.
5 Zhang, J., Albelda, M.T., Liu, Y. and Canary, J.W. (2005) *Chirality*, **17**, 404–420.
6 Lowry, T.M. (1935) *Optical Rotatory Power*, Longmans, Green, London, reprinted by Dover, New York 1964.
7 Eliel, E.L. and Wilen, S.H. (1994) *Stereochemistry of Organic Compounds*, John Wiley & Sons, New York.
8 Gal, J. (2007) *Chirality*, **19**, 89–98.
9 Kelvin, Lord. (1904) *Baltimore Lectures*, C. J. Clay & Sons, London.
10 Barron, L.D. (1986) *Journal of the American Chemical Society*, **108**, 5539–5542.
11 Mason, S.F. (1988) *Chemical Society Reviews*, **17**, 347–359.
12 Bonner, W.A. (1988) *Topics in Stereochemistry*, **18**, 1–96.
13 Feringa, B.L. and van Delden, R.A. (1999) *Angewandte Chemie-International Edition*, **38**, 3418–3438.
14 MacDermott, A.J. (2002) *Chirality in Natural and Applied Science* (eds. Lough, W.J. and Wainer, I.W.), Blackwell Publishing, Oxford, pp. 23–52.
15 Compton, R.N. and Pagni, R.M. (2002) *Advances in Atomic, Molecular, and Optical Physics*, **48**, 219–261.
16 Behr, J-.P.(ed.) (1994) *The Lock and Key Principle*, John Wiley & Sons, New York.
17 Fujii, N. (2002) *Origins Life*, **32**, 103–127.
18 Konno, R., Brückner, H., D'Aniello, A., Fisher, G.H., Fujii, N. and Homma, H.(eds) (2007) *D-Amino Acids: A New Frontier in Amino Acid and Protein Research-Practical Methods and Protocols*, Nova Science Publishers, New York.
19 Westhof, E.(ed.) (1993) *Water and Biological Macromolecules*, CRC Press, Boca Raton.
20 Ball, P. (2005) *Nature*, **436**, 1084–1085.
21 Sandars, P.G.H. (2003) *Origins Life*, **33**, 575–587.
22 Joyce, G.F., Visser, G.M., van Boeckel, C.A.A., van Boom, J.H., Orgel, L.E. and van Westresen, J. (1984) *Nature*, **310**, 602–604.

23 Avetisov, V.A., Goldanskii, V.I. and Kuz'min, V.V. (1991) *Physics Today*, **44**, 33–41.
24 Green, M.M., Peterson, N.C., Sato, T., Teramoto, A., Cook, R. and Lifson, S. (1995) *Science*, **268**, 1860–1866.
25 Green, M.M. (2000) *Circular Dichroism: Principles and Applications* (eds. Berova, N., Nakanishi, K. and Woody, R.W.), Wiley-VCH, New York, pp. 491–520.
26 Klussman, M., Iwamura, H., Mathew, S.P., Wells, D.H., Jr, Pandya, U., Armstrong, A. and Blackmond, D.G. (2006) *Nature*, **441**, 621–623.
27 Plasson, R., Kondepudi, D.K., Bersini, H., Commeyras, A. and Asakura, K. (2007) *Chirality*, **19**, 589–600.
28 Perry, R.H., Wu, C., Nefliu, M. and Cooks, R.G. (2007) *Chemical Communications*, 1071.
29 Fletcher, S.P., Jagt, R.B.C. and Feringa, B.L. (2007) *Chemical Communications*, 2578–2580.
30 Avalos, M., Babiano, R., Cintas, P., Jiménez, J.L. and Palacios, J.C. (2004) *Origins Life*, **34**, 391–405.
31 Cintas, P. (2002) *Angewandte Chemie-International Edition*, **41**, 1139–1145.
32 Rubinstein, I., Eliash, R., Bolbach, G., Weissbuch, I. and Lahav, M. (2007) *Angewandte Chemie-International Edition*, **46**, 3710–3713.
33 Pasteur, L. (1884) *Bulletin de la Société chimique de France*, **41**, 219.
34 Curie, P. (1894) *Journal of Physiology, Paris (3)*, **3**, 393.
35 Barron, L.D. (2000) *Nature*, **405**, 895–896.
36 Wesendrup, R., Laerdahl, J.K., Compton, R.N. and Schwerdtfeger, P. (2003) *Journal of Physical Chemistry A*, **107**, 6668–6673.
37 Barron, L.D. (2002) *Chirality in Natural and Applied Science* (eds. Lough, W.J. and Wainer, I.W.), Blackwell Publishing, Oxford, pp. 53–86.
38 Berestetskii, V.B., Lifshitz, E.M. and Pitaevskii, L.P. (1982) *Quantum Electrodynamics*, Pergamon Press, Oxford.
39 Mislow, K. (1999) *Topics in Stereochemistry*, **22**, 1–82.
40 Zocher, H. and Török, C. (1953) *Proceedings of the National Academy of Sciences of the United States of America*, **39**, 681–686.
41 Tellegen, B.D.H. (1948) *Philips Research Reports*, **3**, 81–101.
42 Lindell, I.V., Sihvola, A.H., Tretyakov, S.A. and Viitanen, A.J. (1994) *Electromagnetic Waves in Chiral and Bi-Isotropic Media*, Artech House, Boston.
43 Weiglhofer, W.S. and Lakhtakia, A. (1998) *AEU-International Journal of Electronics and Communications*, **52**, 276–279.
44 Ghosh, A., Sheridon, N.K. and Fischer, P. (2008) *Small*, **4**, 1956–1958.
45 Wagnière, G.H. and Meir, A. (1982) *Chemical Physics Letters*, **93**, 78–81.
46 Kleindienst, P. and Wagnière, G. (1998) *Chemical Physics Letters*, **288**, 89–97.
47 Rikken, G.L.J.A. and Raupach, E. (1997) *Nature*, **390**, 493–494.
48 Rikken, G.L.J.A., Fölling, J. and Wyder, P. (2001) *Physical Review Letters*, **87**, art. No. 236602.
49 Krstić, V., Roth, S., Burghard, M., Kern, K. and Rikken, G.L.J.A. (2002) *Journal of Chemical Physics*, **117**, 11315–11319.
50 Gottfried, K. and Weisskopf, V.F. (1984) *Concepts of Particle Physics*, **Vol. 1**, Clarendon Press, Oxford.
51 Hegstrom, R.A., Rein, D.W. and Sandars, P.G.H. (1980) *Journal of Chemical Physics*, **73**, 2329–2341.
52 Bouchiat, M.A. and Bouchiat, C. (1997) *Reports on Progress in Physics*, **60**, 1351–1396.
53 Quack, M. (2002) *Angewandte Chemie-International Edition*, **41**, 4618–4630.
54 Barron, L.D. (1994) *Chemical Physics Letters*, **221**, 311–316.
55 Walba, D.M. (2003) *Topics in Stereochemistry*, **24**, 457–518.
56 Avalos, M., Babiano, R., Cintas, P., Jiménez, J.L. and Palacios, J.C. (2004) *Tetrahedron-Asymmetry*, **15**, 3171–3175.
57 Maggiore, M. (2005) *A Modern Introduction to Quantum Field Theory*, Cambridge University Press, Cambridge.
58 Kondepudi, D.K., Kaufman, R. and Singh, N. (1990) *Science*, **250**, 975–976.

59 Pagni, R.M. and Compton, R.N. (2002) *Crystal Growth and Design*, **2**, 249–253.
60 Humbolt, V., Barlow, S.M. and Raval, R. (2004) *Progress in Surface Science*, **76**, 1–19.
61 Ernst, K.-H. (2006) *Topics in Current Chemistry*, **265**, 209–252.
62 Halperin, B.I., March-Russell, J. and Wilczek, F. (1989) *Physical Review*, **B40**, 8726–8744.
63 Arnaut, L.R. (1997) *Journal of Electromagnetic Waves and Applications*, **11**, 1459–1482.
64 Hicks, J.M., Petralli-Mallow, T. and Byers, J.D. (1994) *Faraday Discussions*, **99**, 341–357.
65 Papakostas, A., Potts, A., Bagnall, D.M., Prosvirnin, S.L., Coles, H.J. and Zheludev, N.I. (2003) *Physical Review Letters*, **90**, art. no. 107404.
66 Kuwata-Gonokami, M., Saito, N., Ino, Y., Kauranen, M., Jefimovs, K., Vallius, T., Turunen, J. and Svirko, Y. (2005) *Physical Review Letters*, **95**, art. No. 227401.
67 Avalos, M., Babiano, R., Cintas, P., Jiménez, J.L., Palacios, J.C. and Barron, L.D. (1988) *Chemical Reviews*, **98**, 2391–2404.
68 Jaeger, F.M. (1930) *Optical Activity and High-Temperature Measurements*, McGraw-Hill, New York.
69 Meierhenrich, U.J., Nahon, L., Alcarez, C., Bredehöft, J.H., Hoffman, S.V., Barbier, B. and Brack, A. (2005) *Angewandte Chemie-International Edition*, **44**, 5630–5634.
70 Ribó, J.M., Crusats, J., Sagués, F., Claret, J. and Rubires, R. (2001) *Science*, **292**, 2063–2066.
71 Rubires, R., Farrera, J.-A. and Ribó, J.M. (2001) *Chemistry – A European Journal*, **7**, 436–446.
72 Escudero, C., Crusats, J., Díez-Pérez, I., El-Hachemi, Z. and Ríbo, J.M. (2006) *Angewandte Chemie-International Edition*, **45**, 8032–8035.
73 Dzwolak, W., Loksztejn, A., Galinska-Rakoczy, A., Adachi, R., Goto, Y. and Rupnicki, L. (2007) *Journal of the American Chemical Society*, **129**, 7517–7522.
74 Tomar, S., Green, M.M. and Day, L.A. (2007) *Journal of the American Chemical Society*, **129**, 3367–3375.
75 Rikken, G.L.J.A. and Raupach, E. (2000) *Nature*, **405**, 932–935.
76 Deamer, D.W., Dick, R., Thiemann, W. and Shinitzky, M. (2007) *Chirality*, **19**, 751–763.
77 Sullivan, R., Pyda, M., Pak, J., Wunderlich, B., Thompson, J.R., Pagni, R., Pan, H., Barnes, C., Schwerdtfeger, P. and Compton, R.N. (2003) *Journal of Physical Chemistry A*, **107**, 6674–6680.
78 Lahav, M., Weissbuch, I., Shavit, E., Reiner, C., Nicholson, G.J. and Schurig, V. (2006) *Origins Life*, **36**, 151–170.
79 Barron, L.D. (1987) *Chemical Physics Letters*, **135**, 1–8.
80 Barron, L.D. (1994) *Science*, **266**, 1491–1492.
81 Berova, N., Nakanishi, K. and Woody, R.W.(eds) (2000) *Circular Dichroism: Principles and Applications*, 2nd edn, Wiley-VCH, New York.
82 Urbanová, M. andMaloň, P.(2007) *Analytical Methods in Supramolecular Chemistry* (ed. Schalley, C.A.), Wiley-VCH, New York, pp. 265–304.
83 Berova, N., Di Bari, L. and Pescitelli, G. (2007) *Chemical Society Reviews*, **36**, 914–931.
84 Glättli, A., Daura, X., Seebach, D. and van Gunsteren, W.F. (2002) *Journal of the American Chemical Society*, **124**, 12972–12978.
85 Barron, L.D., Bogaard, M.P. and Buckingham, A.D. (1973) *Journal of the American Chemical Society*, **95**, 603–605.
86 Holzwarth, G., Hsu, E.C., Mosher, H.S., Faulkner, T.R. and Moscowitz, A.J. (1974) *Journal of the American Chemical Society*, **96**, 251–252.
87 Stephens, P.J. and Devlin, F.J. (2000) *Chirality*, **12**, 172–179.
88 Haesler, J., Schindelholz, I., Riguet, E., Bochet, C.G. and Hug, W. (2007) *Nature*, **446**, 526–529.
89 Macleod, N.A., Johannessen, C., Hecht, L., Barron, L.D. and Simons, J.P. (2006)

International Journal of Mass Spectrometry, **253**, 193–200.

90 Urbanová, M., Setnička, V., Devlin, F.J. and Stephens, P.J. (2005) *Journal of the American Chemical Society*, **127**, 6700–6711.

91 Ducasse, L., Castet, F., Fritsch, A., Huc, I. and Buffeteau, T. (2007) *Journal of Physical Chemistry A*, **111**, 5092–5098.

92 Keiderling, T.A. (2002) *Current Opinion in Chemical Biology*, **6**, 682–688.

93 Barron, L.D. (2006) *Current Opinion in Structural Biology*, **16**, 638–643.

94 Kapitán, J., Zhu, F., Hecht, L., Gardiner, J., Seebach, D., and Barron, L.D. (2008) *Angewandte Chemie – International Edition*, **47**, 6392–6994.

95 Blanch, E.W., Hecht, L., Syme, C.D., Volpetti, V., Lomonossoff, G.P., Nielsen, K. and Barron, L.D. (2002) *The Journal of General Virology*, **83**, 2593–2600.

2
Optically Active Supramolecules
Alessandro Scarso and Giuseppe Borsato

2.1
Introduction to Supramolecular Stereochemistry

Most scientists know the concept of chirality as the property of an object to be nonsuperimposable on its mirror image, but not all of them know that the different odor of lemon and orange arise due to the presence of (S) and (R)-limonene enantiomers, respectively [1]. Such simple but highly explicative observation reminds us that our nose can clearly distinguish between these two molecules that differ only in the stereochemical distribution of the atoms. Almost all living systems are intrinsically chiral and enantiomerically pure because they are all made up of chiral enantiopure building blocks like D-saccharides, D-nucleotides and L-amino acids and such chirality implemented in the biopolymers governs stereoselective recognition of chiral external molecular stimuli – such as the limonene molecules – by means of weak intermolecular forces. If we consider the enormous amount of different chiral natural and artificially synthesized molecules we come into contact with during our life, we understand the pivotal importance of unraveling chiral recognition phenomena. It is a natural consequence that international as well as national organizations like the W.H.O. and the F.D.A., that are devoted to monitoring and maintaining of citizens' health, impart stringent requirements on the commercialization of present and new chiral drugs only after they have been tested and produced as single enantiomers [2].

Molecules are chemical entities that can interact with each other by weak intermolecular forces with association lifetimes that span from $<10^{-9}$ s for statistical encounters to days for elaborate assemblies depending on the number, orientation, strength and nature of the weak interactions. The formation of intermolecular aggregates is known as "supramolecular chemistry" with the term supramolecular referring to "the chemistry beyond the molecules" [3]. This concept is intended as the bottom-up approach that allows nanofabrication of a plethora of thermodynamically controlled functional molecular devices made of several units characterized by different geometries, symmetries and forms displaying particular features like reversibility, self-correction and self-recognition, widening the borders of practical chemical structures [4].

Chirality at the Nanoscale: Nanoparticles, Surfaces, Materials and more. Edited by David B. Amabilino
Copyright © 2009 WILEY-VCH Verlag GmbH & Co. KGaA, Weinheim
ISBN: 978-3-527-32013-4

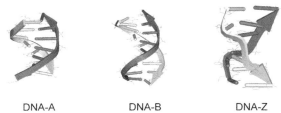

DNA-A   DNA-B   DNA-Z

Figure 2.1 Examples of natural assemblies featuring supramolecular chirality: right-handed double helix of DNA-A and B and left-handed for DNA-Z arising from D-nucleotides as a function of the different base pair sequence, salt concentration, or presence of certain cations.

Chirality is a feature that belongs generally to objects, it is observed in supramolecules as well as in common molecules, with the important difference that while the latter can be chiral only if they do not have symmetry planes, for supramolecular aggregates the resulting assembly can be chiral even if each subunit is intrinsically achiral. This is an important observation that clearly speaks for the importance of studying chirality of self-assembled aggregates. Nature is pervaded by examples of supramolecular chiral structures like helices, ribbons, polyhedral structures and many others [5], where the aggregate is chiral and where the intrinsic chirality implemented in the biopolymeric subunits steer the self-assembled structure to assume, in many cases, highly symmetric [6] enantiomerically pure supramolecules. Examples of such chiral control is evident in double helices of DNA where D nucleotides induce only a right-handed double helix in DNA-A and DNA-B, while left-handed in DNA-Z as a function of the sequence of base pairs, salt concentration, or the presence of certain cations (Figure 2.1).

Supramolecular chirality arises when finite molecular noncovalent assemblies are present in a nonsymmetrical arrangement that exists as two nonsuperimposable mirror images (Figure 2.2) and it is distinguished from "chiral molecular recognition" that refers to one to one host–guest chiral interaction [7].

The topic of supramolecular chirality applied to artificial assemblies in the general sense has been reviewed recently [8–12] and the general phenomenon has been classified into two principal domains as a function of the chiral or achiral nature of the constituent subunits. The first phenomenon is called "noncovalent diastereoselective synthesis" [11] and is observed when the supramolecular aggregate is produced with a certain degree of diastereoselectivity starting from building blocks comprising also chiral moieties and that, in some cases, are characterized by enantioselective self-recognition [13] leading preferentially to *meso* or chiral racemic supramolecules. Conversely, a second class of supramolecular chirality called "noncovalent enantioselective synthesis" [11] arises when the supramolecular aggregate is obtained starting from achiral building blocks and can be produced in an enantiomerically enriched form via the chiral memory effect [9, 14]. The present contribution is aimed at focusing the attention of the reader to the latter kind of chiral assemblies because it represents the ultimate degree of sophistication in supramolecular chirality. In Figure 2.3 are depicted schematic examples of the different approaches applied to a model supra-

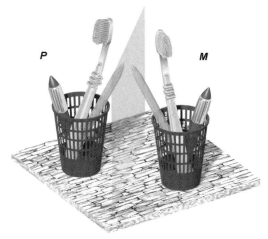

Figure 2.2 Example of a macroscopic chiral racemic object obtained by achiral units: a pen, a marking pen and a toothbrush in a penholder can assume two enantiomeric distributions even if each singular item is achiral.

molecule (A). In the "diastereoselective noncovalent synthesis" an enantiopure chiral residue influences the stereochemical outcome of the supramolecular assembly favoring one diastereoisomer over the other (B), while in the "enantioselective noncovalent synthesis" the chiral external influence to induce the chirality of the supramolecular aggregate favoring one enantiomer by means of weak interactions and, once removed, the supramolecule slowly returns to a racemic mixture (C).

Hydrogen-bonded and metal–ligand coordinated chiral assemblies are covered in the present work as these intermolecular interactions are the most efficient in terms of directionality and binding energy. Emphasis is placed on self-assembled structures characterized by closed topography thus behaving as capsular hosts where a cavity is present and suitable for interaction with neutral or charged molecular guests. Encapsulation is a unique supramolecular phenomenon that markedly influences the behavior of trapped guests giving rise to new forms of isomerism, reactivity, and topology that has reached a high degree of sophistication [15] and that is intrinsically oriented to chiral molecular recognition with the ultimate purpose of Nature mimesis. Moreover, the employment of achiral self-assembling components for the formation of chiral supramolecular structures is a strategy that is much less costly in terms of time required for the synthesis and purification of the components and can positively take advantage of combinatorial and dynamic effects [16].

2.1.1
Survey of Weak Intermolecular Attractive Forces

While the amount of energy involved in a typical covalent chemical bond lies in the range 170–450 kJ/mol, weak intermolecular interactions are typically characterized

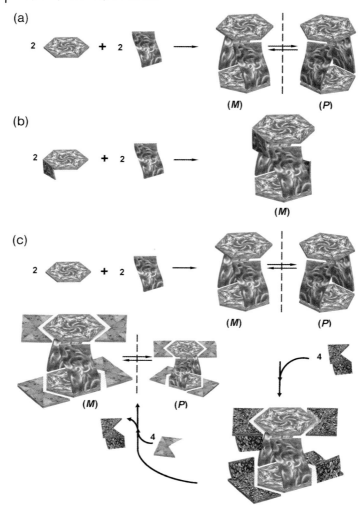

Figure 2.3 (a) The assembly of the four molecular units leads to the formation of a chiral supramolecular aggregate as a racemic mixture of M and P enantiomers; (b) if one of the units is intrinsically chiral, such chirality influences the supramolecular assembly favoring one chiral arrangement over the other, namely supramolecular diastereoselective synthesis; (c) The chiral racemic supramolecular aggregate can interact by means of intermolecular weak forces with a chiral influence that steers the supramolecule towards one stereoisomer over the other. If the aggregate is sufficiently stable, it is possible to displace the chiral influence with an achiral analogue thus leading to an enantioenriched supramolecule made of achiral units. Such a species is not indefinitely stable and slowly racemize.

by much lower energies [17]: in general, the enthalpy involved in intermolecular attractive forces are from one to two orders of magnitude weaker. Intermolecular forces can be classified into several types of interactions that span from the more common hydrogen bonding (8–80 kJ/mol) [18], metal–ligand coordination

(40–130 kJ/mol) [15f, 19] and ion pairing (2–10 kJ/mol) [20], to the much weaker in terms of energy of binding van der Waals dispersive interactions [21], cation-π [22], anion-π [23], CH-π [24] and the recently outlined halogen bond (0.2–3.0 kJ/mol) [25]. The hydrophobic effect can be an additional feature that enhances self-assembly and binding affinity. The latter phenomenon is displayed only in aqueous medium and generally it is caused by entropic effects, involving the typical water network of hydrogen bonds, that makes it overall more favorable to stick together the molecules of an apolar solutes rather than separately solvate each one [26]. It is worth noting that weak interactions between molecular synthons are rarely monotopic, but more frequently multiple complementary interactions take place at the same time with the overall effect of strengthening the interactions between the molecular partners, making intermolecular forces macroscopically relevant in terms of binding energy [27, 28]. Moreover, when multiple interactions are possible, the reciprocal orientation between each couple of attractive residues influences the others if placed in close proximity, with overall results that can be higher or lower in energy than the simple algebraic addition of the single contributions [29].

Hydrogen-bonding prevails among other weak interactions in terms of widespread diffusion in nature, importance and energy of binding: needless to say the pivotal role played by this attractive force in the folding of proteins and many other natural macromolecules, leading to their secondary, tertiary and quaternary structure that are essential to explicate peculiar biological activity. This makes hydrogen bonding one of the favorite tools for the nanofabrication of supramolecular finite aggregates. An illustrative example of a chiral hydrogen-bonded aggregate self-assembled by two achiral counterparts is shown in Figure 2.4. Both molecules, the *bis*-carboxylic acid and the *bis*-aniline are achiral, but the two hydrogen bonds place the two molecules in a reciprocal dissymmetric arrangement, overall a chiral racemic supramolecule.

Metal–ligand coordination is an attractive force characterized by a wide range of properties as a function of the combination of metal and coordinating atom and lies in between covalent bonds and weak interactions. Geometric aspects like distance and directionality between the two counterparts are crucial issues when discussing weak intermolecular forces except in the case of ionic attractions [20]. From this point of view, metal–ligand coordination offers a high level of geometrical control over the assembly together with the possible presence of stereogenic centers directly implemented, or in close vicinity to the metal center. Figure 2.5 shows different chiral structures that arise by coordination of achiral bidentate ligands to metal centers characterized by octahedral or square planar coordination geometry.

2.1.2
Timescale of Supramolecular Interactions and Racemization Processes

Proper docking of subunits for self-assembly of supramolecular chiral objects is a dynamic equilibrium that is favored by the geometric preorganization of the binding moieties imparted by the molecular scaffold. Careful design of counterparts is an important issue to tackle targets like effective self-assembly and long lifetime of supramolecules. It has been demonstrated that increasing the rigidity of the partners

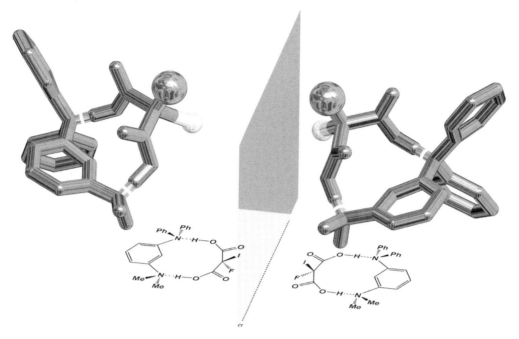

Figure 2.4 Example of a hypothetical hydrogen-bonded chiral supramolecule derived by the association of two achiral units, a dicarboxylic acid and dianiline.

results in an increase of binding free energy mainly because of decreased flexibility, lower number of possible conformers, more negative ΔH of binding (favorable) and consequently more negative entropy ΔS (unfavorable), in agreement with the general observation concerning enthalpy/entropy compensation for association phenomena [30].

Capsular assemblies are generally held together by several single weak attractive interactions between subunits and guest and lifetimes for such association and dissociation phenomena is in the order of 10^{-3}–10^3 s. for hydrogen-bonded systems [31], while the lifetime for metal–ligand assemblies can be up to a few orders of magnitude higher as a function of the intrinsic nature of the aggregate [32].

As far as chiral assemblies made of achiral counterparts are concerned, the lifetime of supramolecular encounters between subunits is strictly related to several aspects of their supramolecular behavior like *in-primis* the activation energy for racemization

Figure 2.5 Chiral complexes arising by coordination of achiral ligands to metal centers: (a) δ and λ enantiomers by ethylenediamine (en) coordination in octahedral [Co(en) Cl$_2$(CO)$_2$]; (b) Δ and Λ enantiomers by oxalate (ox) coordination in octahedral [Fe(ox)$_3$]$^{3-}$; (c) enantiomeric complexes arising by coordination of diamines ligands to square planar [Pt(iso-butylenediamine)(meso-stilbenediamine)]$^{2+}$.

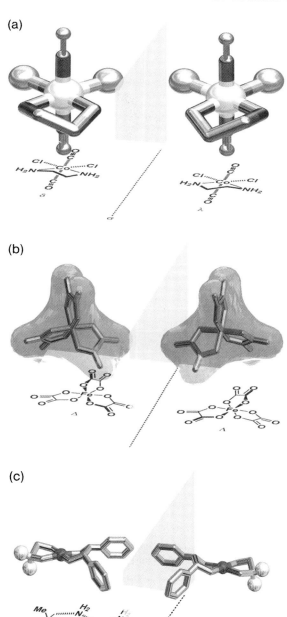

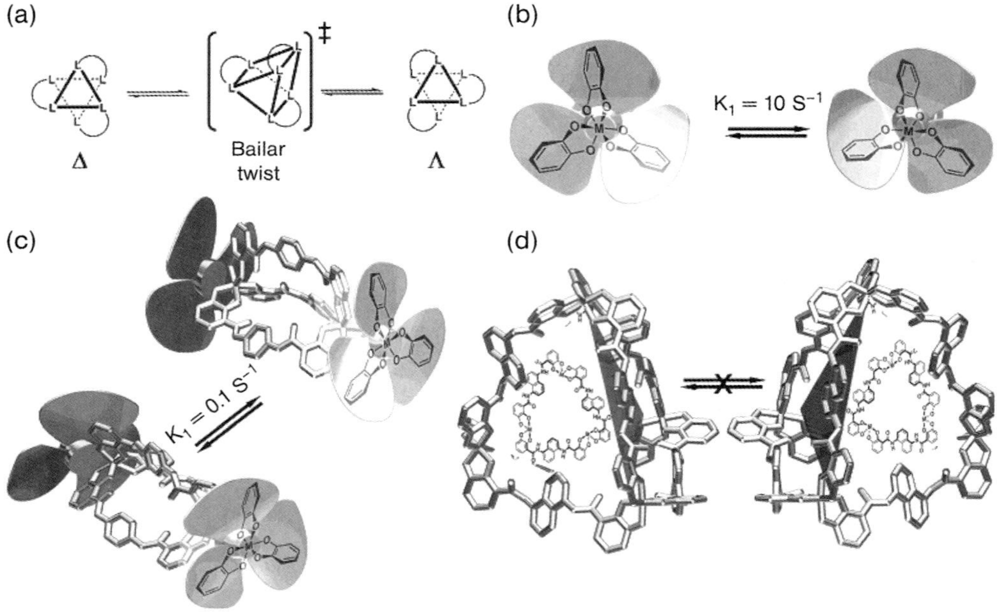

Figure 2.6 Schematic representation of the decreasing racemization rate observed for octahedral metal assemblies comprising bidentate ligands: (a) Bailar twist mechanism for interconversion of Δ and Λ octahedral complexes comprising bidentate ligands; (b) monomeric complexes racemize quickly, while (c) dinuclear systems have lower racemization rates due to mechanical coupling imparted by the rigid ligand; (d) for tetrahedral assemblies with four metal corners racemization is a slow process.

[32], their host–guest chemistry and the mechanism of association–dissociation and in-out exchange of the guest.

While racemization of hydrogen-bonded chiral assemblies necessarily involves rupture of attractive interactions and consequently possible guest release, for metal–ligand assemblies comprising octahedral metal centers with bidentate ligands, another mechanism pathway is possible besides the dissociative one. Such coordination geometry is characterized by a racemization mechanism called Bailar twist (trigonal) typical for small bite angle ligands that involves directly the metal ion and occurs by twisting around one of the pseudo C_3 axes of symmetry [12, 33]. The rigidity of the organic linker that bridge different metal ions plays a crucial role in the mechanical coupling of the stereochemistry of the metal centers (Figure 2.6).

For tightly bonded assemblies the guest can be released without complete disassembly of the multimeric structure, but simply by means of gating through one of the walls of the monomers, thus limiting the enthalpic cost and favoring chiral memory effect.

Encapsulation phenomena are related with solvation and desolvation aspects and the solvent plays a pivotal role both on a thermodynamic point of view, entropically favoring in many examples the encapsulation [34], as well as the kinetic point of view, competing with the guest for binding. If prerequisites like size and shape of the guest are fulfilled, capsular hosts are characterized by high levels of surface buried areas [30] due to the complete surrounding of the guest. This is an important parameter that is correlated to the average binding constants that for nonmetal–ligand assemblies are in the range of $10^{3.4\pm1.6}$ M^{-1}.

2.2
Self-Assembly of Intrinsically Chiral Molecular Capsules

The present section illustrates selected examples of overall racemic supramolecular assemblies ordered in terms of their nature (purely organic or coordination capsules), size of the structure and increasing level of complexity.

2.2.1
Hydrogen-Bonded Assemblies

The employment of concave-shaped molecules endowed with hydrogen-bond donors and acceptors represents one of the easiest methodologies for the construction of chiral supramolecules driven by the self-assembly of the subunits into well-defined aggregates. The kinetic stability of the supramolecular structure influences positively the racemization process that is slow on the NMR timescale and allows for the observation of two distinct species for the encapsulated and free guest.

2.2.1.1 Double Rosettes

Chiral amplification is the phenomenon observed when a chiral unit present in a small initial enantiomeric bias, induces a much higher stereo preference (high de or ee) when introduced into a bigger aggregate. Double-rosette hydrogen-bonded assemblies – whose name comes from their cyclic flower-like form – are typical examples of such a relationship [35]. Racemic self-assembled double rosettes form spontaneously when three calix[4]arene dimelamines units and six achiral barbiturates (BAR) or cyanurate (CYA) are mixed in a 1 : 2 ratio in apolar solvents, as depicted in Figure 2.7. The nine achiral components are held together by 36 hydrogen bonds and the supramolecular aggregate can adopt up to three diastereoisomeric forms with D_3, C_{3h} and C_s symmetry as a function of the conformation of the dimelamine units. More hindered substituents on either dimelamine, barbiturate or cyanurate favor the D_3 isomer which exists as a pair of enantiomers, P and M [36]. Addition of ten equivalents of enantiopure Pirkle's reagent demonstrated the chirality of the assembly. In fact, the interaction between the racemic supramolecular structure and the enantiopure solvating agent caused the splitting of the ^{1}H-NMR proton signals of the bridging methylene of the calix[4]arene into equally populated signals due to diastereoisomeric complexation, but no imbalance between the amounts of the two species was observed and proved the existence of slowly interconverting chiral assemblies in solution [37]. Employing *tris*-imidazoline achiral units and calix[4]

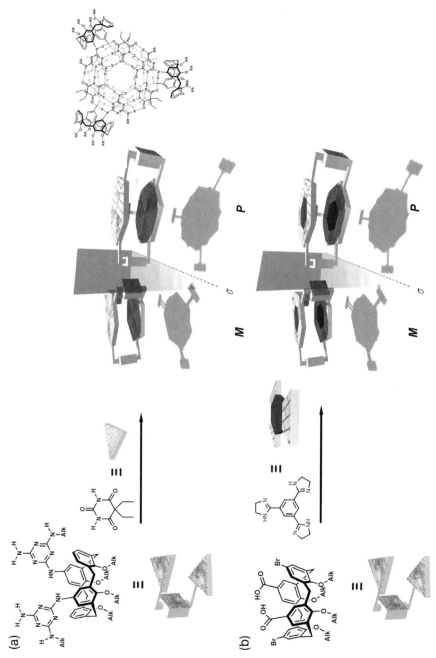

Figure 2.7 Schematic representation of double-rosette assemblies: (a) dimelamine calix[4]arene and barbiturate or cyanurate self-assemble leading to three constitutional isomers with different symmetry. The D_3 symmetric assembly is chiral and exists as a racemic mixture. (b) dicarboxylate calix[4]arene and tris-imidazole self-assemble leading exclusively to a D_3 chiral aggregate.

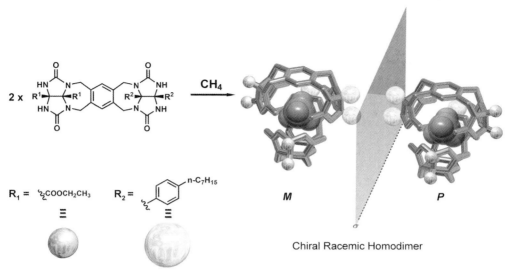

Figure 2.8 Chiral homodimeric tennis ball capsule from an achiral component bearing two different glycoluril residues. The R^1 and R^2 groups have been reduced to spheres for clarity.

arene dicarboxylates in a 2 : 3 ratio, it was possible to achieve the exclusive formation of the D_3 symmetrical self-assembled chiral racemic double rosette due to 12 charged hydrogen bonds in polar protic medium containing up to 50% of deuterated methanol (Figure 2.7) [38].

2.2.1.2 Hydrogen-Bonded Capsules

The smallest known chiral self-assembled host made of achiral building blocks was disclosed and synthesized by Rebek and collaborators modifying the monomeric unit based on the durene scaffold with two different glycoluril units. The so-called *tennis ball* molecular capsule produced by self-assembly of two subunits through formation of eight hydrogen bonds between the glycoluril counterparts offers a small cavity of approximately 60 Å^3 that can accommodate only gaseous guests like methane and ethane [39] (Figure 2.8). Upon dimerization of the monomer, four well-separated N–H signals were observed in the downfield region of the ^{1}H-NMR spectrum arising from the chirality of the assembly, even though it is composed of achiral monomers. This serves as proof of concept to underline how asymmetry can be sensed at a magnetic level more easily then at a binding level. Unfortunately, the small volume of the cavity did not allow encapsulation and investigation of chiral guests.

More elongated monomers based on a rigid polycyclic scaffold endowed with two different glycoluril moieties at the extremities allowed the formation of a chiral *softball* molecular capsule. The association of two subunits through a seam of hydrogen bonds provides a chiral capsule that maintains a C_2 symmetry axis but no longer the planes of symmetry featured by the monomer (Figure 2.9) [40]. The

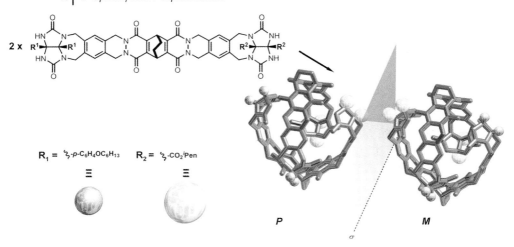

Figure 2.9 Chiral homodimeric softball capsule from an achiral component bearing two different glycoluril residues. The R^1 and R^2 groups have been reduced to spheres for clarity.

cavity has a volume of approximately 240 Å^3 and a spherical shape that enables encapsulation of complementary guests like (R)-(+)-camphor, thus leading to the formation of two diastereoisomeric host–guest assemblies characterized by separate signals but with equal intensity because of the semispherical symmetry of the cavity.

To provide selectivity in the binding as a result of stereorecognition, closer interactions and tighter binding are pivotal prerequisites. Moreover, the asymmetry of the surfaces has to extend into the cavity, rather than be limited to the exterior. A step forward is presented by the evolution of the chiral *softball* monomer units. Each piece contains the same glycoluril functional group at the extremities but is characterized by different lengths of the two spacers between the hydrogen-bonding extremities and the core of the monomer that remains achiral due to the presence of a plane of symmetry. Dimerization provides a chiral dimeric capsule characterized by a dissymmetric inner space [41] that, contrarily to what was observed with the first softball, easily sensed chiral guests not only at the magnetic level but also at the binding level providing different amounts of diastereoisomeric complexes. A total of six different glycoluril monomers with various spacers in terms of size and electronic properties on the two sides of the bicyclic six-membered unit were prepared in order to deeply investigate the phenomenon, and the relative homodimeric capsules were studied with a total of twenty chiral enantiopure guests (Figure 2.10) [42].

Cavity volume varies from 190 to 390 Å^3 and suitable guests were molecules whose structure is related to camphor and pinane derivatives, with average size of the guest in the range 145–225 Å^3. Proper matching of volumes and shapes of both supramolecular hosts as well guests was crucial to ensure reciprocal sensing and preferential stereoselective binding of one enantiomer of the capsule giving rise to diastereose-

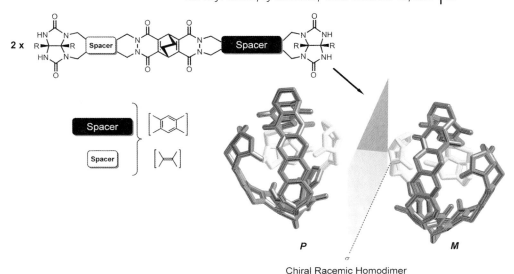

Figure 2.10 Second generation of chiral dissymmetric softball capsules with the same glycoluril residues at the two extremities of the monomer but with different spacers. The R groups have been removed for clarity.

lective host–guest assemblies. This survey on chiral recognition in encapsulation processes focused the attention to the importance of the presence of attractive interactions between host and guest counterparts in order to enhance affinity as well as selectivity and that simple steric requirements are, most of the time, insufficient for the achievement of enantioselective binding.

Calix[4]arene units decorated at the upper rim with hydrogen-bonding moieties, usually urea residues, provide a suitable semirigid concave scaffold for the construction of dimeric capsules with a cavity of about 200 Å^3 (Figure 2.11a) [43]. If the calix[4]arene unit contains two different groups appended on different aromatic residues either on the lower or on the upper rim, it is possible to observe the formation of homodimeric chiral capsules held together by a seam of hydrogen bonds that, behaving like a zip, seal the equatorial region of the supramolecular structure (Figure 2.11b). As far as homodimeric chiral structures are concerned, these form when calix[4]arenes of general structure ABBB, AABB and ABAB self-assemble [44], regardless of the orientation of the hydrogen-bond seam, and the effects of this supramolecular chirality are more evident for calix[4]arenes modified on the upper rather than on the lower rim. For ABAB monomers, only one chiral racemic assembly is possible while for ABBB and AABB monomers, two possible regioisomeric dimers are observed with a diastereoselective ratio that, for the ABBB, is a function of the solvent (CDCl$_3$ 0%; C$_6$D$_{12}$ 71%). For AABB, the relative amount of the regioisomers spans from 1 : 1 to 2 : 1, increasing the steric difference between A and B from n-hexyl to adamantyl residues. Moreover, the equilibrium between the regioisomeric forms could be shifted entirely to one side if two adjacent A and two B groups are covalently connected.

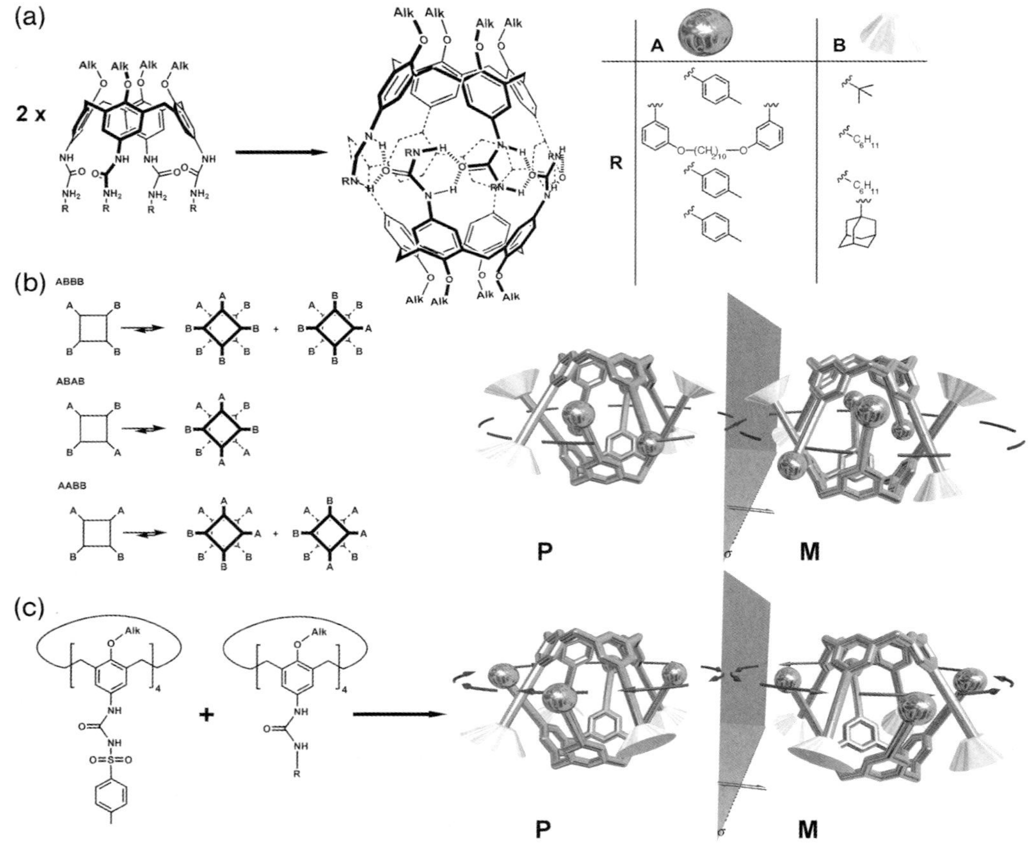

Figure 2.11 (a) Self-assembly of calix[4]arene monomers endowed with urea residues leading to dimeric hydrogen-bonded capsules; (b) Examples of homodimeric self-assembly of achiral units leading to chiral racemic capsules; (c) Heterodimeric self-assembly between tetra-tosyl and tetra-tolyl urea hydrogen-bonding moieties leading to chiral capsules as a consequence of the directionality of the hydrogen-bonding belt.

For AAAA monomers homodimerization provides a single species where the two halves of the capsule are identical. In the presence of chiral enantiopure suitable guests like (+)-nopinone or (−)-myrtenal, encapsulation occurs and causes differentiation between the north and south pole of the supramolecular assembly as a result of restricted tumbling of the trapped guest that induces asymmetry to the host as it becomes chiral because of the directionality of the hydrogen bonds [45]. Combination of tetra-aryl AAAA and tetra-tosyl monomers BBBB provides preferentially heterodimeric rather than homodimeric assemblies. Such capsular assemblies are chiral because of the hydrogen-bond network that is localized in the equatorial region and that can be either clockwise or counter-clockwise [46] (Figure 2.11c).

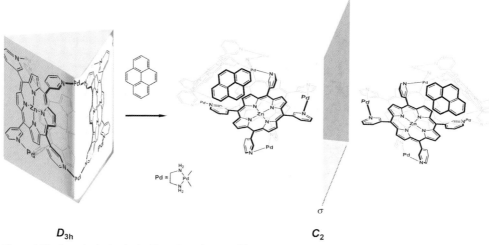

Figure 2.12 Guest inclusion in the D_{3h} prismatic assembly causes a transition into a C_2 symmetry structure that is chiral and exists as a pair of enantiomers.

2.2.2
Metal–ligand Assemblies

In the literature there is a plethora of examples, recently reviewed thoroughly [12], concerning metal–ligand supramolecular chiral aggregates spanning from helicates, to trigonal pyramids, to more complicated bipyramids, octahedral and cuboctahedra structures, all possible by combinations of different stoichiometries between metal and ligand. Herein, we report clear-cut examples of chiral closed structures where the host–guest chemistry plays a crucial role in the asymmetry of the assembly. An elegant example of a guest-induced conformational change into a chiral aggregate is provided by the supramolecular prism formed upon coordination of three face bridging Zn(II)-porphyrin ligands to six Pd(II) ions. While the assembly displayed D_{3h} symmetry, in the presence of a suitable elongated aromatic guest like pyrene, it turned into a C_2 symmetry chiral structure triggered by apical to equatorial flipping of two pyridine–Pd–pyridine hinges placed at diagonal positions (Figure 2.12).

Tetrahedral assemblies are the most represented among chiral supramolecular structures due to the extensive work pursued by Raymond and collaborators. As introduced in Figure 2.6, rigid, bidentate moieties are suitable chelating ligands for octahedral cations and the chirality implemented in the metal center can be coupled to another metal through the ligand. They are extremely effective in imparting unique features to the assembly; size and rigidity of the spacer are key points that dictate the geometrical preference of the supramolecular chiral aggregate. As far as the size of the ligand is concerned, it is worth noting that, as reported in Figure 2.13, the smaller naphthyl ligand provided directly the M_4L_6 tetrahedral assembly independently of the presence of a templating guest. Conversely, with larger ligands the capsular assembly

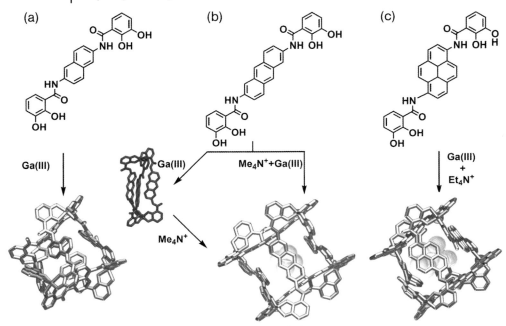

Figure 2.13 Effect of the size of the ligand on the host–guest chemistry of metal–ligand coordination self-assembled structures: (a) small ligand leads to chiral tetrahedral assembly without the aid of templating guests while larger ligands (b) and (c) provide chiral assemblies only in the presence of suitable cationic guests.

formed only upon addition of a suitable guest like in the cases of the anthryl derivative that required tetramethylammonium guest and of the pyryl ligand that necessitated the larger tetraethylammonium guest [47].

The rigidity of the ligand plays a crucial role in determining the final structure of the self-assembled aggregate, in fact considering *tris*-bidentate chelating ligands based on the catechol moiety, it was observed that those bearing a rigid aromatic nuclei provided chiral tetrahedral M_4L_4 structures that showed guests binding for tetra-methyl and tetra-ethyl ammonium cations, while ligands containing conformational more flexible methylene spacers did not self-assemble into tri-dimensional structures but gave simple 1:1 metal–ligand interaction (Figure 2.14) [48].

As an example of further supramolecular complexity, it is worth describing the system formed from six end-capped Pd(II) diamine units bridged by four pyridyl ligands that, in the presence of suitable guests like neutral aromatic units, self-assemble into the chiral octahedral structure illustrated in Figure 2.15. The role of the guest is crucial in order to impart the proper assembly and to induce the dissymetrization that makes the capsule chiral.

2.2 Self-Assembly of Intrinsically Chiral Molecular Capsules | 45

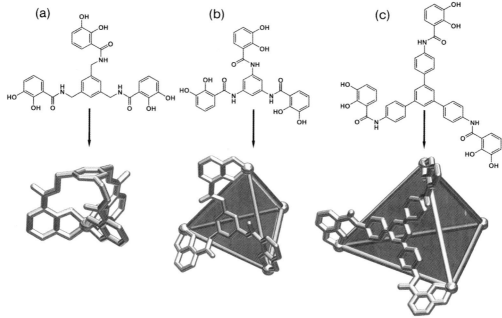

Figure 2.14 The rigidity of the ligand is crucial to define the geometry of the assembly: (a) more flexible ligand bearing methylene units provides only the ML monomeric octahedral complex, while for the rigid aromatic *tris*-bidentate chelating ligands (b) and (c) only the chiral metal–ligand M_4L_4 assembly is observed.

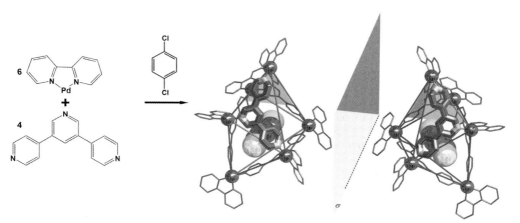

Figure 2.15 Example of octahedral chiral structure obtained by combination of Pd(II) and C_{2v}-symmetric pyridyl ligands. The structure self-assembles only in the presence of suitable aromatic guests.

2.3
Chiral Induction in the Formation of Supramolecular Systems

Self-assembly of achiral building blocks leads to the formation of chiral enantiomeric structures in equimolar amounts (racemate), because of the absence of a source of symmetry breaking. As will be discussed in Section 2.4, sensing of chirality between supramolecular chiral hosts and chiral guests is possible with different levels of stereorecognition and leads to the formation of diastereoisomeric assemblies based on the complementarity between the two, as well as the strength of the interactions. Such species are in thermodynamic equilibrium under dynamic exchange of the building blocks at a rate that depends on the rate of association and dissociation of the host. Quick removal of the guest restores the original racemic composition unless host dissociation into the original building blocks is a much slower phenomenon than in–out guest exchange. In such a case, it is possible to prepare a host–guest combination with a certain degree of diastereoselectivity and replace the enantiopure templating partner with an achiral guest maintaining the preference for one enantiomeric host assembly. This phenomenon is called the chiral memory effect [9, 14] and basically it occurs when the slow disassembly of the supramolecular host allows it to retain a preferred enantiomeric structure even if the enantiopure guest that is responsible for the chiral induction has been removed. Simply speaking, it is like walking on a shoreline: the print of the left foot on the sand is chiral. After the foot is lifted, the print persists for a while but slowly fades due to the leveling off of the sand grains. The retaining of enantioselectivity achieved by such a method is simply due to the kinetic stability of the host and it can persist long enough to ensure full characterization, even though it inexorably undergoes slow racemization by dissociation and reassembly until reaching the isoenergetic racemic composition. The structure thus obtained is chiral with a certain degree of enantiopurity even though it is made out of achiral components.

2.3.1
Chiral Memory Effect in Hydrogen-Bonded Assemblies

One of the earliest examples of the chiral memory effect by a supramolecular assembly is based on a hydrogen-bonded double-decker rosette that is characterized by a high number of directional hydrogen bonds that lead to a half-lifetime of several hours for the assembly disassembly process. Reinhoudt and collaborators exploited this peculiarity preparing a double rosette from achiral calix[4]arene dimelamine and chiral enantiopure (R)-BAR, obtaining prevalently the M isomer in 98% de (Figure 2.16). This species was then equilibrated with the stronger binding achiral monomer cyanurate that resulted in replacement of the (R)-BAR from the original chiral assembly without complete dissociation of the structure and yielded a chiral double rosette made of achiral components with 90% ee. The kinetic stability of the assembly allowed complete replacement of the chiral inductor BAR preserving the M chirality of the rosette as memory of the native form of the assembly, despite the fact that it no longer contains any chiral components [49, 50].

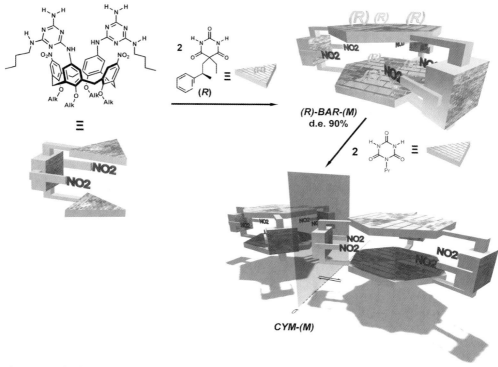

Figure 2.16 Chiral memory effect on double rosette where kinetic stability of the hydrogen-bonded assembly allows retention of configuration even after replacement of the chiral subunit (R)-BAR with the achiral cyanurate.

The same result was obtained with an external chiral inductor, as in the case illustrated in Figure 2.17 where the racemic rosette composed of achiral calix[4]arene bearing dimelamine moieties with pyridine functionalities and an achiral cyanurate was initially enantioenriched by interaction with enantiopure D-dibenzoyl tartaric acid via two-point hydrogen-bonding interactions providing predominantly the *P* diastereoisomeric supramolecular complex with 90% de [51]. Treatment with ethylenediamine led to precipitation of the chiral inductor leaving the enantioenriched chiral double-rosette made of achiral components with 90% ee. In this case the memory effect observed is even stronger than in the former example with activation energy towards racemization as high as 119 kJ/mol and a half-lifetime of one week at room temperature.

Aiming at demonstrating chiral memory effect in a *softball* capsule, Rebek and collaborators endowed the *softball* monomers with extra phenolic groups in order to provide four extra hydrogen bonds, thus increasing the strength of the supramolecular assembly [52]. The capsule forms when suitable guests are provided and with enantiopure (+)-pinanediol two diastereoisomeric complexes in 2:1 ratio were

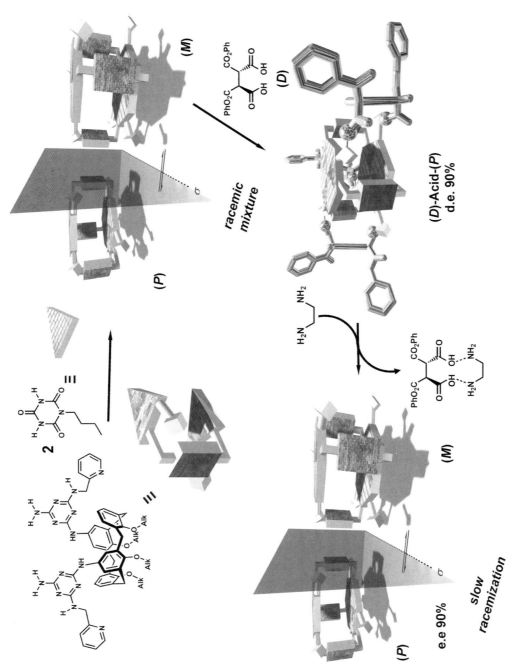

Figure 2.17 Chiral memory effect on double rosette mediated by enantiopure chiral inductor D-dibenzoyl tartaric acid. The favored (P) configuration of the rosette was retained even after removal by precipitation of the external chiral inductor.

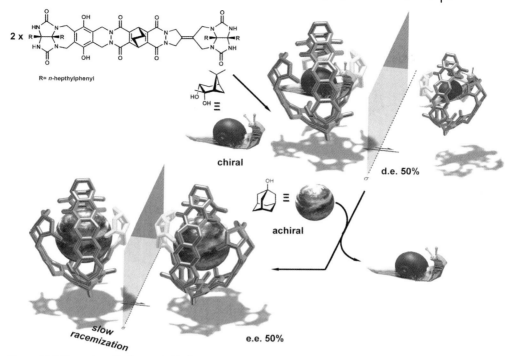

Figure 2.18 Racemic softball capsule binds an enantiopure guest leading preferentially to one diastereoisomer. The (+)-pinanediol can be exchanged with an achiral one without affecting the stereochemistry of the supramolecular assembly due to strong chiral memory effect.

observed after several hours of equilibration. Subsequent exchange of the encapsulated guest with an excess of achiral 1-adamantanol and again with the enantiomer (−)-pinanediol showed the same de (50%) in favor of the less stable diastereoisomeric complex (Figure 2.18). These experiments demonstrated that slow exchange and relative kinetic inertness of the capsule result in the chiral memory effect, even in the presence of the wrong enantiomeric guest. Eventually, in a few days, the system slowly equilibrated to a 2 : 1 diastereoisomeric ratio in favor of the (−)-enantiomer complex.

2.3.2
Chiral Memory Effect in Metal–Ligand Assemblies

Chiral racemic hosts held together by means of metal–ligand interactions are ideal targets for the exploitation of the chiral memory effect because of the generally slower dissociation rate of the coordination bonds compared with hydrogen bonds. Beautiful examples of elegant approaches to enantioselective noncovalent syntheses were

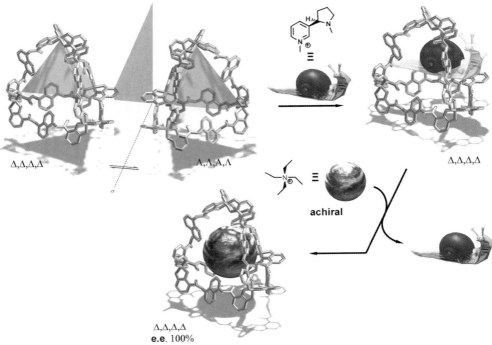

Figure 2.19 Chiral memory effect of a tetrahedral metal–ligand capsule allows for the isolation of an enantiopure host by precipitation and subsequent substitution of the chiral guest with an achiral one leaving unaffected the Δ,Δ,Δ,Δ stereochemistry of the structure.

disclosed by Raymond. In Figure 2.19, the molecular capsule $[Ga_4L_6]^{12-}$ is described with L = 1,5-*bis*(2,3-dihydroxybenzamido)naphthalene) that can exist in two enantiomeric forms (Δ,Δ,Δ,Δ and Λ,Λ,Λ,Λ) due to the strong mechanically interlocked configuration of the coordinated ligands around each octahedral metal cation present on the corner of the assembly. The host can easily accommodate into the cavity a chiral organic cation like *N*-methylnicotinium (Nic), thus leading to the formation of two diastereoisomeric host–guest complexes that are characterized by different relative energies as well as solubility. The Δ,Δ,Δ,Δ-[(Nic)⊂Ga$_4$L$_6$]$^{11-}$ form is the less soluble of the two and it was isolated, characterized in the solid state as well as in solution and when placed in contact with an achiral cation like Et$_4$N$^+$, having a higher binding affinity, led to the formation of the enantiopure Δ,Δ,Δ,Δ-[(Et$_4$N)⊂Ga$_4$L$_6$]$^{11-}$ species that was stable and retained the homoconfigurational cluster structure without racemization for at least eight months [53]. This stability is a further consequence of the slow ligand exchange rate of gallium cations coupled with the mechanical stiffness provided by the interconnection of the metal centers in a rigid tridimensional tetrahedral arrangement.

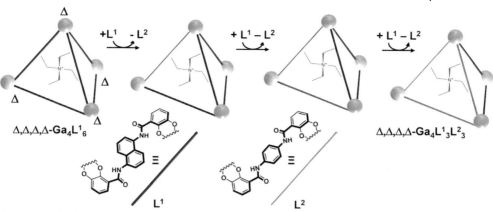

Figure 2.20 Chiral memory effect of a tetrahedral metal–ligand capsule allows substitution of several ligand molecules that make up the edges of the tetrahedron leaving unaffected the overall Δ,Δ,Δ,Δ stereochemistry of the structure.

Analogously, the same self-assembled chiral enantiopure host Δ,Δ,Δ,Δ-[(Et$_4$N) ⊂ Ga$_4$L$^1_6$]$^{11-}$ comprising L^1 naphthyl *bis*-catecolamide ligand was treated with a different one, L^2 phenyl *bis*-catecolamide, which prefers to provide M$_2$L$^2_3$ arrangements rather than the M$_4$L$^2_6$ structure (Figure 2.20). Ligand exchange on the capsule occurred in a stepwise fashion leading to several species Δ,Δ,Δ,Δ-[(Et$_4$N) ⊂ Ga$_4$L$_n$L$^2_{6-n}$]$^{11-}$ arising from substitution of the edges of the tetrahedral assembly but the stereochemistry of the assembly remained unaffected as confirmed by CD measurements as indication of the extreme chiral memory present in the progenitor species [54].

2.4
Chiral Spaces for Chiral Recognition

Simple confinement of a chiral guest within a chiral pocket usually leads to a low level of stereoselectivity because selection on the basis of size, shape and repulsive steric imbalance of the guest is intrinsically limited by effects related to the packing coefficient. In fact, it is well established that in solution as well as in encapsulation phenomena packing coefficients [55], expressed as the volume of the guest over the volume of the cavity (gases ∼0.35, solids 0.7–0.9, liquids 0.5), are usually in the range 0.55 ±0.09, which means that about half of the available space remains unfilled and resides in between host and guest. Conversely, the extremely high levels of stereo-recognition observed in enzymes for substrate and transition-state binding are achieved by concomitant presence of complementary surfaces as well as weak attractive interaction between guest and host, thus docking in close proximity the counterparts and enabling reciprocal effective sensing [30].

2.4.1
Enantioselective Recognition within Chiral Racemic Self-Assembled Hosts

This section deals with chiral racemic self-assembled hosts made by achiral building blocks and their interaction with chiral, racemic or enantiopure molecular guests. The formation of two diastereoisomeric supramolecular host–guest assemblies is generally an equilibrium reaction and thus a far higher level of diastereoselectivity, albeit only in a few cases, has been achieved.

2.4.1.1 Hydrogen-Bonded Hosts

Dimeric calix[4]arenes are the smallest chiral supramolecular aggregates that have displayed enantioselective binding. Interaction between the heterodimeric assembly composed by a tetra-tolyl and a tetra-tosyl urea reported in Figure 2.10 and a racemic guest characterized by a rigid bicyclic structure like norcamphor resulted in encapsulation of the guest and provide two sets of diastereoisomeric assemblies in a 1.3 : 1 ratio (de ~ 13%). The same level of stereoselectivity was observed with (R)-(+)-3-methylcyclopentanone indicating a small level of stereorecognition exploiting only steric interactions [56].

Another clear-cut example of enantioselective binding was observed within the cavity of chiral *softball* molecular assemblies (Figure 2.9). This structure was extensively investigated with several suitable chiral guests and small but substantial differences were observed, like in the case of the alcohol (+)-pinanediol that holds a hydrogen-bond donor residue and produced a de 32%, while the corresponding ketone (+)-nopinone provided scarce results (de 0%). It is worth noting that in a few cases the level of diastereoselectivity achieved was above 50%, but never greater than 60% [42] likely due to the average distance present between the capsule and the chiral guest. This is a consequence of the packing coefficients of approximately 0.5, which means that half of the cavity's volume is not filled and this space surrounds and is interposed between the guests, unless attractive interactions between the two counterparts allow reciprocal docking and sensing.

A step further towards high stereoselective binding was observed by implementing attractive interactions with steric interactions, like in the saccharides enantioselective recognition displayed by hydrogen-bonded rosette assemblies. Tethering of two calix[4]arene dimelamine units led, in the presence of barbiturate or cyanurate counterparts, to expanded tetrarosette assemblies endowed with a spacious cavity where medium-sized guests can be accommodated. Figure 2.21 shows the structure of the assembly that exists as a racemic mixture of *P* and *M* enantiomers. Interaction of this racemic host with enantiopure saccharides – like n-octyl β-D-glucopyranoside (β-D-gluc) – caused the shift and splitting of some ^{1}H-NMR resonances of the tetrarosette due to hydrogen bonding of the guest within the cavity, which associates with a binding constant of approximately $20\,M^{-1}$. The dynamic equilibrium between the two enantiomers of the host was shifted towards the preferred formation of the *P* enantiomer as proved by the induced negative Cotton effect observed in the CD spectrum [57], while a mirror effect was observed employing β-L-gluc that favored the *M* enantiomer. Such enantioselective recognition represents a model example of

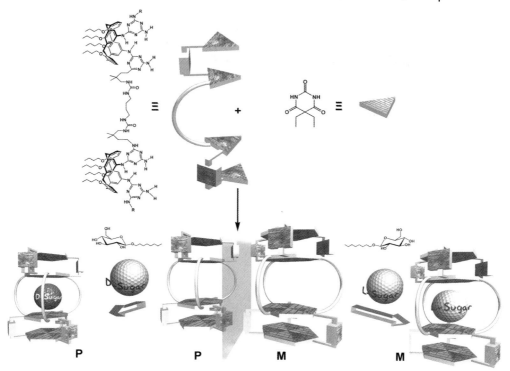

Figure 2.21 Dimeric calix[4]arene dimelamines self-assemble in the presence of barbiturate to provide a chiral racemic tetrarosette assembly. In the presence of chiral enantiopure saccharide guests the host equilibrates into an enantiomerically pure structure as a remarkable example of enantioselective supramolecular recognition.

enantioselective amplification of the best host from a racemic dynamic mixture induced by the templating effect imparted by the guest.

2.4.1.2 Metal–Ligand Hosts

Diastereoselective formation of host–guest assemblies comprised of chiral capsules and chiral guests is an emerging topic in supramolecular chemistry, equally distributed among hydrogen-bonded and metal–ligand assemblies. One interesting example of the latter species is provided by the chiral supramolecular box composed of self-assembled Zn-porphyrin units *ZnP* whose chirality arises from the reciprocal orientation of the porphyrin units along the poly-alkyne spacer (Figure 2.22) [58]. Racemization of the assembly is rather slow ($t_{1/2} \sim 11.5$ h at $20\,^\circ$C) due to the presence of eight strong Zn(II)-pyridine coordination bonds allowing enantiomer separation by means of chiral chromatographic methods. Interaction of the racemic supramolecular box with enantiopure (*R*)-limonene as the solvent showed the formation of strong CD active absorptions in the visible region, indicative of selective binding

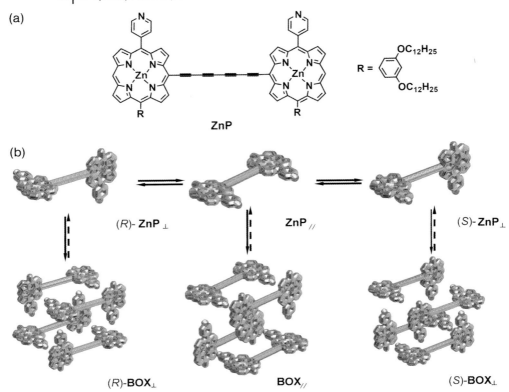

Figure 2.22 (a) Schematic representation of rotamers of Zn-porphyrin units and (b) their self-assembly into a chiral supramolecular box.

of one of the two enantiomers of the host with the enantiopure guest. The degree of diastereoselectivity was rather low (3%) and linear with the enantiopurity of limonene but the high molecular ellipticity of the system allows the chiral supramolecular box to be exploited as a chiroptical sensor of chiral hydrocarbons.

For metal–ligand assemblies the earliest examples of chiral recognition dealt with dimeric aromatic structures held together by means of pyridine-Pd(II) interactions (Figure 2.23). The first one is particularly interesting because the shape of the guests steers the formation of an achiral self-assembled host when the pseudospherical $CBrCCl_3$ guest is added, while 1,3,5-benzentricarboxylic acid induces the formation of a chiral host. Addition of (R)-mandelic acid proved the formation of two enantiomeric assemblies but stereoselective binding was observed only with (S)-1-acetoxyethylbenzene in a 3:2 diastereoselective ratio (20% de) [59].

A higher level of diastereoselectivity was observed with the homo-oxacalix[3]aryl esters sealed into a dimeric structure by coordination between three Pd(II) ions and m-substituted pyridine moieties. The assembly is chiral as depicted in Figure 2.24 and the cavity was able to sense chiral ammonium guests like (S)-2-methylbuty-

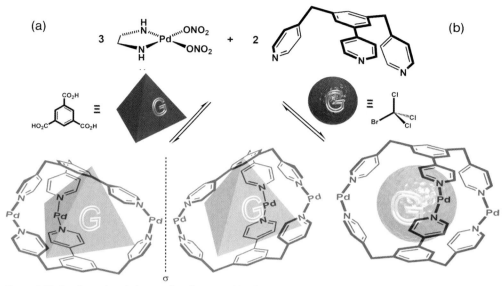

Figure 2.23 Pyridine-Pd(II) linkages allow for assembly of (a) chiral racemic host templated by a planar guest; (b) achiral host templated by a spherical guest.

lammonium triflate that caused a marked imbalance between the enantiomeric host structures with a diastereoselectivity of up to 70% [60].

The extent of chiral recognition increases with supramolecular hosts characterized by closer contact with the chiral guest. One elegant example is represented by the tetrahedral cagelike structure $[Ga_4L_6]^{12-}$ (L = 1,5-bis(2,3-dihydrobenzamido)naphthalene, where the chirality is present in the homochiral $\Delta,\Delta,\Delta,\Delta$ or $\Lambda,\Lambda,\Lambda,\Lambda$ configuration with respect to each metal center (Figure 2.13a) [15h]. The hydrophobic cavity of approximately 300–350 Å^3 can encapsulate a variety of monocationic species, either organic or organometallic ones. For the latter, a high level of stereoselective binding was achieved employing chiral organometallic complexes like tetrahedral Ru (II) complexes bearing a sterically hindered cyclopentadienyl residue and a diene ligand. The interaction between the racemic host and guest leads to the formation of four stereoisomeric products, basically two diastereoisomers each composed of a pair of enantiomers (Figure 2.25a). The level of diastereoselectivity achieved is remarkable for supramolecular assemblies and is a function of the steric hindrance of the ligands of the Ru(II) complex, with de of up to 70% for 2-ethyl-1,3-butadiene [61]. As far as organic guests are concerned, chiral phosphonium cations arising by the reaction between ketones and highly nucleophilic phosphines under controlled acidity in water cannot survive in solution for more than one hour; however, in the presence of the metal–ligand self-assembled host described above, they are encapsulated and survive for several days [62]. The nanoscopic environment provided by the capsule is chiral and racemic and can sense the formation of a pair of enantiomers of the phosphonium cation when ketones bearing different alkyl residues like

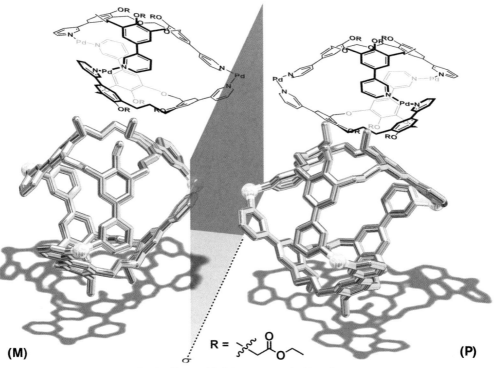

Figure 2.24 A chiral self-assembled dimeric capsule driven by metal–ligand coordination between Pd(II) and pyridine moieties of homo-oxacalix[3]aryl esters ligand.

2-butanone are employed leading to two diastereoisomeric pairs of enantiomers. Diastereoselectivity is positively influenced by the overall steric bulkiness of the cationic guest and for 2-butanone ranges between 24% for trimethyl phosphine up to 50% for triethyl phosphine (Figure 2.25b), but tends to decrease with time due to the equilibration process that takes place after the initial encapsulation of the chiral guest.

2.4.2
Interguests Chiral Sensing within Achiral Self-Assembled Hosts

The asymmetric influence of a chiral object on the surrounding space can also occur within achiral assemblies, analogously to what happens in a shoe box containing only a left shoe. The space around the shoe is chiral and a different right shoe can be accommodated better then another left one due to the small size of the container that forces the two guests to sterically interact with each other (Figure 2.26). This trivial example can be extended to a molecular level where achiral molecular self-assembled structures characterized by medium-large cavities can accommodate a chiral molecule, thus influencing, on a stereochemical level, the encapsulation of further molecules. Chirality is a feature related to the spatial distribution of objects, therefore

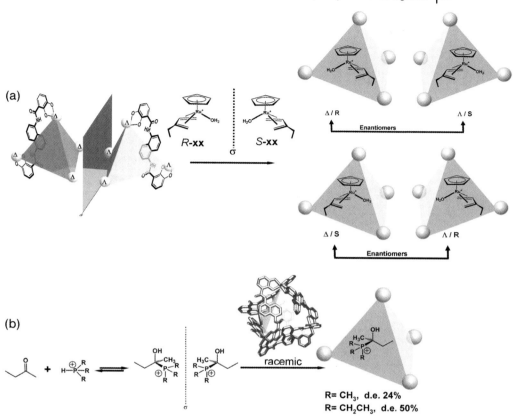

Figure 2.25 Schematic representation of the chiral tetrahedral host structure [Ga$_4$L$_6$]$^{12-}$ and suitable chiral guests with the different levels of stereoselective binding for (a) Ru(II) organometallic complexes and (b) phosphonium cation.

any symmetric closed space becomes chiral if a chiral object is placed in it. This reflects directly into the mutual electronic and steric interaction between guests and, as a consequence, in their stereorecognition. Examples of this approach to chiral supramolecular recognition are very recent, with few examples both for metal–ligand assemblies as well as for hydrogen-bonded supramolecular structures. This approach represents the ultimate level of enantioselective recognition where a molecule plays both roles, cause and effect, of the enantiomeric recognition.

2.4.2.1 Hydrogen-Bonded Hosts

Cylindrical hydrogen-bonded self-assembled achiral structures provide important nanosized molecular containers where concepts like interguest enantioselective recognition can be easily studied because the cylindrical space limits the number of possible reciprocal orientations of two guests compared to a spherical space because of the constrictions imposed by the geometry of the cavity (Figure 2.26). In

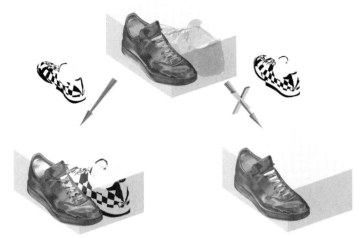

Figure 2.26 A right shoe in a shoe box leaves a chiral space that can more easily accommodate a left shoe rather than another right shoe because of the reciprocal sensing of chirality imposed by the small space available.

addition, the polarity of the cavity can favor certain spatial arrangements among all the possible ones. The elongated shape of the cavity, of average volume of 425 Å^3, hampers free tumbling for many guests and allows a better control over their reciprocal position. It is worth noting that coencapsulation of species results in an effective concentration of ∼4 M in the cavity for lifetimes as high as 1 s.

The simplest example of enantioselective interguest recognition was reported by the double encapsulation of (±)-*trans*-1,2-cyclohexanediol whose adequate size, shape and polarity allowed accommodation in pairs within the cavity, giving two diastereoisomeric complexes, the homochiral couple (R)-(R)/(S)-(S) and the heterochiral (R)-(S) combination in a <10% de in favor of the homochiral combination [63]. An analogous concept was observed with chiral α-halo or α-hydroxy carboxylic acids [64] that are accommodated in pairs within the cavity, but only a few of them give rise to diastereoselective binding, in particular substitution at the β position hinders the guests and results in diastereoselective ratios up to 1.6 : 1 (23% de). For bromo and hydroxy acids, a heterochiral combination is the most stable for (±)-3-methyl-2-hydroxy-butyric acid, while the opposite was observed for (±)-3-methyl-2-bromo-butyric acid. A role is also played by the belt of imide hydrogen bonds that participate in guest binding when a high packing coefficient of the guests is present, while low packing coefficients allow linear carboxylic acid dimer formation.

Coencapsulation can occur between different chiral molecules with enantioselective binding observed for gusts endowed with hydrogen-bonding moieties. Encapsulation of (S)-mandelic acid in the presence of (±)-2-butanol gives rise to two diastereoisomeric complexes in a 1.1 ratio (5% de) at 303 K and 1.3 (13% de) at 283 K in favor of the combination (S)-mandelic acid with (R)-2-butanol as a consequence of

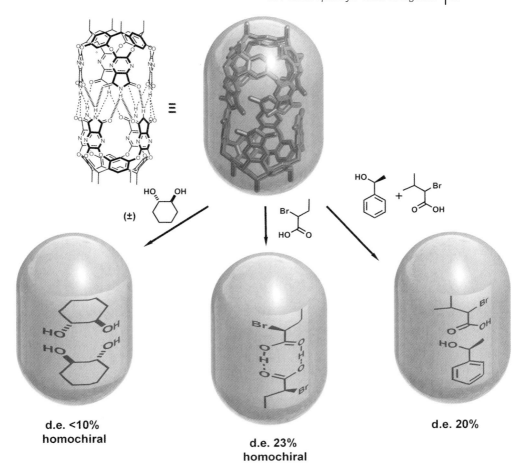

Figure 2.27 Diastereoselective coencapsulation of chiral guests within an achiral self-assembled host. The restricted space available induces reciprocal stereoselective sensing of the guests that leads to preferential homochiral coencapsulation of 1,2-cyclohexandiol and 2-bromo-butyric acid. Coencapsulation is possible for different molecules where the primary enantiopure guest (S)-1-phenyl-ethanol) preferentially selects one of the enantiomer of a secondary guest ((±)-3-methyl-2-bromo-butyric acid).

weak intermolecular attractive interactions between the two guests that are instructed and favored by the shape and size of the cavity of the molecular capsule. Diastereoselective ratio increased up to 1.5 (20% de) for the couple (S)-1-phenylethanol and (±)-3-methyl-2-bromo-butyric acid (Figure 2.27). It is worth noting that only the presence of weak attractive interactions like hydrogen bonds ensures stereoselective interactions, while simple steric requirements are not sufficient for enantioselective recognition within achiral molecular capsules as a consequence of the upper limit of the packing coefficients [55].

The methodology can be further extended to triple coencapsulation. Three molecules of enantiopure (R)-propylene sulfide can be accommodated within the

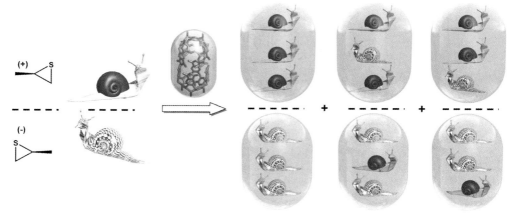

Figure 2.28 Coencapsulation of three chiral molecules of propylene sulfide into an achiral cylindrical capsule leads to the formation of three racemic diastereoisomeric complexes. Their reciprocal ratio depends on the enantiopurity of the guest.

cavity, without exchanging their position on the NMR chemical shift timescale, and they give rise to different resonances. Decreasing gradually the enantiomeric purity of the guest caused the appearance of other diastereoisomeric capsules that were assigned on the basis of the relative statistical abundance indicating no enantioselective binding. The pattern of signals observed is a direct consequence of the enantiomeric distribution of the guest (Figure 2.28). It is therefore possible to determine the enantiopurity of a certain species without the use of any chiral method. Simple encapsulation in an achiral container offers chiral self-sensing of the guest and provides, through formation of diastereoisomeric species, an output that is directly related to the optical purity of the guest [65].

2.4.2.2 Metal–Ligand Hosts

Recently, Fujita applied the peculiar features of the bowl-shaped coordinating self-assembled host composed of six Pd(II) corners and four *tris*-(3-pyridyl)triazine ligands to the enantiomeric self-recognition of binaphthol derivatives as guests in the container in water (Figure 2.29). The hydrophobic cavity of the host can accommodate two molecules of (*R*)-BINOL within the coordination cage. When the assembly is placed in contact with a hexane solution of racemic ($\pm$)-H$_8$BINOL, a new species with a dimeric capsular structure arose, characterized by the inclusion of two homochiral molecules, one of (*R*)-BINOL and one of (*S*)-H$_8$BINOL, while the enantiomeric (*R*)-H$_8$BINOL was excluded from the supramolecular assembly due to steric interactions and remained in the solution from where it was isolated and its enantiopurity degree assessed. The chirality imparted by the two homochiral molecules of (*R*)-BINOL in the constricted space of the apolar cavity of the capsular assembly was sufficient to promote enantioselective binding of a secondary guest of opposite chirality with an ee as high as 87% [66].

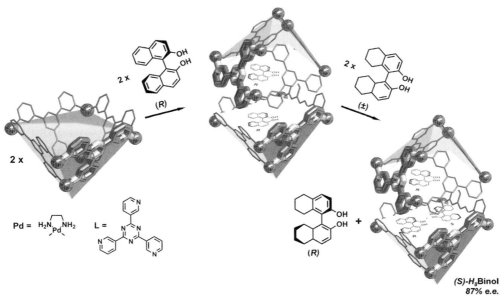

Figure 2.29 Heterobimolecular recognition of (S)-H$_8$BINOL by a chiral cavity provided by two (R)-BINOL guests included into a self-assembled capsule held together by a metal–ligand coordination network.

2.5
Conclusion and Outlook

The study and preparation of supramolecular capsular assemblies is experiencing a transition from being initially a serendipitous discovery, to becoming potentially suitable for several applications, *in primis* drug delivery and, more probably, catalysis. In fact, use of self-assembled capsular catalysts would complement classical organometallic catalysis as well as the emerging new field of organocatalysis for molecular transformations. Some outstanding examples of this powerful approach are already available, characterized by unique features like stabilization of reactive intermediates [67], peculiar substrates, as well as product selectivities [68]. The opportunity provided by supramolecular chirality to control the space that surrounds an encapsulated guest on a stereochemical level theoretically allows fine construction of chiral nanoreactors able to promote asymmetric reactions, with the aim of following the path of Nature in the evolution that ended up with enzymes as the most active and selective known catalysts.

At present, supramolecular chiral hosts, originated by self-assembly of subunits, provide receptive environments that offer low to moderate stereoselectivity, with few exceptions, if compared to that available using conventional reagents or chiral catalysts [69]. However, the observed selectivities are encouraging considering that they rely only on weak attractive and repulsive intermolecular interactions. Stereochemical matching between host and guest surfaces is not sufficient in order to achieve very high stereoselective binding because a large part of the empty space that remains tends to reside between the partners of the encapsulation process [55]. A possible solution in order to tackle this issue would be the implementation of attractive interactions that will complement the steric ones and favor closer approach between host and guest. This would have positive cooperative effects as observed with enzymes and natural receptors that tend to modify their tridimensional structure in order to better complement the noncovalent interactions with the guest [30, 70].

Acknowledgments

The authors wish to thank Dr. Trevor Dale and Prof. Julius Rebek Jr for reviewing this manuscript.

Abbreviations and Symbols

BAR	barbiturate
BINOL	1,1′-bi(2-naphthol)
CD	circular dichroism
CYA	cyanurate
de	diastereoisomeric excess
ee	enantiomeric excess
I	ionic strength
L	ligand
H_8BINOL	5,5′,6,6′,7,7′,8,8′-Octahydro(1,1′binaphthalene)-2,2′-diol
M	metal
PC	packing coefficient
^{1}H-NMR	proton nuclear magnetic resonance

References

1 Feringa, B.L. and van Delden, R.A. (1999) *Angewandte Chemie-International Edition*, **38**, 3418–3438.

2 (a) Blaser, H.U., Spindler, F. and Studer, M. (2001) *Applied Catalysis A*, **221**, 119–143; (b) Blaser, H.U. (2003) *Chemical Communications*, 293–296.

3 Lehn, J.-M. (1995) *Supramolecular Chemistry: Concempts and Perspectives*, VCH, Weinheim.

4 Lawrence, D.S., Jiang, T. and Levett, M. (1995) *Chemical Reviews*, **95**, 2229–2260.

5 Philp, D. and Stoddart, J.F. (1996) *Angewandte Chemie-International Edition*, **35**, 1154–1196.

6 Caulder, D.L. and Raymond, K.N. (1999) *Accounts of Chemical Research*, **32**, 975–982.

7 (a) Chen, H., Weiner, W.S. and Hamilton, A.D. (1997) *Current Opinion in Chemical Biology*, **1**, 458–466; (b) Fitzmaurice, R.J., Kyne, G.M., Douheret, D. and Kilburn, J.D. (2002) *Journal of the Chemical Society-Perkin Transactions 1*, 841–864.

8 Gibb, B.C. (2002) *Journal of Supramolecular Chemistry*, **2**, 123–131.
9 Scarso, A. and Rebek, J., Jr (2006) Chiral Spaces in Supramolecular Assemblies, in Topics, in *Current Chemistry Volume 265: Supramolecular Chirality* (eds. Crego-Calama, M. and Reinhoudt, D.N.), Springer-Verlag, Berlin, Heidelberg, pp. 1–46.
10 Scarso, A., Shivanyuk, A., Hayashida, O. and Rebek, J., Jr (2004) Chiral Spaces in Encapsulation Complexes, Chapter 22, in *Progress in Biological Chirality* (eds G. Palyi, C. Zucchi and L. Caglioti), Elsevier, Oxford (GB), pp. 261–270.
11 Mateos-Timoneda, M.A., Crego-Calama, M. and Reinhoudt, D.N. (2004) *Chemical Society Reviews*, **33**, 363–372.
12 Seeber, G., Tiedemann, B.E.F. and Raymond, K.N. (2006) Supramolecular Chirality in Coordination Chemistry in Topics, in *Current Chemistry Volume 265: Supramolecular Chirality* (eds. Crego-Calama, M. and Reinhoudt, D.N.), Springer-Verlag, Berlin, Heidelberg, pp. 147–183.13.
13 Example of chiral enantioselective self-recognition of racemic building blocks: (a) Claessens, C.G. and Torres, T. (2002) *Journal of the American Chemical Society*, **124**, 14522–14523; (b) Kim, H.-J., Moon, D., Lah, M.S. and Hong, J.-I. (2002) *Angewandte Chemie-International Edition*, **41**, 3174–3177.
14 Seeber, G., Tiedemann, B.E.F. and Raymond, K.N. (2006) Supramolecular Chirality in Coordination Chemistry in Topics, in *Current Chemistry Volume 265: Supramolecular Chirality* (eds. Crego-Calama, M. and Reinhoudt, D.N.), pp. 147–183.
15 (a) Conn, M.M. and Rebek, J., Jr (1997) *Chemical Reviews*, **97**, 1647–1668; (b) Rebek, J., Jr (1996) *Chemical Society Reviews*, 255–264; (c) Rebek, J., Jr (2000) *Chemical Communications*, **8**, 637–643; (d) Rebek, J., Jr (1999) *Accounts of Chemical Research*, **32**, 278–286; (e) Rebek, J., Jr (2005) *Angewandte Chemie-International Edition*, **44**, 2068–2078; (f) Fujita, M. (1998) *Chemical Society Reviews*, **27**, 417–425; (g) Leininger, S., Olenyuk, B. and Stang, P.J. (2000) *Chemical Reviews*, **100**, 853–908; (h) Fiedler, D., Leung, D.H., Bergman, R.G. and Raymond, K.N. (2005) *Accounts of Chemical Research*, **38**, 349–358.
16 Lehn, J.-M. (2007) *Chemical Society Reviews*, **36**, 151–160.
17 Goshe, A.J., Steele, I.M., Ceccarelli, C., Rheingold, A.L. and Bosnich, B. (2002) *Proceedings of the National Academy of Sciences of the United States of America*, **99**, 4823–4829.
18 Prins, L., Reinhoudt, D.N. and Timmerman, P. (2001) *Angewandte Chemie-International Edition*, **40**, 2382–2426, and references therein.
19 Pitt, M.A. and Johnson, D.W. (2007) *Chemical Society Reviews*, **9**, 1441–1453.
20 Hossain, M.A. and Schneider, H.-J. (1999) *Chemistry – A European Journal*, **5**, 1284–1290.
21 Meyer, E.A., Castellano, R.K. and Diederich, F. (2003) *Angewandte Chemie-International Edition*, **42**, 1210–1250.
22 Meyer, .E.A., Castellano, R.K. and Diederich, F. (2003) *Angewandte Chemie-International Edition*, **42**, 1210–1250.
23 Schottel, B.L., Chifotides, H.T. and Dunbar, K.R. (2008) *Chemical Society Reviews*, **37**, 68–83.
24 Nishio, M. (2005) *Tetrahedron*, **61**, 6923–6950.
25 (a) Metrangolo, P. and Resnati, G. (2001) *Chemistry – A European Journal*, **7**, 2511–2519; (b) Auffinger, P., Hays, F.A., Westhof, E., Ho, O.S. (2004) *Proceedings of the National Academy of Sciences, USA*, **101**, 16789–16794.
26 Blokzijl, W. and Engberts, J.B.F.N. (1993) *Angewandte Chemie-International Edition*, **32**, 1545–1579.
27 Mulder, A., Huskens, J. and Reinhoudt, D.N. (2004) *Organic and Biomolecular Chemistry*, **2**, 3409–3424.
28 Mammen, M., Choi, S.-K. and Whitesides, G.M. (1998) *Angewandte Chemie-International Edition*, **37**, 2754–2794.

29 (a) Jorgensen, W.L. and Pranata, J. (1990) *Journal of the American Chemical Society*, **112**, 2008–2010; (b) Murray, T.J. and Zimmerman, S.C. (1992) *Journal of the American Chemical Society*, **114**, 4010–4011.

30 Houk, K.N., Leach, A.G., Kim, S.P. and Zhang, X. (2003) *Angewandte Chemie-International Edition*, **42**, 4872–4897.

31 Palmer, L.C. and Rebek, J., Jr (2004) *Organic and Biomolecular Chemistry*, **2**, 3051–3059.

32 Pluth, M.D. and Raymond, K.N. (2007) *Chemical Society Reviews*, **36**, 161–171.

33 Heard, P.J. (2007) *Chemical Society Reviews*, **36**, 551–569.

34 Kang, J. and Rebek, J., Jr (1996) *Nature*, **382**, 239–241.

35 Prins, L.J., Timmerman, P. and Reinhoudt, D.N. (2001) *Journal of the American Chemical Society*, **123**, 10153–10163.

36 Prins, L.J., Jolliffe, K.A., Hulst, R., Timmerman, P. and Reinhoudt, D.N. (2000) *Journal of the American Chemical Society*, **122**, 3617–3627.

37 Prins, L.J., Hulst, R., Timmerman, P. and Reinhoudt, D.N. (2002) *Chemistry – A European Journal*, **8**, 2288–2301.

38 Lee, H.Y., Moon, D., Lah, M.S. and Hong, J.-I. (2006) *The Journal of Organic Chemistry*, **71**, 9225–9228.

39 Szabo, T., Hilmersson, G. and Rebek, J., Jr (1998) *Journal of the American Chemical Society*, **120**, 6193–6194.

40 Tokunaga, Y. and Rebek, J., Jr (1998) *Journal of the American Chemical Society*, **120**, 66–69.

41 Rivera, J.M., Martín, T. and Rebek, J., Jr (1998) *Science*, **279**, 1021–1023.

42 Rivera, J.M., Martín, T. and Rebek, J., Jr (2001) *Journal of the American Chemical Society*, **123**, 5213–5220.

43 Rebek, J., Jr (2000) *Chemical Communications*, 637–643.

44 Pop, A., Vysotsky, M.O., Saadioui, M. and Böhmer, V. (2003) *Chemical Communications*, 1124–1125.

45 Castellano, R.K., Kim, B.H. and Rebek, J., Jr (1997) *Journal of the American Chemical Society*, **119**, 12671–12672.

46 Thondorf, I., Rudzevich, Y., Rudzevich, V. and Böhmer, V. (2007) *Organic and Biomolecular Chemistry*, **17**, 2775–2782.

47 (a) Caulder, D.L., Powers, R.E., Parac, T.N. and Raymond, K.N., (1998) *Angewandte Chemie-International Edition*, **37**, 1840–1843; (b) Scherer, M., Caulder, D.L., Johnson, D.W. and Raymond, K.N. (1999) *Angewandte Chemie-International Edition*, **38**, 1587–1592; (c) Johnson, D.W. and Raymond, K.N. (2001) *Inorganic Chemistry*, **40**, 5157–5171.

48 Weitl, F.L. and Raymond, K.N. (1979) *Journal of the American Chemical Society*, **101**, 2728–2731.

49 Prins, L.J., De Jong, F., Timmerman, P. and Reinhoudt, D.N. (2000) *Nature*, **408**, 181–184.

50 Prins, L.J., Verhage, J.J., de Jong, F., Timmerman, P. and Reinhoudt, D.N. (2002) *Chemistry – A European Journal*, **8**, 2302–2313.

51 Ishi,-I, T., Crego-Calama, M., Timmerman, P., Reinhoudt, D.N. and Shinkai, S. (2002) *Journal of the American Chemical Society*, **124**, 14631–14641.

52 Rivera, J.M., Craig, S.L., Martín, T. and Rebek, J., Jr (2000) *Angewandte Chemie-International Edition*, **39**, 2130–2132.

53 Terpin, A.J., Ziegler, M., Johnson, D.W. and Raymond, K.N. (2001) *Angewandte Chemie-International Edition*, **40**, 157–160.

54 Ziegler, M., Davis, A.V., Johnson, D.W. and Raymond, K.N. (2003) *Angewandte Chemie-International Edition*, **42**, 665–668.

55 Mecozzi, S. and Rebek, J., Jr (1998) *Chemistry – A European Journal*, **4**, 1016–1022.

56 Castellano, R.K., Nuckolls, C. and Rebek, J., Jr (1999) *Journal of the American Chemical Society*, **121**, 11156–11163.

57 Ishi,-I, T., Mateos-Timoneda, M.A., Timmerman, P., Crego-Calama, M., Reinhoudt, D.N. and Shinkai, S. (2003) *Angewandte Chemie-International Edition*, **42**, 2300–2305.

58 Aimi, J., Oya, K., Tsuda, A. and Aida, T. (2007) *Angewandte Chemie-International Edition*, **46**, 2031–2035.

59 Hiraoka, S. and Fujita, M. (1999) *Journal of the American Chemical Society*, **121**, 10239–10240.

60 Ikeda, A., Udzu, H., Zhong, Z., Shinkai, S., Sakamoto, S. and Yamaguchi, K. (2001) *Journal of the American Chemical Society*, **123**, 3872–3877.

61 Fiedler, D., Leung, D.H., Bergman, R.G. and Raymond, K.N. (2004) *Journal of the American Chemical Society*, **126**, 3674–3675.

62 Brumaghim, J.L., Michels, M. and Raymond, K.N. (2004) *European Journal of Organic Chemistry*, 4552–4559.

63 Heinz, T., Rudkevich, D.M. and Rebek, J., Jr (1999) *Angewandte Chemie-International Edition*, **38**, 1136–1139.

64 (a) Scarso, A., Shivanyuk, A., Hayashida, O. and Rebek, J., Jr (2003) *Journal of the American Chemical Society*, **125**, 6239–6243; (b) Palmer, L.C., Zhao, Y.-L., Houk, K.N. and Rebek, J., Jr (2005) *Chemical Communications*, **29**, 3667–3669.

65 Yamanaka, M. and Rebek, J., Jr (2004) *Chemical Communications*, 1690–1691.

66 Yoshizawa, M., Tamura, M. and Fujita, M. (2007) *Angewandte Chemie-International Edition*, **46**, 3874–3876.

67 Rebek, J., Jr (2006) *Nature*, **444**, 557, and references therein.

68 Vriezema, D.M., Comellas Aragonès, M., Elemans, J.A.A.W., Cornelissen, J.J.L.M., Rowan, A.E. and Nolte, R.J.M. (2005) *Chemical Reviews*, **105**, 1445–1489.

69 Koblenz, T.S., Wassenaar, J. and Reek, J.N.H. (2008) *Chemical Society Reviews*, **37**, 247–262.

70 Williams, D.H., Stephens, E., O'Brien, D.P. and Zhou, M. (2004) *Angewandte Chemie-International Edition*, **43**, 6596–6616.

3
Chiral Nanoparticles
Cyrille Gautier and Thomas Bürgi

3.1
Introduction

The future development of nanotechnology is likely to rely on the construction of new devices by self-assembly of nanoscale building blocks. This strategy is largely inspired by life, which relies on self-assembly of molecules, and that is dominated by homochirality, as emphasized in Chapter 1. While the origin of homochirality remains unclear, the transcription and amplification of chirality from small chiral molecules such as *L*-amino acids to large biological materials like proteins inevitably played a crucial role in its evolution. Obviously, chirality of nanoparticles (NPs), which are promising building blocks for the bottom-up approach, will become an important parameter in the miniaturization race. The properties and structures of inorganic solids are usually catalogued without reference to their size. However, when dimensions of bulk materials are shrinking down to the nanoscale, intrinsic properties and geometry are turned upside down. New behaviors appear and the structure of nanomaterials may significantly differ from the highly symmetric structure found in bulk materials and in some cases, low-symmetry chiral structures become more stable. Future technologies and integration of NPs in new devices will rely on these size effects, which are briefly introduced in Section 3.2.1. NPs behaviors are easily and finely tunable by adjusting parameters such as composition, size, shape, organization and functionalization. During the last twenty years, new characterization tools such as electron and scanning probe microscopes have been developed that allow the investigation of materials with a resolution close to the atomic level. In parallel, a large variety of synthetic routes for the preparation of NPs have been developed, as will be described in Section 3.2.2. After a brief overview of general preparation routes special emphasis will be given to pathways yielding NPs that are optically active, which are asymmetric catalysts or that are simply observed to be chiral in pure or racemic forms by X-ray diffraction or microscopy. Ensembles of NPs produced by the best available techniques still suffer from inhomogeneous broadening due to impurities or to size and shape distribution. Therefore Section 3.2.3 deals with size and shape separation, which are essential in

Chirality at the Nanoscale: Nanoparticles, Surfaces, Materials and more. Edited by David B. Amabilino
Copyright © 2009 WILEY-VCH Verlag GmbH & Co. KGaA, Weinheim
ISBN: 978-3-527-32013-4

order to understand the chiroptical properties. Manifestation and origins of optical activity in metal-based electronic transitions (MBETs) of monolayer protected NPs are discussed in Section 3.3. The next section deals with optically active coordination clusters, which correspond to the inferior limit of optically active metallic NPs and that can serve as models for a better comprehension of larger particles. The very new and few studies of chiral organic nanoparticles (ONPs) are overviewed in Section 3.5. Despite their novelty, this type of nanomaterial seems to have a great potential in a wide range of applications.

NPs play an important role in catalysis. For example, Haruta has discovered that gold NPs dispersed across the surfaces of certain oxides are active catalysts for carbon monoxide oxidation at room temperature, whereas bulk gold is totally inactive [1]. Actually, the use of NPs in catalysis is in constant progress and recently, few examples of enantioselective catalysis with NPs have been reported and are enumerated in Section 3.6. This section also discusses the potential applications of chiral NPs in various fields such as enantiodiscrimination but also to their possible integration and organization in liquid crystals (LCs). Finally, as will become evident by reading this chapter, the study of chiral NPs is still in its discovery phase. The last section will describe the perspectives of these types of materials.

3.2
Nanoparticle Properties and Synthesis

3.2.1
Nanoparticle Properties

This section gives a short introduction on NP properties in general, in order to facilitate the understanding of asymmetric NP properties. Several books and reviews describe this subject extensively [2–5]. NPs can be understood as intermediate between molecular (atomic) and bulk matter and their physical properties are neither those of molecules nor those of bulk material. From a physical point of view, most of the properties of the inorganic NPs are related to three phenomena: surface plasmon (SP) resonance, quantum size effects (QSEs) for metallic and semiconducting NPs, respectively, and reduction in melting temperature (T_m) of small metallic and semiconducting particles with the decrease in the crystal size. For example, the band gap of CdS can be tuned between 2.5 and 4 eV and the T_m varies from 1600 down to 400 °C when the size is reduced [6].

As illustrated in Figure 3.1a, SP resonance arises from the interaction between surface charges of metallic NPs and an electromagnetic field [7]. SPs are waves that propagate along the surface of the conducting NPs. The free electrons respond collectively by oscillating in resonance with light giving rise to absorption and scattering of light. The resulting red color of gold NPs has been used for centuries in coloration of stained-glass windows. On the other hand, the QSEs refer to electron confinement in semiconducting NPs. Both SP resonance and QSEs of NPs are easily tunable. Important modifications of electrical and optical properties can be observed,

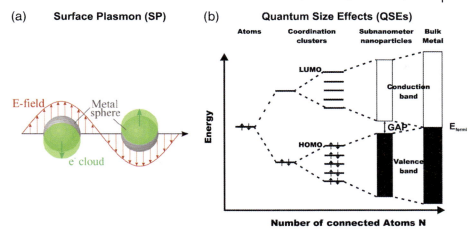

Figure 3.1 (a) A schematic representation of the plasmon oscillation for a sphere, showing the displacement of the conduction electron charge cloud relative to the nuclei. Reprinted with permission from Ref. [7]. Copyright (2003) American Chemical Society. (b) Schematic illustration of the electronic energy levels as a function of the number of connected atoms. Some bulk metals such as gold become semiconducting when their size is reduced to the nanoscale. From left to right, evolution of energy levels by binding more and more metal atoms together. The discrete energy levels observed in coordination clusters merge into energy bands for small NPs (1–3 nm in diameter) with a band gap conferring semiconducting properties. The band gap progressively decreases as the size of the NPs increases and the NPs are becoming metallic as the corresponding bulk metal.

which may be used for new applications such as fabrication of transistors, biosensors and catalysts. The physical behavior is dictated by the size, the shape, the nature, the composition, and even the organization of NPs. These physical characteristics can be combined with chemical and biological properties for hybrid NPs, which are surrounded with organic molecules or biological materials.

As illustrated in Figure 3.2, the most striking demonstration of QSEs is the variation of color with size. This is shown in Figure 3.2 (A) for gold NPs and is further evidenced by their systematic bathochromic shift in absorption spectra as their size is increasing (B). For semiconducting NPs the emission is a function of their size (C). Section 3.3 shows that the chiroptical properties of NPs are also dictated by QSEs.

When the size of metallic and semiconducting particles is reduced, their dispersion, that is their fraction of surface atoms, increases, and so does the surface to volume ratio. At the nanoscale, the large contribution of the surface energy to the overall system is responsible for the large change in thermodynamic properties as for example the diminution of T_m. This also has important consequences on the structure of NPs and could be a driving force for inducing chirality in the latter. The shape is also an important factor for the tuning of physical properties. For example, the SP resonance band of spherical NPs splits into different bands for anisotropic NPs such as nanorods (NRDs) [8]. Colloidal semiconducting NRDs such as CdSe rods emit linearly polarized light that is highly desirable in a wide variety of

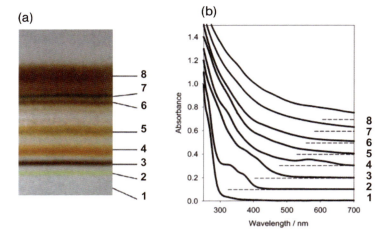

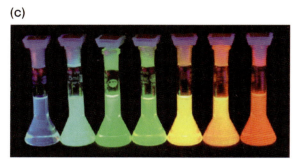

Figure 3.2 (a) Gold NPs separated according to their size by polyacrylamide gel electrophoresis (PAGE). NPs are numbered according to their increasing size. (b) UV-vis spectra of the size-separated gold NPs. A clear red shift of the absorption onset is observed as the size of NPs increases. A and B are reprinted with permission from Ref. [38]. Copyright (2006) American Chemical Society. (c) Fluorescence induced by exposure to ultraviolet light of vials containing various sized cadmium selenide (CdSe) quantum dots.

applications, whereas emission of similar spherical quantum dots is not polarized [9]. The surface of nanocrystals exposes different crystallographic facets, which can grow at different rates or be selectively functionalized [10, 11]. This allows the preparation of NPs with a large range of shapes such as NRDs, nanotripods and nanostars. Such NPs are ideal "building blocks" for the bottom-up preparation of new nanodevices through self-assembling of NPs directed by functional groups within their organic shell [10, 12]. Recently, chiral single-walled nanotubes of Pt and Au were observed by ultrahigh vacuum transmission electron microscopy [13, 14]. They consist of five or six atomic rows that coil helically around the axis of the tube.

Organization of metallic NPs has a large impact on their electronic and optical properties. For example, arrays of pairs of parallel NRDs have been shown to have a negative refractive index at the optical communication wavelength. Such metamaterials have been predicted to act as perfect lenses [15]. Due to a coupling of the SP resonance the formation of a polymeric network of gold NPs in solution is observable through a color change from red to blue. This can be used for colorimetric sensing of biomolecules [10, 12, 16]. The controlled coupling of SPs of well-ordered NP lattices is promising for preparing optoelectronic devices [17]. Experiments and calculations have shown that arrays of chiral micro- and nanometallic particles with subwavelength period can rotate the polarization plane azimuth of the transmitted wave at normal incidence [18, 19]. This behavior is similar to that observed for chiral LCs and opens the possibility to create planar polarization-sensitive devices via self-assembly of NPs with a chiral shape or a chiral organization.

ONPs also show a very broad range of interesting properties like intensive color, fluorescence, photoconductivity and photochemical behavior. As for their inorganic analogues, the optical properties of ONPs depend on their size. In particular, ONPs of polycyclic aromatic hydrocarbons are considered as "transition" materials and share many properties with conventional semiconductors. Their excitonic transitions are responsible for the size effect on the absorption and emission of the ONPs. For example, charge-transfer transitions between the stacked molecules in the 1-phenyl-3-((dimethylamino)styryl)-5-((dimethylamino)phenyl)-2-pyrazoline (PDDP) NPs as well as $\pi-\pi^*$ transitions are shifted to the high-energy side with decreasing NP size [20]. The photoluminescence properties of organic molecules upon formation of ONPs can either be quenched or largely increased depending on the molecular shapes, conformational flexibility, intramolecular movements, packing structure and organizational morphology. Molecules that are very weakly luminescent in solution can become highly fluorescent upon formation of ONPs due to aggregate-induced enhanced emission (AIEE) in ONPs. These properties offer promising potential use for ONPs in optoelectronics and biology. Very recently, Park *et al.* reported photoswitchable ONPs using a molecule with a fluorescent unit covalently bound to a photochromic moiety [21]. The fluorescent unit is virtually nonfluorescent in the solution but it shows an AIEE in the ONPs. This type of high-contrast ON/OFF fluorescence switching can potentially be used in rewritable high-density optical data storage.

3.2.2
Preparation, Purification and Size Separation

3.2.2.1 Preparation

The literature dealing with inorganic NPs synthesis is uncountable and a wide variety of techniques have been reported using precursors from liquid, solid and gas phases. Metallic NPs can be prepared by the top-down (physical) approach for example by mechanical subdivision of metallic aggregates or by lithography techniques. NPs can also be prepared by the opposite approach called bottom-up (chemical). The top-down approach easily allows control of the organization of NPs in 1, 2 and 3 dimensions. However, the bottom-up approach is considerably cheaper and allows the preparation of very small particles.

The chemical approach is based on the nucleation and growth of metallic atoms in liquid media or in gas phase. The literature describes five general methods for the chemical preparation of transition metal NPs. (1) Chemical reduction of metal salts, (2) thermal, photochemical or sonochemical decomposition, (3) electrochemical reduction, (4) metal-vapor synthesis and (5) ligand reduction and displacement from organometallic complexes.

Among the huge number of methods listed in the literature, only a few have been used for the preparation of optically active metal NPs. A few examples of micro- and nanoparticle arrays (not necessarily of chiral NPs) with chiral shapes [18] or chiral organizations [19, 22] have been studied theoretically [19] or prepared using top-down methods such as electron-beam lithography and lift-off techniques.

Most optically active metallic NPs have been prepared by the bottom-up approach and more specifically by the chemical reduction of metal ions or of precursor complexes in solution, which is illustrated in Figure 3.3 [23]. The reduction of the metal salts is often performed in the presence of chiral stabilizers such as thiols, phosphines and amines (DNA and alkaloid) in a one-pot procedure (direct syntheses). (A) NPs are prepared in a single phase when the metallic salt and the ligand are both soluble in the same solvent such as water, alcohols, acetic acid, tetrahydrofuran (THF) or a mixture of these. (B) If their solubilities are not compatible, the Brust–Schiffrin synthesis [24] provides a ready access in a biphasic reaction taking advantage of phase-transfer compounds, generally tetraoctylammonium bromide (TOAB), to shuttle ionic reagents to an organic phase where particle nucleation, growth and passivation occur. (C) When the functional groups are not compatible with the reducing agent, the optical activity can be induced via a postsynthetic modification of the ligand shell like a ligand exchange reaction (indirect syntheses). This two-step method presents some advantage in both purification and separation. Indeed, some NPs are very well characterized and both their separation and purification processes are well established. The gold NPs prepared by the citrate method of Turkevich et al. [25] and those developed by Schmid [26], which are protected by triphenylphosphine (TPP), are surely the more frequently used NPs for the ligand exchange reaction. This strategy avoids a time-consuming search of efficient parameters for purification and separation of each new type of NPs. However, the postsynthetic modifications such as ligand exchange in an excess of ligand often modify the size distribution of the NPs [27–30].

ONPs are generally prepared using simple methods such as reprecipitation [31], evaporation or formation of microemulsion [32]. However, many other methods have been reported such as laser ablation of organic crystal in a liquid [33] and vapor-driven self-assembly (VDSA) [34]. In reprecipitation, the most commonly used process, a molecularly disperse solution of the organic compounds in a water-miscible solvent is mixed vigorously with an aqueous phase, which induces the nucleation and growth of ONPs. The size is controlled by the growth parameters such as the monomer and the solvent concentrations. The stability of ONPs prepared by the reprecipitation method can be increased by the formation of a protective layer using a surfactant or a water-soluble polymer as for example gelatin or poly(vinyl alcohol) (PVA). The size of ONPs can be tailored by adding the stabilizer at different aging times [35].

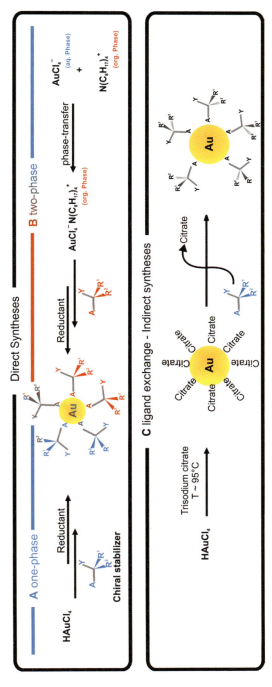

Figure 3.3 Different methods exploited for the syntheses of optically active NPs with chiral ligands.

3.2.3
Purification and Separation of Nanoparticles

The physical properties of NPs are strongly size and shape dependent, as emphasized in the last section. Fine tuning of properties for a specific purpose often requires a perfect control of these two parameters as well as a high control of functionalization and purity. Procedures for the preparation of NPs are in constant progress but till now only few methods yield monodisperse particles with high purity. However, this fine tuning can be reached after appropriate purification and separation processes. Current methods for purification of NPs samples, that is removal of free ligand and reducing agent, involve centrifugation, precipitation, washing, dialysis, chromatography or extraction to remove impurities. Size selection can be based on fractional crystallization, size exclusion chromatography (SEC), electrophoresis and membrane based methods such as ultrafiltration or diafiltration. These techniques are particularly appropriate for the size separation of water-soluble NPs but are generally time consuming and only viable for production on a small scale. The size separation of NPs soluble in organic media suffers from a lack of methods compatible with organic solvents. The fundamental studies of the size-dependent chiroptical properties have been mainly performed with water-soluble NPs separated according to their size and charge by high-density polyacrylamide gel electrophoresis PAGE (see Figure 3.2) [36–40]. The only size selection of chiral NPs soluble in organic media has been realized using SEC [41].

3.3
Chiroptical Properties of Inorganic Nanoparticles

Due to their organic shell, monolayer-protected NPs can be dissolved in various solvents and are thus amenable to chiroptical techniques such as electronic circular dichroism (ECD) and vibrational circular dichroism (VCD). The former has demonstrated its aptitude for the study of protein secondary and tertiary structures, whereas the latter has been used for the determination of conformation and absolute configuration of organic molecules in solution [42, 43]. Recently, these complementary techniques have been applied to gold NPs covered with different chiral organic ligands. VCD in the infrared region selectively probes molecular vibrations located in the organic shell, whereas ECD in the UV-visible region is sensitive towards electronic transitions that may be located in the inorganic core.

3.3.1
Vibrational Circular Dichroism

VCD spectroscopy was recently applied to gold NPs of about 2 nm core diameter in order to study the conformation of adsorbed chiral cysteine derivatives [38, 44]. In order to extract structural information the measured spectrum has to be compared with the calculated ones for different conformers. Figure 3.4 shows IR and VCD

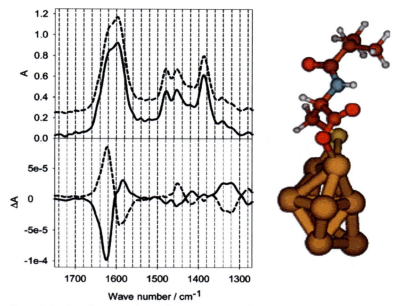

Figure 3.4 Infrared (top) and VCD (bottom) spectra of NILC protected gold NPs. The calculated VCD spectrum of the conformer on the right on a Au_8 cluster fits best the experimental spectrum. Reproduced with permission from Ref. [38]. Copyright (2006) American Chemical Society.

spectra of gold NPs covered by the two enantiomers of N-isobutyryl-cysteine (NIC). Whereas the IR spectra are identical for the two enantiomers, the VCD spectra show a mirror-image relationship. Density functional theory (DFT) calculations show that the structure of the underlying gold cluster does not have a large effect on the simulated VCD spectra, whereas the conformation of the adsorbed thiol has a large influence. The calculated VCD spectrum of one stable conformer of NIC adsorbed on a small gold cluster matches well with the experimental data. This conformation is characterized by an interaction of the carboxylate with the gold cluster (see Figure 3.4). Thus the carboxylate group seems to be a second anchoring point beside the strong Au–S bond. This two-point interaction may influence the optical activity of the NPs as discussed in Section 3.3.3.

3.3.2
Circular Dichroism

It is tempting to assume that the structure of metallic NPs corresponds to a fragment of the highly symmetric bulk crystal lattice. However, in 1996 calculations performed by Wetzel and DePristo, and experimental observation of Riley *et al.* indicated that naked Ni_{39} clusters prefer a lower (D_5) symmetry chiral structure [45, 46]. At the same time, Whetten and coworkers experimentally observed optical activity in the MBETs for a Au_{28} cluster covered with L-glutathione (GSH), a chiral tripeptide [36]. They, furthermore, isolated similar well-defined clusters with different mass by PAGE. The three smaller isolated clusters with a core mass between 4 and 8 kDa showed a strong optical activity, whereas neither the crude mixture nor the higher molecular weight components possess such a strong optical activity. The optical activity in the near-infrared (NIR), visible and near-ultraviolet is clearly size dependent and its amplitude is comparable to the signals observed for intrinsically chiral conjugated systems like chiral fullerenes or larger helicenes.

Recently, optical activity in MBETs has been reported for gold, silver and palladium nanoclusters having a more or less well-defined size and an organic shell composed of different chiral molecules [23, 37–39, 47–49]. Figure 3.5 summarizes the chemical structures of the different molecules that have been shown to induce optical activity in NPs as well as their size and the maximum of amplitude of the anisotropy factor ($\Delta\varepsilon/\varepsilon$) measured for gold or silver NPs.

Pairs of metal clusters with similar inorganic core, size or size distribution and the same number and type of ligands but with opposite absolute configuration exhibit mirror image ECD spectra as common chiral molecules do. No optical activity is observed for the clusters prepared with a racemic mixture of ligand. However, in the case of silver particles covered by penicillamine it was reported that the UV-vis spectra of the separated NPs are slightly different from the homochiral ones [37]. These silver NPs are not as stable as gold NPs. In all cases where optical activity in size-selected NPs was observed, the ECD signals change with the size of the NPs. In some cases the anisotropy factor gradually increases with a decrease in the mean cluster diameter [37–39]. Thus, it seems that the subnanometer and the nanometer classes of nanoclusters are the best candidates for displaying optical activity. This may be related to the fact that in this range scale, most of the metal atoms reside at the surface of the core and thus interact directly with the chiral ligands. Another explanation for the tendency of decreasing optical activity with increasing particle size is simply the increased conformational space for larger particles (larger number of gold atoms and ligands) and thus the increased probability of multiple energy minima on the potential energy surface. An increasing number of conformers leads to a decreased observable optical activity as positive and negative bands of different conformers average out.

In contrast, optical activity was also reported by Park and coworkers for considerably larger penicillamine or cysteine capped silver particles (23.5 nm) [48] and also for silver nanocrystals grown on a double-stranded DNA scaffold [49]. However, these particles were not size separated, and it cannot be excluded that the observed optical activity is due to a fraction of small particles. Recently, this hypothesis was verified by Kimura and coworkers who have separated silver NPs according to their size and observed optical activity only for NPs in the nanometer range [39].

When comparing the characteristics of all the ligands able to impart optical activity in the MBETs, it becomes evident that most of them and especially cysteine derivatives are able to perform hydrogen-bond-mediated self-assembly. This property was proposed to be a crucial parameter for inducing optical activity [48, 50]. However, the atropisomeric bidentate ligands BINAS and BINAP do not display such behavior,

3.3 Chiroptical Properties of Inorganic Nanoparticles

Name	Structure	METAL $\Delta\varepsilon/\varepsilon$ max. SIZE
N-acetyl-L-cysteine (NALC)	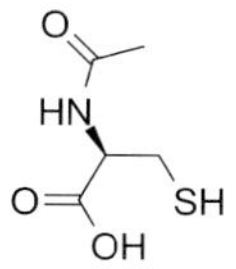	
N-isobutyryl-L-cysteine (NILC)	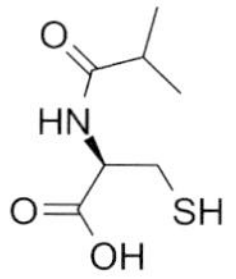	Au 0.5×10^{-3} (1 to 2 nm)
L-cysteine	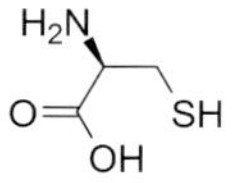	
L-cysteine methyl ester	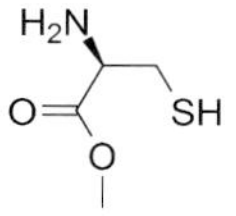	Au 0 (20 to 25 nm)
L-penicillamine (L-pen)	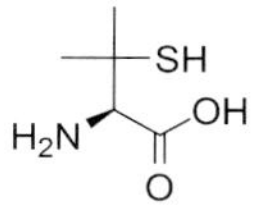	Au $\quad$ Ag 3×10^{-4} to 1×10^{-4} $\quad$ 1×10^{-3} to 1×10^{-5} (0.57 to 1.75 nm) $\quad$ (1 to 1.9 nm)
L-glutathione (GSH)	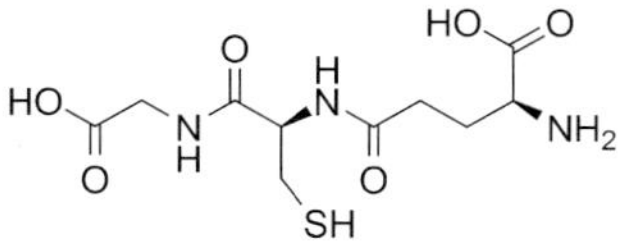	Au 2×10^{-4} to 1×10^{-3} (0.7 to 1 nm)
(R)-BINAS	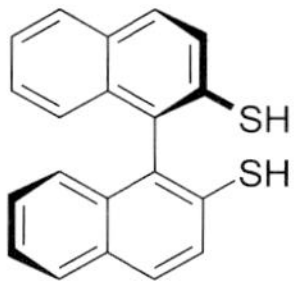	Au 4×10^{-3} (1 to 2 nm)
(R)-BINAP	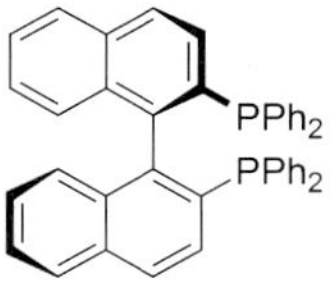	Au 1.4×10^{-3} (1 to 2 nm)

Figure 3.5 Chemical structures of chiral molecules inducing optical activity to metal NPs.

despite the fact that they are particularly well suited to impart optical activity [23, 47]. On the other hand, a common feature is that all the ligands can interact with the cluster surface by at least two functional groups, as emphasized in Section 3.3.1. Furthermore, it was demonstrated that by blocking the acid group (anchoring point) in cysteine derivatives the optical activity was lost [48]. In addition, Hegmann and coworkers studied the optical activity of small gold NPs covered by the chiral naproxen and observed that this molecule, which only possesses one interaction point, does not induce optical activity in MBETs [51]. Recently, CdS quantum dots capped with D- and L-penicillamine have been prepared. These quantum dots are strongly white-emitting and show strong optical activity with an almost identical mirror-image relationship in the range 200–390 nm [52]. The luminescence of such chiral quantum dots does not result in circularly polarized light.

In all of the cases described above, the optical activity in the NIR, visible and UV can be attributed neither to the metallic precursors nor to the organic species and its origin remains unclear due to the lack of structural information and to the very few examples of well-defined nanoclusters.

3.3.3
Origin of Optical Activity in Metal-Based Transitions

The observed optical activity in MBETs of small metal particles protected with chiral thiols can be attributed to two opposite and one intermediate model (see Figure 3.6). In the first one, the optical activity arises from an intrinsically chiral inorganic core (A). In the presence of chiral ligands one of the two possible enantiomers of the core is favored. Such behavior is found, as discussed in Section 3.4, for coordination clusters with a chiral framework. In the second one, the inorganic core can be achiral and the optical activity is induced by a chiral environment due to the chiral organic shell through a vicinal effect or through a chiral electrostatic field (B). Both models have

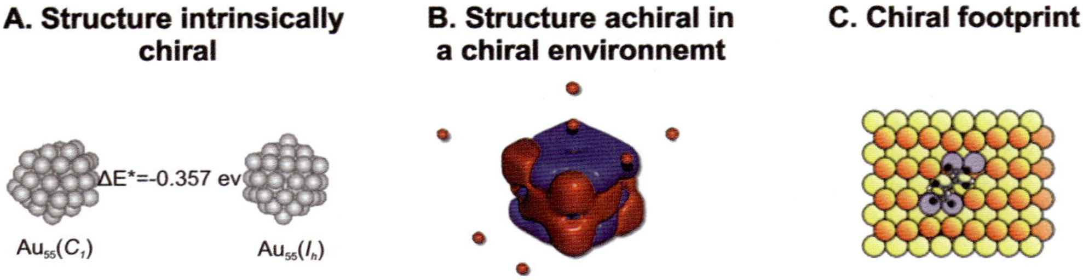

A. Structure intrinsically chiral

$\Delta E^* = -0.357$ ev

$Au_{55}(C_1)$ $Au_{55}(I_h)$

B. Structure achiral in a chiral environnemt

C. Chiral footprint

Figure 3.6 Possible origins of optical activity observed for metal particles. (a) The calculated chiral structure (C_1) of bare Au_{55} is more stable than the highly symmetric structure (I_h). Reprinted from Ref. [53] with kind permission of The European Physical Journal (EPJ). (b) Chiral distribution of electron density in Au_{28} gold clusters induced by a chiral point-charge system. Red points correspond to negative point charges and red and blue surfaces in the core represent, respectively, regions of high and low electron density. Reproduced from Ref. [55] by permission of the PCCP Owner Societies. (c) Chiral footprint imparted by bitartrate on Ni(100) surface. Reproduced with permission from Ref. [56]. Copyright (2002) American Chemical Society.

support from theoretical calculations [53–55]. Garzón *et al.* have predicted that small metal particles such as Au_{28} or Au_{55} prefer low-symmetry chiral over high-symmetry nonchiral structures [53, 54]. Goldsmith and coworkers have demonstrated that optical activity could arise from an achiral metal core perturbed by a dissymmetric field originating from the chiral organic shell [55]. Trends in the electronic transition frequencies and amplitudes with cluster size observed experimentally are qualitatively in agreement with both models. It is likely that the two mechanisms concurrently impart optical activity to the core and the key question whether the core is chiral or not remained unanswered at that point. In the intermediate model, the grand core can be achiral but the relaxation of the surface atoms involved in the adsorption of the chiral ligand creates a chiral "footprint" similarly as observed for tartaric acid adsorption on Ni surfaces (C) [56]. This is favored for ligands that possess at least two anchoring points on the surface. Such double interactions have indeed been documented on surfaces [57–60] and seem to be a common characteristic of the ligand able to induce optical activity in MBETs as discussed in Section 3.3.2. DFT calculations have shown that the ligands are not only playing the role of passivating molecules, but they also distort the metal cluster structure [53]. Very recently, the first total structure determination of a small gold-thiolate nanocluster composed of 102 gold atoms and 44 *p*-mercaptobenzoic acids (*p*-MBA) has been published [61]. As shown in Figure 3.7, the particles are chiral and formed as a racemic mixture in the

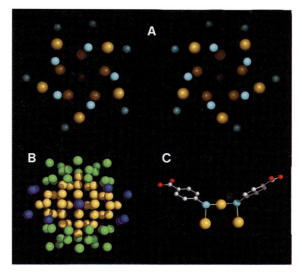

Figure 3.7 X-ray crystal structure determination of the $Au_{102}(p\text{-MBA})_{44}$ NPs. (a) View down the cluster axis of the two enantiomers (gold atoms in yellow and sulfur atoms in cyan). (b) Packing of gold atoms in the core. Marks decahedron (MD) (2,1,2) in yellow, two 20-atom "caps" at the poles in green, and the 13-atom equatorial band in blue. (c) Sulfur–gold interactions in the surface of the NPs. Example of two *p*-MBAs interacting with three gold atoms in a bridge conformation. Gold atoms are yellow, sulfur atoms are cyan, oxygen atoms are red, and carbon atoms are gray. Reprinted from Jadzinsky, Pablo D. *et al.*, Ref. [61], with permission from AAAS.

crystal (A). However, the central gold atoms are packed in a Marks decahedron (MD) that is as highly symmetric as the fcc structure of bulk gold (B). The chirality arises from the geometry of equatorial atoms on the surface. The deviations in local symmetry may reflect the interaction of the equatorial atoms with the *p*-MBA monolayer. Furthermore, most sulfur atoms bonded to two gold atoms in staple motifs are stereogenic centers (C).

Evidently, we can not assume that the structure of the center part of NPs considerably smaller than 102 gold atoms resembles the bulk metal structure. Indeed, Häkkinen and coworkers, using DFT calculations, predicted that the structure of a $Au_{38}(SCH_3)_{24}$ cluster consists of ring-like $(AuSCH_3)_4$ units protecting a central Au_{14} core [62]. The three models described above can still act concomitantly but their individual contribution remained undetermined. Distortion of surface atoms involved in the adsorption of capping agent, is now proved to be a major element in chirality of NPs. Bidentate chiral ligands have demonstrated their ability to induce optical activity in MBETs probably by controlling enantioselectively the geometry of the surface atoms and the absolute configuration of the thiolate-stereogenic centers. Recently, we have shown that the optical activity of Au_{15} and Au_{18} clusters protected with one enantiomer of NIC is reversed with a perfect mirror image relationship upon exchange of the absorbed thiol for its opposite enantiomer. This experiment clearly demonstrates that the optical activity is dictated by the absolute configuration of the absorbed thiols. This optical isomerization implies that if the clusters exhibit intrinsically chiral structures, these structures are not stable enough in one enantiomeric form to withstand a switch of the absolute configuration of the passivating thiol [63].

3.4
Optically Active Coordination Clusters

As NPs and organometallic complexes, coordination clusters have stimulated great interest in the field of catalysis and especially for asymmetric catalysis [64, 65]. In addition to the classical asymmetric induction originating from the chirality of the organic ligands, clusters are expected to be able to induce chirality through their chiral metallic framework.

The development of intrinsically chiral organometallic clusters is closely related to methods for their rational synthesis with a desirable molecular geometry. The work of Beurich [66] on metal-exchange reaction as well as the multistep addition and substitution reactions by systematic incorporation of organometallic units have laid the basis on cluster design [67–71]. Actually, various types of chiral clusters such as hetero- or homometallic tetrahedrane-type clusters [72, 73] or pentanuclear wing-type butterfly heterotrimetallic clusters [74] are accessible by systematic synthetic procedures. Such pathways inevitably lead to a racemic mixture of the chiral metallic framework. The mixture can be enriched in one of its enantiomeric forms when a chiral ligand is used [75]. Tunik *et al.* have reported that the reaction of the ligand (*S*)-BINAP [76] with $H_4Ru_4(CO)_{12}$ or (*S*,*S*)-BOTPHOS [77] with $CpRhRu_3(H)_3(CO)_{10}$

produces clusters with up to 100% stereoselectivity. However, this type of quantitative stereoselectivity is still uncommon and, furthermore, the separation into enantiomers remains one of the major challenges in order to use this material as asymmetric catalyst. The traditional method consists in the derivatization of the racemic mixture with a chiral auxiliary before separation of the diastereoisomers by chromatography or fractional crystallization. This method allows the efficient separation of enantiomers of clusters that are optically active even after removal of the auxiliary [78, 79]. However, this pathway is often time consuming or inefficient and removal of the chiral auxiliary can lead to damaging or racemizing the cluster [75, 80]. High-pressure liquid chromatography is an efficient alternative mild method that does not require derivatization when a chiral stationary phase is used. Nevertheless, very few clusters have been separated by this method [81]. The optical activity of enantiopure chiral clusters is very strong and depends on the composition of the metallic framework as much as on the type of the capping ligand. The presence of a chiral auxiliary modulates the ECD curves but does not change the general appearance of the signal [72].

Metal clusters have become especially important in a wide variety of heterogeneous catalytic reactions such as for example, oxidation, isomerization of alkenes, carbonylation and reduction of multiple bonds [82]. This list is not exhaustive and industrial reactions, such as oxidation of butane to maleic anhydride, can enlarge this panel [83]. However, despite this rich catalytic activity of organometallic clusters, the envisioned breakthrough in the field of asymmetric catalysis using clusters with framework chirality has not been reported yet. This situation is probably due to on the one hand the very small number of enantiopure clusters known today, but also on the other hand due to the racemization of the metallic framework under the catalytic conditions such as CO pressure, UV irradiation or heating [72]. Furthermore, the question whether organometallic complexes and clusters act as catalyst or whether they are precursor for the formation of NPs that are the real catalyst remains open for many reactions. Recently, Mario Barberes and coworkers published the catalytic cyclopropanation of styrene with ethyl diazoacetate using a chiral heterodimetallic cubane-type cluster with 25% enantiomeric excess (ee) but this small ee cannot be attributed only to the chiral metallic framework since the ligands are also chiral [84]. The concept of asymmetric catalysis using a metallic framework still remains to be demonstrated.

3.5
Nanoparticles of Chiral Organic Compounds

In recent years ONPs of various compounds have been the focus of interest of researchers due to their potential use in the fields of optoelectronics, nonlinear optics, photonics, sensing [85], recognition [86], and DNA delivery [87]. ONPs have been extensively described in the book of Masuhara [88]. ONPs bridge the gap between isolated molecules and the bulk crystal. As for their inorganic analogues, the properties of ONPs strongly depend on their size, their shape and also on their

3 Chiral Nanoparticles

assembly. However, in the case of ONPs, the electronic and optical properties are fundamentally different from those of metal and semiconductor NPs. This has to do with the much weaker interaction between the constituents in ONPs, which are typically of the van der Waals type. In the case of ONPs the size-dependent properties arise from aggregation, increased intermolecular interaction (many-body interaction) and surface effects [20, 89]. For example emission spectra are tunable by alteration of the size of the ONPs. Besides this, additional great diversity of properties can be obtained through variation of organic molecular structures. Park and coworkers have shown that ONPs of 1-cyano-*trans*-1,2-bis-(4′-methylbiphenyl)ethylene (CN-MBE) exhibit an enhanced emission when compared to the weak fluorescence of the monomers in solution. The fluorescence switches off in the presence of organic vapor, which is interesting for sensing applications [85]. This unusual enhancement of emission in CN-MBE NPs is attributed to the synergetic effect of intramolecular planarization and *J*-type aggregate formation that restrict excimer formation to NPs.

The optical activity of chiral ONPs is also size dependent. Recently, different ONPs from chiral auxiliary such as (*R*)-(+)-1,1′-bi-2-naphthol dimethyl ether (BNDE), (*R*)-di-2-naphthylprolinol (DNP) and (*R*)-(−)-2,2′-bis-(*p*-toluenesulfonyloxy)-1,1′-binaphthalene (*R*-BTBN) have been prepared and studied by Yao and coworkers [90–92]. Their studies show that the optical activity of the chiral ONPs follows the same trend as the absorption. The exciton chirality peaks evolves to the low-energy side with increase in particle size. Surprisingly, the ECD spectra of the ONPs of BNDE and DNP are completely opposite to ECD spectra of the dilute monomers as shown in Figure 3.8. For the *R*-BTBN NPs, the intensity ratio of the first to the third Cotton effects increases as the size of the ONPs increases to 60 nm. This trend is explained by the more effective excimer formation between the two chromophores in adjacent

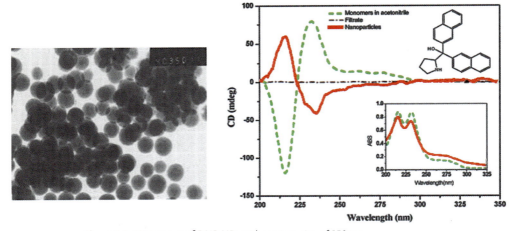

Figure 3.8 TEM image of DNP NPs with average size of 150 nm (left). CD and UV (inset) spectra of DNP monomer (dashed green line), NPs (solid red line) and the filtrate (dash dot black line). The structure of (*R*)-di-2-naphthylprolinol is shown as the inset. Reprinted from ref [91] with permission from RSC.

molecules due to the change of dihedral angle as the ONPs grow in size. These works successfully demonstrate the manipulation of exciton chirality by simply changing the size of the ONPs. This degree of chirality control might be of practical value, for example, for the use of the particles as active components in optically switchable devices and in asymmetric syntheses via photochemistry in the solid state with the ionic chiral auxiliary approach [93]. Cho et al. have demonstrated the ability of chiral hexablock copolymeric NPs for the chiral recognition of bilirubin [86].

3.6
Applications

NPs are of considerable interest because of their wide variety of potential application in various fields such as biosensing [94], optics [95], electronics, photonics [17], catalysis [96], nanotechnology [5] and drug or DNA delivery [87, 97]. Chiral NPs are particularly interesting for asymmetry amplification at different length scales, which means as chiral catalysts for chemical synthesis and chiral-selective membranes or chiral dopant in liquid crystals (LCs). Further applications may be envisaged for the detection of chirality or may be related to the optical activity.

3.6.1
Asymmetric Catalysis

Metallic NPs are potentially able to be active in catalysis. NPs can also be used as catalyst support. In this case the properties of homogeneous and heterogeneous catalysis are combined. For example, the catalytic properties of complexes supported on NPs can be influenced by the neighboring chiral ligands [98]. An advantage of such a catalyst is the rapid separation from products and substrates by using precipitation, membrane-based techniques, SEC, or even magnetic fields. Recently, Hu and coworkers reported the immobilization of preformed Ru catalysts on magnetite NPs. These systems catalyzed the asymmetric hydrogenation of aromatic ketones with an ee up to 98%. The latter chiral catalyst, which can be removed magnetically, was easily recycled up to 14 times without loss of activity and enantioselectivity [99].

The intrinsic catalytic behavior of NPs has been widely examined. Many NPs proved to be efficient and selective catalysts not only for reactions that are also known to be catalyzed by molecular complexes, such as olefin hydrogenation or C–C coupling [2], but, moreover, for reactions that are not or are poorly catalyzed by molecular species, such as aromatic hydrogenation [100]. Unambiguous distinction between colloidal and molecular catalysis is, however, often very difficult to make [101, 102]. As shown in Section 3.3, metallic NPs may be intrinsically chiral and, as discussed for organometallic clusters, optically active NPs can potentially combine the chirality of their core and of their organic shell in order to induce enantioselectivity to prochiral substrates. Despite the impressive progress in catalysis; however, only a few NPs systems have been shown to be efficient in asymmetric catalysis till today [47, 96, 103–110].

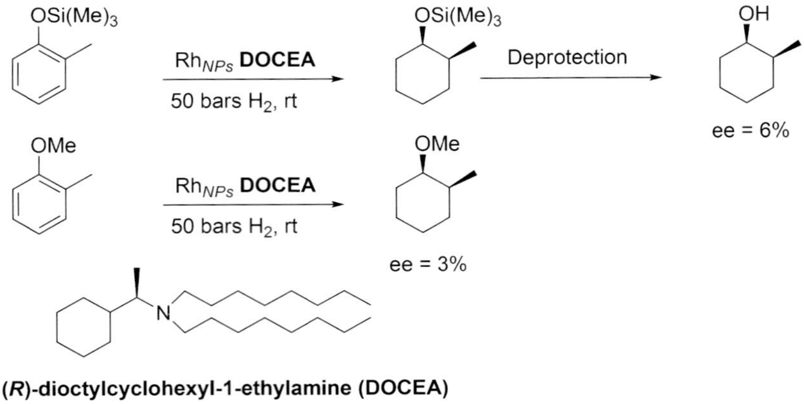

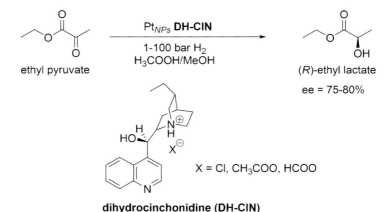

Figure 3.9 Enantioselective reduction of o-cresol trimethylsilyl ether and 2-methylanisol.

The first example of an asymmetric reaction shown in Figure 3.9 was reported in 1994 by Lemaire and coworkers who used rhodium NPs stabilized by a chiral amine (R)-dioctylcyclohexyl-1-ethylamine for the hydrogenation of o-cresol trimethylsilyl ether (top) or 2-methylanisole (bottom) with poor but significant ee values [111].

As shown in Figure 3.10, the most relevant systems involve Pt and Pd NPs stabilized by cinchonidine, in the hydrogenation of pyruvate derivatives. This system was first reported by Orito and coworkers [112] for supported Pt particles and since then much effort has been devoted to unravel its mechanism [113]. Bönneman and coworkers extended this concept to free metal particles in 1996. They reported the hydrogenation of ethyl pyruvate into (R)-ethyl lactate with 75–80% ee using platinum particles stabilized with dihydrocinchonidine salt (DH-CIN) in the liquid phase or immobilized on charcoal or silica [103, 105].

Figure 3.10 Enantioselective reduction of ethyl pyruvate.

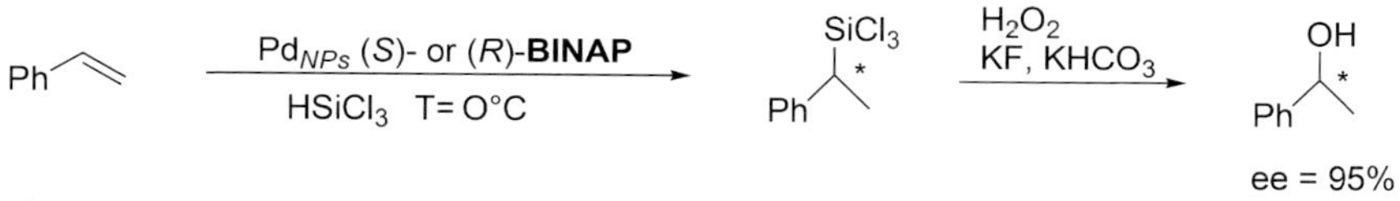

Figure 3.11 Stereoselective hydrosilylation of styrene with trichlorosilane in the presence of chiral Pd catalyst.

The preparation of the catalyst was more recently revisited and Pt NPs stabilized with PVP and modified with cinchonidine have allowed improving the ee up to 95–98% [106–108]. The sense of enantioselectivity can be reversed with respect to the cinchonidine Pt NPs system either by changing the chiral modifier to cinchonine (CN) or by changing Pt for Pd [114]. The latter switch is not explained but it reveals the significant role of the metal in the enantioselectivity. Interestingly, also the size of the NPs seems to influence enantioselectivity.

In 2003, Tamara et al. reported that Pd NPs (2 nm) optically active in MBETs and stabilized by BINAP, catalyze the hydrosilylation of styrene with trichlorosilane with an ee of 95% (see Figure 3.11). Interestingly, palladium complexes coordinated with BINAP are not active for this reaction [47].

Chaudret and coworkers performed the enantioselective allylic alkylation of rac-3-acetoxy-1,3-diphenyl-1-propene with dimethyl malonate catalyzed by palladium NPs (4 nm) in the presence of a chiral xylofuranoside diphosphite. This reaction, shown in Figure 3.12, gives rise to 97% ee for the alkylation product and a kinetic resolution of the substrate recovered with 90% ee [96].

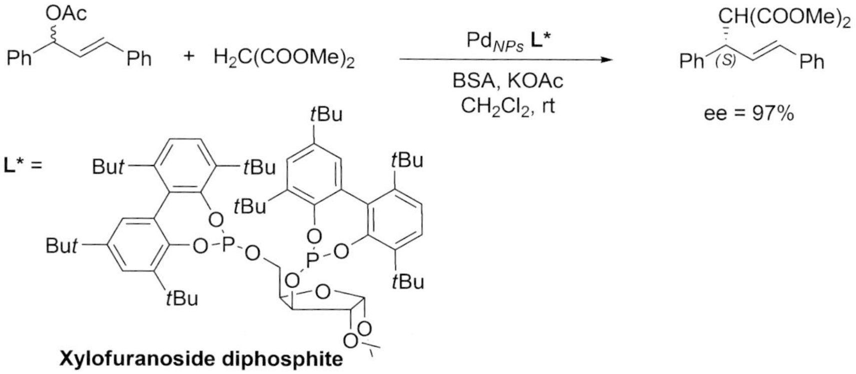

Figure 3.12 Enantioselective allylic alkylation catalyzed by palladium NPs.

3.6.2
Nanoparticles in Liquid-Crystal Media

A challenge in nanotechnology remains the control of the organization of NPs in order to integrate them in high-tech devices such as waveguides, band gap materials, light-scattering devices, flat panels and perfect lenses by avoiding the use of the top-down

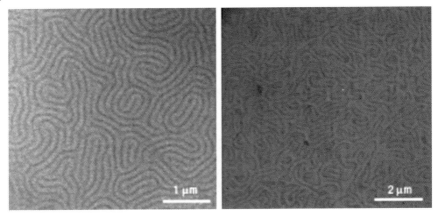

Figure 3.13 TEM micrographs of the fingerprint cholesteric texture for the pure (left) and doped with achiral NPs (right) chiral liquid-crystalline material. Reprinted with permission from Macmillan Publishers Ltd: Nature Materials, Ref. [116], copyright (2002).

approach. Liquid crystalline (LC) materials appear as perfect candidates for the control of the crucial parameters size, shape and self-assembly of nanoscale materials in a one-pot process. The use of LCs in syntheses and self-assembly has recently been reviewed by Hegmann and coworkers [115]. Organization of nanomaterials in LCs can, furthermore, respond to external stimuli such as an electric (magnetic) field or temperature. This opens the possibility to use such composites as electrical actuators.

Figure 3.13 shows that achiral NPs coated with a zwitterionic surfactant when dispersed in a cholesteric LC can order in accordance to the chiral helical structure of the chiral phase. The platinum NPs form periodic ribbons, which mimic the well-known cholesteric texture. De Guerville et al. also demonstrated that the NPs not only decorate the pristine structure but create a novel structure characterized by a larger periodicity [116, 117]. They also observed that this periodicity, that is the distance between the ribbons, can be tuned by varying the molar fraction of chiral mesogens present in the pure cholesteric host.

Recently, Hegmann demonstrated that gold NPs protected by a chiral ligand, a Naproxen functionalized thiol, can be used to induce chirality to a nonchiral nematic LC phase [51]. The use of gold NPs protected with chiral ligand as chiral dopant in LC phase was furthermore confirmed using induced CD studies of a nematic LC (N-LC) doped with three different NPs. This experiment has revealed a transfer of chirality from the NPs to the N-LC phase that resulted in the formation of a chiral nematic phase with opposite helical sense in comparison to the pure chiral ligand dispersed in the same N-LC host [118]. The formed stripe texture might be tunable according to the size of the NPs or their functionality. This opens a pathway for improving LC mixtures for a variety of applications. For example, N-LCs doped with gold nanoclusters can be aligned and electrically reoriented at lower threshold voltages than the pure N-LCs [119].

3.6.3
Chiral Discrimination

Self-assembled monolayers of chiral thiols on metal surfaces have proved their potential for enantiodiscrimination [60]. On account of their higher surface area compared with monolayers, chiral NPs should be even more interesting for discrimination. Kong and coworkers have already shown that gold NPs are efficient sensors with the capability of probing chiral amino acids at the subpicomolar level. The signal was monitored by differential potential voltammetry using a glassy carbon electrode modified with bovine serum albumine, which is a chiral modifier, and amplified by silver atoms anchored on the gold NPs [120]. Similarly, Rotello *et al.* have clearly demonstrated by calorimetric studies that the NPs bearing enantiomeric and diastereoisomeric end groups afford distinctly different binding affinities towards a protein target [121]. This reveals that chirality on NPs is an important prerequisite for the specific recognition required for therapeutic applications.

3.7
Outlook

Nanoscience is still in the discovery phase, and this is particularly true for chiral NPs. Only very few examples of well-defined optically active NPs have been synthesized and only on a small scale (milligrams). However, examples of applications described in the last section show the very large potential of chiral NPs, for example in enantioselective catalysis, in LC displays and in chiral recognition. Progresses in enantioselective synthesis and in resolution of chiral NPs are expected in the near future. Integration of NPs in larger systems may allow the preparation of nanoscale optical components [22] and planar polarization-sensitive devices [18]. The merging of optical activity with SPs may pave the way for new types of "chirophotonic devices." NPs are already starting materials or catalysts for the preparation of larger anisotropic nanomaterials such as nanorods. Perhaps chiral NPs will allow the preparation of larger chiral nanomaterials. Finally, chiral NPs are also of fundamental interest as nanometer-size analogues of extended chiral metal surfaces and serve as models for the better understanding of interactions between surfaces and organic molecules.

References

1 Haruta, M. (1997) *Catalysis Today*, **36**, 153.
2 Roucoux, A., Schulz, J. and Patin, H. (2002) *Chemical Reviews*, **102**, 3757.
3 Schmid, G. (2004) *Nanoparticles From Theory to Application*, Wiley-VCH, Weinheim.
4 Dahl, J.A., Maddux, B.L.S. and Hutchison, J.E. (2007) *Chemical Reviews*, **107**, 2228.
5 Daniel, M.C. and Astruc, D. (2004) *Chemical Reviews*, **104**, 293.
6 Goldstein, A.N., Echer, C.M. and Alivisatos, A.P. (1992) *Science*, **256**, 1425.

7 Kelly, K.L., Coronado, E., Zhao, L.L. and Schatz, G.C. (2003) *The Journal of Physical Chemistry. B*, **107**, 668.

8 Park, S., Pelton, M., Liu, M., Guyot-Sionnest, P. and Scherer, N.F. (2007) *The Journal of Physical Chemistry C*, **111**, 116.

9 Hu, J., Li, L.-s., Yang, W., Manna, L., Wang, L.-w. and Alivisatos, A.P. (2001) *Science*, **292**, 2060.

10 Caswell, K.K., Wilson, J.N., Bunz, U.H.F. and Murphy, C.J. (2003) *Journal of the American Chemical Society*, **125**, 13914.

11 Hu, X., Cheng, W., Wang, T., Wang, E. and Dong, S. (2005) *Nanotechnology*, **16**, 2164.

12 Chang, J.Y., Wu, H., Chen, H., Ling, Y.C. and Tan, W. (2005) *Chemical Communications*, 1092.

13 Oshima, Y., Koizumi, H., Mouri, K., Hirayama, H., Takayanagi, K. and Kondo, Y. (2002) *Physical Review B-Condensed Matter*, **65**, 121401.

14 Oshima, Y., Onga, A. and Takayanagi, K. (2003) *Physical Review Letters*, **91**, 205503.

15 Pendry, J.B. (2000) *Physical Review Letters*, **85**, 3966.

16 Elghanian, R., Storhoff, J.J., Mucic, R.C., Letsinger, R.L. and Mirkin, C.A. (1997) *Science* **277**, 1078.

17 Barnes, W.L., Dereux, A. and Ebbesen, T.W. (2003) *Nature*, **424**, 824.

18 Tuomas, V., Konstantins, J., Jari, T., Pasi, V. and Yuri, S. (2003) *Applied Physics Letters*, **83**, 234.

19 Yuri, S., Nikolay, Z. and Michail, O. (2001) *Applied Physics Letters*, **78**, 498.

20 Fu, H.B. and Yao, J.N. (2001) *Journal of the American Chemical Society*, **123**, 1434.

21 Lim, S.-J., An, B.-K., Jung, S.D., Chung, M.-A. and Park, S.Y. (2004) *Angewandte Chemie*, **43**, 6346.

22 Canfield, B.,Kujala, S., Laiho, K., Jefimovs, K., Turunen, J. and Kauranen, M. (2005) Plasmonics: Metallic Nanostructures and Their Optical Properties III (ed. Stockman, M.I.), Proceedings of the SPIE, **5927**, 62.

23 Yanagimoto, Y., Negishi, Y., Fujihara, H. and Tsukuda, T. (2006) *The Journal of Physical Chemistry. B*, **110**, 11611.

24 Brust, M., Walker, M., Bethell, D., Schiffrin, D.J. and Whyman, R. (1994) *Journal of the Chemical Society. Chemical Communications*, 801.

25 Turkevich, J., Stevenson, P. and Hillier, J. (1951) *Discussions of the Faraday Society*, **11**, 55.

26 Schmid, G. (1990) *Inorganic Syntheses*, **27**, 214.

27 Song, Y., Huang, T. and Murray, R.W. (2003) *Journal of the American Chemical Society*, **125**, 11694.

28 Shichibu, Y., Negishi, Y., Tsukuda, T. and Teranishi, T. (2005) *Journal of the American Chemical Society*, **127**, 13464.

29 Balasubramanian, R., Guo, R., Mills, A.J. and Murray, R.W. (2005) *Journal of the American Chemical Society*, **127**, 8126.

30 Shichibu, Y., Negishi, Y., Tsunoyama, H., Kanehara, M., Teranishi, T. and Tsukuda, T. (2007) *Small*, **3**, 835.

31 Denkov, N.D., Velev, O.D., Kralchevsky, P.A., Ivanov, I.B., Yoshimura, H. and Nagayama, K. (1993) *Nature*, **361**, 26.

32 Debuigne, F., Jeunieau, L., Wiame, M. and Nagy, J.B. (2000) *Langmuir*, **16**, 7605.

33 Tamaki, Y., Asahi, T. and Masuhara, H. (2000) *Applied Surface Science*, **168**, 85.

34 An, B.-K., Kwon, S.-K. and Park, Soo Y. (2007) *Angewandte Chemie*, **46**, 1978.

35 Xie, R., Xiao, D., Fu, H., Ji, X., Yang, W. and Yao, J. (2001) *New Journal of Chemistry*, **25**, 1362.

36 Schaaff, T.G., Knight, G., Shafigullin, M.N., Borkman, R.F. and Whetten, R.L. (1998) *The Journal of Physical Chemistry. B*, **102**, 10643.

37 Yao, H., Miki, K., Nishida, N., Sasaki, A. and Kimura, K. (2005) *Journal of the American Chemical Society*, **127**, 15536.

38 Gautier, C. and Bürgi, T. (2006) *Journal of the American Chemical Society*, **128**, 11079.

39 Nishida, N., Yao, H., Ueda, T., Sasaki, A. and Kimura, K. (2007,) *Chemistry of Materials*, **19**, 2831.

40 Yao, H., Fukui, T. and Kimura, K. (2007) *The Journal of Physical Chemistry C*, **111**, 14968.

41 Gautier, C., Taras, R., Gladiali, S. and Bürgi, T. (2008) *Chirality*, **20**, 486.

42 Nafie, L.A. (1997) *Annual Review of Physical Chemistry*, **48**, 357.
43 Freedman, T.B., Cao, X., Dukor, R.K. and Nafie, L.A. (2003) *Chirality*, **15**, 743.
44 Gautier, C. and Bürgi, T. (2005) *Chemical Communications*, **43**, 5393.
45 Wetzel, T.L. and DePristo, A.E. (1996) *Journal of Chemical Physics*, **105**, 572.
46 Parks, E.K., Kerns, K.P. and Riley, S.J. (1998) *Journal of Chemical Physics*, **109**, 10207.
47 Tamura, M. and Fujihara, H. (2003) *Journal of the American Chemical Society*, **125**, 15742.
48 Li, T., Park, H.G., Lee, H.S. and Choi, S.H. (2004) *Nanotechnology*, **15**, S660.
49 Shemer, G., Krichevski, O., Markovich, G., Molotsky, T., Lubitz, I. and Kotlyar, A.B. (2006) *Journal of the American Chemical Society*, **128**, 11006.
50 Bovet, N., McMillan, N., Gadegaard, N. and Kadodwala, M. (2007) *The Journal of Physical Chemistry. B*, **111**, 10005.
51 Qi, H. and Hegmann, T. (2006) *Journal of Materials Chemistry*, **16**, 4197.
52 Moloney, M.P., Gun'ko, Y.K. and Kelly, J.M. (2007) *Chemical Communications*, 3900.
53 Garzón, I.L., Beltran, M.R., Gonzalez, G., Gutierrez-Gonzalez, I., Michaelian, K., Reyes-Nava, J.A. and Rodriguez-Hernandez, J.I. (2003) *European Physical Journal D*, **24**, 105.
54 Garzón, I.L., Reyes-Nava, J.A., Rodríguez-Hernández, J.I., Sigal, I., Beltrán, M.R. and Michaelian, K. (2002) *Physical Review B-Condensed Matter*, **66**, 073403.
55 Goldsmith, M.R., George, C.B., Zuber, G., Naaman, R., Waldeck, D.H., Wipf, P. and Beratan, D.N. (2006) *Physical Chemistry Chemical Physics*, **8**, 63.
56 Humblot, V., Haq, S., Muryn, C., Hofer, W.A. and Raval, R. (2002) *Journal of the American Chemical Society*, **124**, 503.
57 Bieri, M. and Bürgi, T. (2005) *The Journal of Physical Chemistry. B*, **109**, 22476.
58 Bieri, M. and Bürgi, T. (2005) *Langmuir*, **21**, 1354.
59 Bieri, M. and Bürgi, T. (2006) *Physical Chemistry Chemical Physics*, **8**, 513.
60 Bieri, M., Gautier, C. and Bürgi, T. (2007) *ChemPhysPhysChem*, **9**, 671.
61 Jadzinsky, P.D., Calero, G., Ackerson, C.J., Bushnell, D.A. and Kornberg, R.D. (2007) *Science*, **318**, 430.
62 Häkkinen, H., Walter, M. and Gronbeck, H. (2006) *The Journal of Physical Chemistry. B*, **110**, 9927.
63 Gautier, C. and Bürgi, T. (2008) *Journal of the American Chemical Society*, **130**, 7077.
64 Bladon, P., Pauson, P.L., Brunner, H. and Eder, R. (1988) *Journal of Organometallic Chemistry*, **355**, 449.
65 Adams, R.D. (1998) *Catalysis by Di- and Polynuclear Metal Cluster Complexes*, John Wiley and Sons, New York.
66 Beurich, H. and Vahrenkamp, H. (1978) *Angewandte Chemie-International Edition*, **17**, 863.
67 Mani, D. and Vahrenkamp, H. (1986) *Chemische Berichte*, **119**, 3639.
68 Müller, M. and Vahrenkamp, H. (1983) *Chemische Berichte*, **116**, 2311.
69 Müller, M. and Vahrenkamp, H. (1983) *Chemische Berichte*, **116**, 2322.
70 Gusbeth, P. and Vahrenkamp, H. (1985) *Chemische Berichte*, **118**, 1746.
71 Gordon, F. and Stone, A. (1984) *Angewandte Chemie-International Edition*, **23**, 89.
72 Vahrenkamp, H. (1989) *Journal of Organometallic Chemistry*, **370**, 65.
73 Vieille-Petit, L., Suss-Fink, G., Therrien, B., Ward, T.R., Stoeckli-Evans, H., Labat, G., Karmazin-Brelot, L., Neels, A., Bürgi, T., Finke, R.G. and Hagen, C.M. (2005) *Organometallics*, **24**, 6104.
74 Gubin, S.P., Galuzina, T.V., Golovaneva, I.F., Klyagina, A.P., Polyakova, L.A., Belyakova, O.A., Zubavichus, Y.V. and Slovokhotov, Y.L. (1997) *Journal of Organometallic Chemistry*, **549**, 55.
75 Eckehart Roland, H.V. (1984) *Chemische Berichte*, **117**, 1039.
76 Tunik, S.P., Pilyugina, T.S., Koshevoy, I.O., Selivanov, S.I., Haukka, M. and Pakkanen, T.A. (2004) *Organometallics*, **23**, 568.

77 Koshevoy, I.O., Tunik, S.P., Poe, A.J., Lough, A., Pursiainen, J. and Pirila, P. (2004) *Organometallics*, **23**, 2641.
78 Felix Richter, H.V. (1982) *Chemische Berichte*, **115**, 3243.
79 Müller, M. and Vahrenkamp, H. (1983) *Chemische Berichte*, **116**, 2748.
80 Blumhofer, R. and Vahrenkamp, H. (1986) *Chemische Berichte*, **119**, 683.
81 Wang, X., Li, W.Z., Zhao, Q.Y., Li, Y.M. and Chen, L.R. (2005) *Analytical Sciences*, **21**, 125.
82 Moiseev, I.I. and Vargaftik, M.N. (1998) *New Journal of Chemistry*, **22**, 1217.
83 Busca, G. and Centi, G. (1989) *Journal of the American Chemical Society*, **111**, 46.
84 Feliz, M., Guillamón, E., Llusar, R., Vicent, C., Stiriba, S.E., Pérez-Prieto, J. and Barberis, M. (2006) *Chemistry – A European Journal*, **12**, 1486.
85 An, B.K., Kwon, S.K., Jung, S.D. and Park, S.Y. (2002) *Journal of the American Chemical Society*, **124**, 14410.
86 Chung, T.W., Cho, K.Y., Nah, J.W., Akaike, T. and Cho, C.S. (2002) *Langmuir*, **18**, 6462.
87 Kommareddy, S. and Amiji, M. (2007) *Nanomedicine*, **3**, 32.
88 Masuhara, H., Nakanishi, H. and Sasaki, K. (2003) *Single Organic Nanoparticles*, Springer, Verlag.
89 Fu, H., Loo, B.H., Xiao, D., Xie, R., Ji, X., Yao, J., Zhang, B. and Zhang, L. (2002) *Angewandte Chemie-International Edition*, **41**, 962.
90 Xiao, D., Yang, W., Yao, J., Xi, L., Yang, X. and Shuai, Z. (2004) *Journal of the American Chemical Society*, **126**, 15439.
91 Xi, L., Fu, H., Yang, W. and Yao, J. (2005) *Chemical Communications*, 492.
92 Zhang, Y., Peng, A., Wang, J., Yang, W. and Yao, J. (2006) *J Photochem Photobiol, A*, **181**, 94.
93 Scheffer, J.R. and Xia, W. (2005) *Organic Solid State Reactions*, Springer, Berlin/Heidelberg, p. 233.
94 Rosi, N.L. and Mirkin, C.A. (2005) *Chemical Reviews*, **105**, 1547.
95 Sun, S., Murray, C.B., Weller, D., Folks, L. and Moser, A. (2000) *Science*, **287**, 1989.
96 Jansat, S., Gomez, M., Philippot, K., Muller, G., Guiu, E., Claver, C., Castillon, S. and Chaudret, B. (2004) *Journal of the American Chemical Society*, **126**, 1592.
97 Hong, R., Han, G. Fernandez, J.M., Kim, B.j., Forbes, N.S. and Rotello, V.M. (2006) *Journal of the American Chemical Society*, **128**, 1078.
98 Belser, T., Stohr, M. and Pfaltz, A. (2005) *Journal of the American Chemical Society*, **127**, 8720.
99 Hu, A., Yee, G.T. and Lin, W. (2005) *Journal of the American Chemical Society*, **127**, 12486.
100 Widegren, J.A. and Finke, R.G. (2003) *Journal of Molecular Catalysis A-Chemical*, **191**, 187.
101 Davies, I.W., Matty, L., Hughes, D.L. and Reider, P.J. (2001) *Journal of the American Chemical Society*, **123**, 10139.
102 Widegren, J.A. and Finke, R.G. (2003) *Journal of Molecular Catalysis A-Chemical*, **198**, 317.
103 Bönnemann, H. and Braun, G.A. (1996), *Angewandte Chemie-International Edition*, **35**, 1992.
104 Studer, M., Blaser, H.U. and Exner, C. (2003) *Advanced Synthesis and Catalysis*, **345**, 45.
105 Bönnemann, H. and Braun, G.A. (1997) *Chemistry – A European Journal*, **3**, 1200.
106 Zuo, X., Liu, H., Guo, D. and Yang, X. (1999) *Tetrahedron*, **55**, 7787.
107 Köhler, J.U. and Bradley, J.S. (1997) *Catalysis Letters*, **45**, 203.
108 Köhler, J.U. and Bradley, J.S. (1998) *Langmuir*, **14**, 2730.
109 Mevellec, V., Mattioda, C., Schulz, J., Rolland, J.P. and Roucoux, A. (2004) *Journal of Catalysis*, **225**, 1.
110 Jansat, S., Picurelli, D., Pelzer, K., Philippot, K., Gomez, M., Muller, G., Lecante, P. and Chaudret, B. (2006) *New Journal of Chemistry*, **30**, 115.
111 Nasar, K., Fache, F., Lemaire, M., Beziat, J.C., Besson, M. and Gallezot, P. (1994) *Journal of Molecular Catalysis*, **87**, 107.
112 Orito, Y., Imai, S., Nawa, S. and Nguyeng, I. (1979) *Journal of Synthetic Organic Chemistry Japan*, **37**, 173.

113 Bürgi, T. and Baiker, A. (2004) *Accounts of Chemical Research*, **37**, 909.
114 Collier, P.J., Iggo, J.A. and Whyman, R. (1999) *Journal of Molecular Catalysis A-Chemical*, **146**, 149.
115 Hegmann, T., Qi, H. and Marx, V.M. (2007) *Journal of Inorganic and Organometallic Polymers and Materials*, **17**, 483.
116 Mitov, M., Portet, C., Bourgerette, C., Snoeck, E. and Verelst, M. (2002) *Nature Materials*, **1**, 229.
117 Mitov, M., Bourgerette, C. and de Guerville, F. (2004) *Journal of Physics: Condensed Matter*, **16**, S1981.
118 Qi, H., O'Neil, J. and Hegmann, T. (2008) *Journal of Materials Chemistry*, **18**, 374.
119 Qi, H., Kinkead, B. and Hegmann, T. (2008) *Advanced Functional Materials*, **18**, 212.
120 Wang, Y., Yin, X., Shi, M., Li, W., Zhang, L. and Kong, J. (2006) *Talanta*, **69**, 1240.
121 You, C.-C., Agasti, S.S. and Rotello, V.M. (2008) *Chemistry – A European Journal*, **14**, 143.

4
Gels as a Media for Functional Chiral Nanofibers
Sudip Malik, Norifumi Fujita, and Seiji Shinkai

4.1
A Brief Introduction to Gels

4.1.1
Introduction

Gels, one of the most promising materials in the twenty-first century, have generated tremendous research interest nowadays not only for the better basic scientific understanding of gelation phenomena but also to design novel kinds of functional materials based on the self-assembly of molecules. At the advent of nanotechnology where the molecules are assembled into nanometer-scale architectures, the thrust of gel chemistry has been clearly visualized as it is possible to correlate directly between the molecular self-assembly at the nanoscale and the macroscopic material properties, visible to the naked eye. As a benefit of gel chemistry, gels formed by, for instance, dilute solutions of inorganic substances such as silica or clays, polymers and proteins have been studied and are extensively commercialized in the form of toothpaste, soap, shampoos, cosmetics, puddings, contact lenses, drugs, gel pens, clearing agents, and so on in our everyday lives. Within two decades, there has been rapidly growing attention to low molecular weight gelators (LMWG), which is mainly motivated by the well-defined structure–property relationship of LMWGs as well as the many potential applications of them. Gels are usually constructed from the self-assembly of LMWGs including steroid derivatives, peptides, nucleobases, saccharides and even more complex organic structures like dendrimers. In contrast to macromolecular and inorganic counterparts, the network of LMWG is formed solely as a result of noncovalent interactions like solvophobic effects, π–π interactions, van der Waals interaction and hydrogen bonds. The self-assembly process starting from a single molecule leads to the formation of an interpenetrated or entangled network structure that is responsible for preventing the flow of the bulk solvent, which is by far the majority component. The formation of gel, which is completely reversible and at the same time efficient as most LMWGs form gels at concentration well below 0.1% wt/v, is perhaps the finest example of a supramolecular self-assembly process.

Chirality at the Nanoscale: Nanoparticles, Surfaces, Materials and more. Edited by David B. Amabilino
Copyright © 2009 WILEY-VCH Verlag GmbH & Co. KGaA, Weinheim
ISBN: 978-3-527-32013-4

However, most of the organogelators uncovered up to now have been discovered by serendipity rather than design, which is the reason why this task is still difficult as well as challenging. Over the years there have been significant contributions towards the design of different gelators and the understanding of gelation phenomena from the different groups including those led by Weiss [1], Hanabusa [2], Feringa [3], Hamilton [4], Terech [5], and Shinkai [6]. From the available literature [1–9], the following broad conclusions can be drawn about whether a molecule may be expected to gelate or not: (a) The molecules must have *noncovalent interactions* among themselves. Typically, these interactions include π–π interactions, van der Waal interaction, solvophobic effects and hydrogen bonds. (b) These noncovalent interactions should be *directional*, leading to formation of an anisotropic network of LMWGs. (c) The molecule must be *sparingly soluble* in the solvent of choice. The molecule should not be too soluble, otherwise it results in simple dissolution, whereas it should not be too insoluble, otherwise it will result in the simple precipitation of a solid of the material out of the solution.

4.1.2
Definition of Gels

Although everyone knows what a gel is, scientifically it is not easy to define one, as none of the available definitions of gel covers all of the properties. As described by Lloyd, the gel is easier to recognize than to define [10]. In 1961 Ferry explained a gel as a substantially diluted system that exhibits no steady-state flow [11]. Thus, a gel is a state of matter, neither liquid nor solid, or may be termed a semisolid. To correlate between the macroscopic and microscopic properties of a gel, Flory defined a gel in a more comprehensive way [12]. A gel has continuous structures with macroscopic dimensions that are permanent on the time scale of analytical experiment and are solid-like in its rheological behavior below certain stress limit. One important thing common to all gels is that they are comprised of two components, one of which is liquid at the temperature under consideration, and the other of which is a solid, for example LMWGs. The gels themselves have the mechanical properties of solids and under any mechanical stress the gels show the phenomenon of strain. This rule is obeyed in a large part even today, although all networks do not need to be solid and more than two components may construct a gel. Nowadays, the rheological description about the gel is believed to be more justified as well as being a quantitative one.

4.1.3
Classification of Gels

Gels can be classified by taking into account various considerations (Figure 4.1), depending on the source, medium, constitution, and the nature of interactions that are responsible for the creation of the three-dimensional gel network. They are divided into macromolecular gels and small molecular gels (low molecular weight gels). Macromolecular gels are formed from either chemically crosslinked (covalent bonds) or physical interactions (noncovalent bonds).

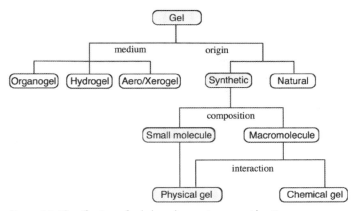

Figure 4.1 Classification of gels based on various considerations.

The gels derived from polyethylene, polyvinyl alcohol, polyamide, and so on [13–15] are examples of chemically crosslinked gels where the networks result from strong chemical bonding and are thermally irreversible. The gels of polyacrylate, polystyrene [14, 16], polyaniline [17] poly(3-alkylthiophenes) [18], and so on are formed by weak physical interactions and they are thermoreversible gels.

According to their history, organogelators can be classified as (a) first-generation gelators that were found by serendipity and primarily studies on only gelation properties rather than functional properties [19] and (b) second-generation gelators that were successfully designed for the gelation as well as simultaneously showing the functional or material properties using various kinds of secondary interactions after gelation [20, 21].

4.1.4
Chirality in Gels

When we think about chirality in gels, most of us imagine that is how stereogenic centers make an effect on the helicity in the gel fibers. Among numerous varieties of assemblies appearing in the gel phase, helical assemblies receive particular attention because of the morphological similarity with naturally occurring helical molecular architectures. In 1969 – note that it was less than 20 years after the discovery of the double-helical structure of DNA – Tachibana et al. [22, 23] reported helical aggregates of 12-hydroxystearic acid and its lithium salts. After dissolving hydroxystearic acids in solvents at high temperature, it forms gels upon cooling to room temperature. As it is an optically active fatty acid, depending on the absolute configuration of the enantiomer, it forms different kinds of helical fibers in the gels. For example, the D-acid produces the left-handed helix and the L-acid shows the right-handed helix (Figures 4.2a and b). This morphological diversity is a clear difference from naturally occurring helical assemblies as generally, the biological system adopts homohelicity in those higher-order molecular structures. Interestingly, the racemic mixture of

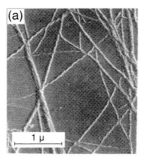 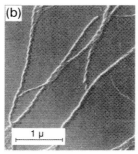 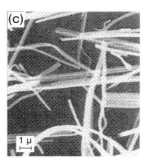

Figure 4.2 TEM pictures of gel fibers prepared from (a) 12D-hydroxystearic acid, (b) 12-L-hydroxystearic acid and (c) the racemate of the lithium salt.

12-hydroxystearic acids produces objects with a plate-like structure and the lithium salt of 12-hydroxystearic acids generates entirely opposite helicity in the fibers to the corresponding acid. The racemic mixture of the salt shows fibers without helicity (Figure 4.2c).

4.2
Chiral Organogels

4.2.1
Steroid-Based Chiral Gelators

Steroids that are versatile building blocks found abundantly in plants and animals play crucial roles to inspire the biochemical activities in living creatures. The chemical structures of steroids consist of a fused tetracyclic androstane skeleton (a, b, c, and d in Figure 4.3) [24]. It creates the complicated ring structures in which skeletons take either a *cis*- or *trans*-configuration at each stereogenic center.

However, the all-*trans*-configuration of a steroid skeleton is commonly observed in the biological systems. One of the derivatives of steroids, cholesterol, which is found in the cell membrane of the tissues in all animals, has a tendency to generate a liquid-crystalline phase (cholesteric phase) where steroid–steroid stacking, which obviously is a result of van der Waals interactions, is operating to create one dimensional (1D) helical structures that are the prerequisite condition for the gelation.

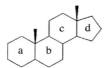

Figure 4.3 Common chemical structure of steroids.

Cholesterol-based gelators have been extensively studied by several groups such as Weiss's group [25–31], Shinkai's group [32–37], and Whitten's group [38–40]. A number of aromatic components such as anthracene [25–29], anthraquinone [27, 30], azobenzene [32, 33], porphyrin [41–43], nucleobase [35], stilbene [40], oligothiophenes [37] have been covalently attached with cholesterol moieties at their 3-position to create highly ordered 1D structures in the oraganogels derived from them. When the cholesterol moieties form 1D helical columnar stacking, the attached aromatic groups are oriented so as to give the face-to-face-type interaction and stabilize the 1D aggregates.

In 1987, Weiss et al. first published the gelation properties of anthracene-appended cholesterol derivatives (**1~3**) [25] that can form a gel with a wide range of organic solvents, from low polarity solvents, such as hydrocarbons, to polar ones such as alcohols, and amines. Even less than 1 wt% of **1** forms the gel with these organic liquids. Later, a series of structural analogues of **1** were devised and also synthesized to investigate the gelation properties systematically. For example, **1** and **2** can gelate several solvents, whereas **3** does not form a gel (Figure 4.4).

Azobenzene-appended cholesterol gelators have been comprehensively studied by Shinkai and his coworkers (Figure 4.5) [33]. Besides the steroid–steroid stacking of

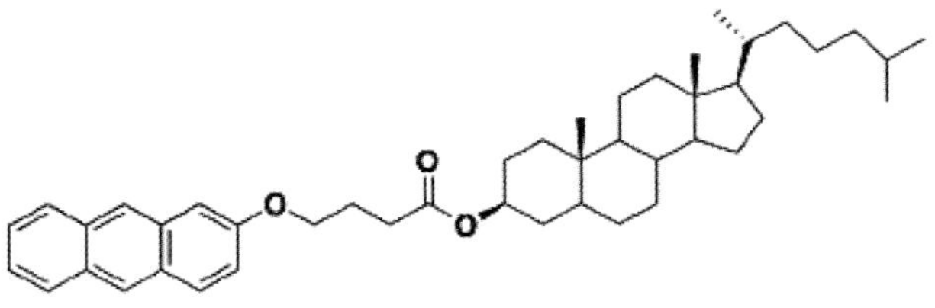

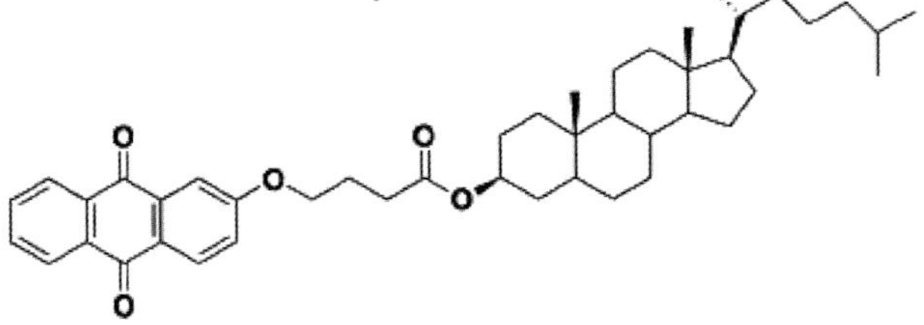

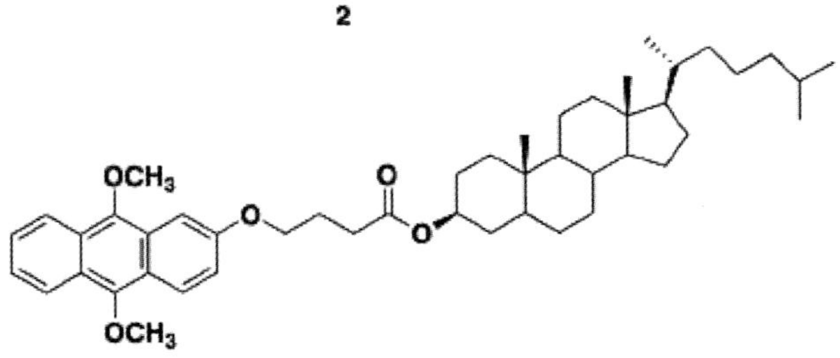

Figure 4.4 Chemical structure of anthracene derivatives based cholesterol gelators.

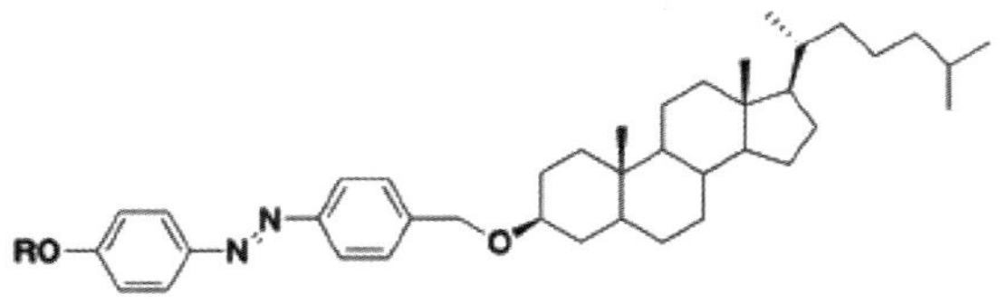

4a: R = CH₃
4b: R = nPr
4c: R = nBu

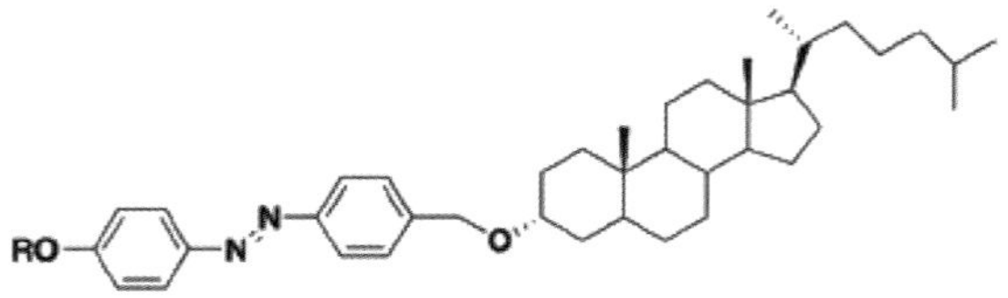

5a: R = CH₃
5b: R = CH₃CH₂

Figure 4.5 Chemical structure of azobenzene-appended cholesterol gelators: note and **5** are epimers at C-3.

the gelators, the spectroscopic properties of the azobenzene moieties afford useful information on how the molecules self-assemble and how the gels are formed. Two series of compounds were synthesized and intensive circular dichroism (CD) spectroscopic investigations were carried out to propose the mechanism of gelation. Generally, azobenzene moieties are attached at the C-3 position of cholesterol. Reaction between an acid chloride and cholesterol produces the natural (S)–configuration at C-3 through the esterification of cholesterol at 3-OH group. The inverted (R)–configuration has been synthesized by treating cholesterol with an acid in the presence of triphenylphosphine and diethyl azobenedicarboxylate. Both the conformers form a gel in many nonpolar and polar organic solvents, but the small structural difference at C-3 of cholesterol causes the dramatic changes in the nature of gel networks.

The azobenzene chromophoric group, having electron absorption in the visible region, produces strong CD active signals upon the formation of gels of these **4** and **5**, whereas in the sol state these are totally CD silent. This observation indicates that in the gel phase the cholesterol moieties aggregate in a specific chiral direction, which prompts the azobenzene chromophores to interact in an asymmetric manner. Compound **4a**, in the gel it forms in 1-butanol, produced a positive exciton coupling band of (R)–chirality, whereas **5b** in its methanol gel showed a negative exciton coupling band of (S)–chirality (Figure 4.6). These results clearly show that the transition dipole moments in the azobenzene moieties are interacting as well as oriented either in clockwise, that is, (R)–chirality or anticlockwise, that is, (R)–chirality. Another unusual investigation is that of the inversion of the CD signal in the case of gels formed by **4b** and **4c**. Among the several gel-preparation conditions,

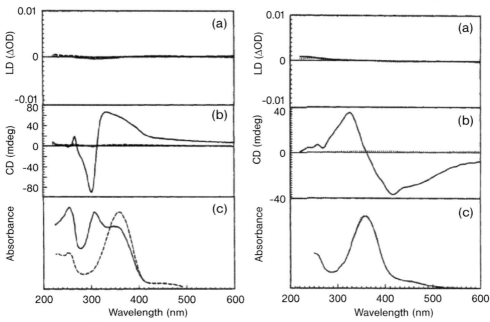

Figure 4.6 (a) LD spectra, (b) CD spectra and (c) absorption spectra; the gel state was observed at 25 °C (solid line) and the solution state was observed at 60 °C (dotted line). *Left*: **4a** (0.2 wt%) in n-butanol and *Right*: **5b** (0.4 wt%) in methanol.

the cooling speed from sol to gel is the most responsible factor associated with this phenomenon. As a result of kinetic influences, a less stable (metastable) aggregate is formed preferentially and immobilized by the gel phase. It is important to note that both the left-handed and right-handed gel networks could be produced from the same sol solution by slow and fast cooling, respectively as confirmed by SEM experiments.

The effect of the absolute configuration at C-3 of the steroid on the gelation was investigated thoroughly. It has been found that **4a** (with (S)-chirality) forms excellent thermally reversible gels with organic solvents, whereas **5a** with (R)-chirality fails to form a gel with the same set of solvents as used in the previous case. The influence of chirality on the gelation can be explained by the structural differences between the (R)- and (S)-isomers as observed in energy minimized structures (Figure 4.7) obtained from the semiempirical molecular orbital calculations. Compound **4a** adopts an extended linear conformation that enables it to stack efficiently into 1D structures. Though the primary driving force comes from steroid stacking, there is an additional stabilizing force coming from the face-to-face orientation of azobenzene groups. On the other hand, **5a** takes an L-shaped or bent structural form in which 1D stacking is energetically less favored because the additional stabilization from the stacking of azobenzene moieties is somewhat restricted here.

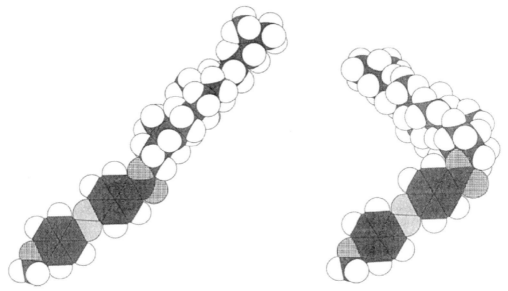

Figure 4.7 Energy minimized structures of **4a** (left) and **5a** (right).

Porphyrin-appended cholesterol gelators were reported by our group [41–43]. In this series the first compounds designed and synthesized are **6a** with natural (*S*)-chirality at the 3-position and **6b** with (*R*)-chirality at the 3-position (Figure 4.8). Gelator **6a** forms an organogel with cyclohexane and methylcyclohexane. On the other hand, gelator **6b** does not form a gel in these solvents. As explained above, in the aggregated state, the porphyrin moieties in **6a** with (*S*)-chirality can interact with each other to stabilize 1D packing, whereas such packing in **6b** is more difficult because of steric factors originating from the inverted (*R*)-chirality. In the gel state, porphyrin moieties are packed in a J-type aggregation fashion confirmed by UV-vis absorption and CD studies.

Later, a series of Zn(II) porphyrin-appended cholesterol derivatives were designed and synthesized by varying the length of the spacer [$(CH_2)_n$] between the steroid moiety and Zn(II) porphyrin [43] and the gelation abilities have shown the dependence on the length of spacer. Gelators with an even number of carbon atoms in the spacer moiety (**7a** and **7c**) form a gel with aromatic solvents like *p*-xylene, toluene and benzene, whereas those with an odd number of carbon atoms in the spacer (**7b** and **7d**) do not form a gel.

In the chiral gel state, gelator molecules are stacked in a 1D helical manner through the steroid–steroid interaction and the additional interactions among Zn(II) porphyrin moieties play a crucial role in stabilizing the aggregates rather than hydrogen-bonding interactions between amide and carbamate groups. These aggregates are highly sensitive toward the number of carbon atoms in the spacer. Only in the case of an even number of carbon atoms do both steroid–steroid interactions as well as Zn

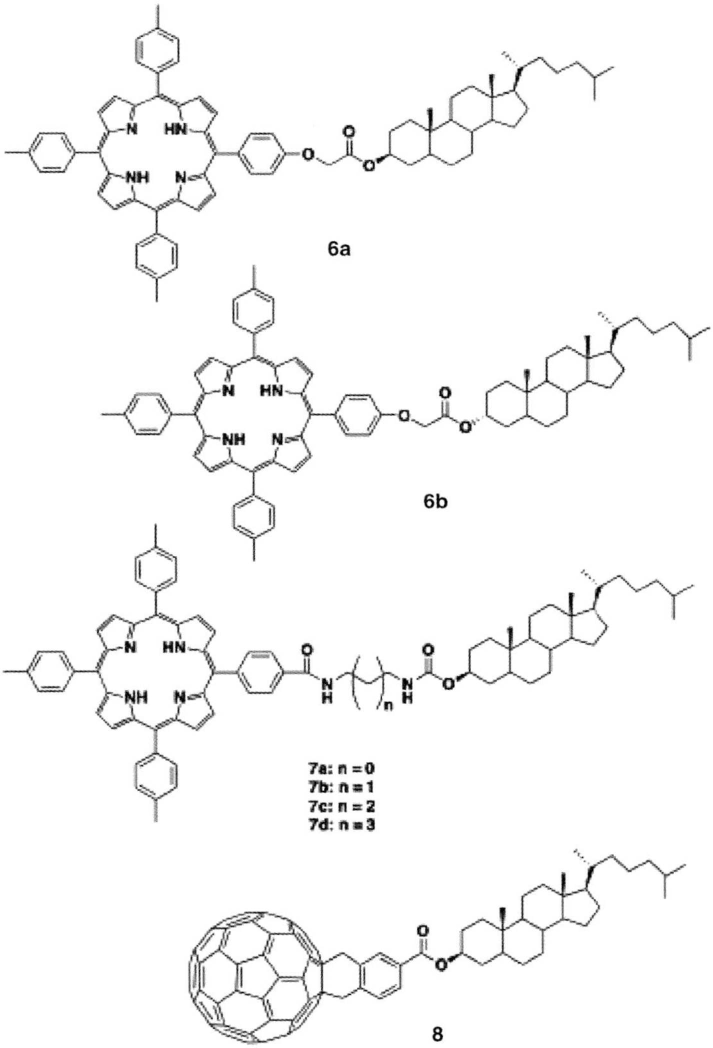

Figure 4.8 Molecular structures of porphyrin and [60] fullerene-based cholesterol gelators.

(II) porphyrin–Zn(II) porphyrin interactions (Figure 4.9, left) cooperate effectively to construct the 1D helical superstructure of gelator molecules.

Interestingly, the influence of [60]fullerene on the gelling ability of the Zn(II) porphyrin-appended cholesterol derivatives has been studied. The gel-to-sol phase-transition temperature of **7a** and **7c** has been significantly increased with increasing equivalent of added [60]fullerene. This result indicates the reinforcement of gel aggregates caused by the intermolecular Zn(II) porphyrin – [60]fullerene interactions that is reflected in UV-vis and CD spectroscopic investigations (Figure 4.9, right). The

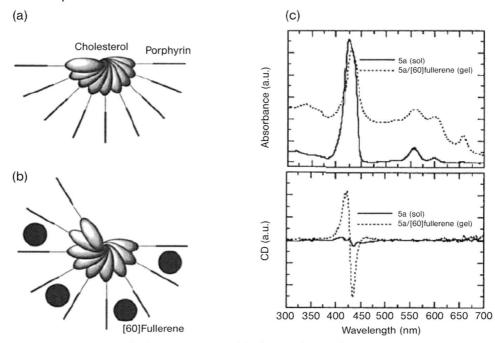

Figure 4.9 Left side: 1D aggregate models of **7** (a) in absence, (b) in presence of [60] fullerene in the gel phase. Right side: (c) UV-vis and absorption spectra **7a** (full line) and **7a**/[60] fullerene complex (2 : 1 complex, dotted line) in toluene ($2.55 \times 10^{-4}\,\mathrm{mol/dm^3}$ for **7a**) at $20\,^\circ\mathrm{C}$.

Soret band of the porphyrin shifts to longer wavelengths and the CD spectra corresponding to the Soret band shows a strong Cotton effect upon gelation of **7a** in the presence of [60]fullerene only. The proposed aggregation mode is to orient the left-handed arrangement of porphyrins through the formation of a sandwich-like complex (Figure 4.9, left).

Later, [60]fullerene-appended gelators were reported from the same group [44] and only (S)-chirality of **8** forms a gel in dichloromethane and CD studies indicate that fullerene moieties are oriented chirally in the gel phase. The influence of Zn(II) porphyrin is tested by adding Zn(II) porphyrin in the gel of **8**. However, the gel is rather destabilized. The result is in contrast to the behavior of **7a** in the presence of fullerene.

A *trans*-stilbene-appended cholesterol gelator was successfully designed and synthesized by Whitten *et al.* (Figure 4.10) [40]. A series of time transient atomic force microscopy (AFM) images of **9** in 1-octanol were recorded in order to monitor the sol-to-gel phase transition onto a hydrophobic graphite surface. At the initial stage, the dewetting process starts from the surface and the gelator–solvent interaction is enhanced, leading to formation of condensed droplets that eventually transform into

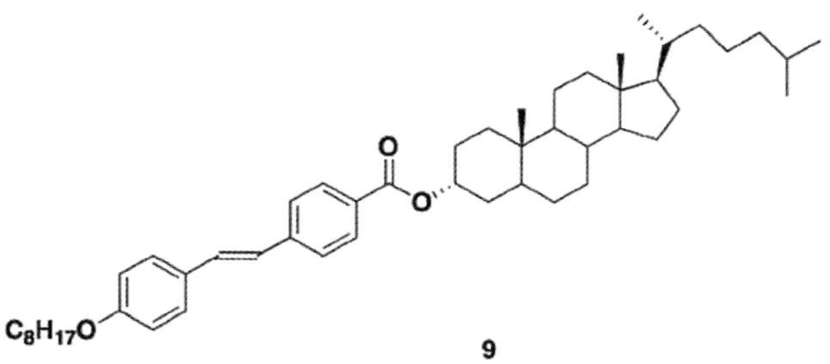

9

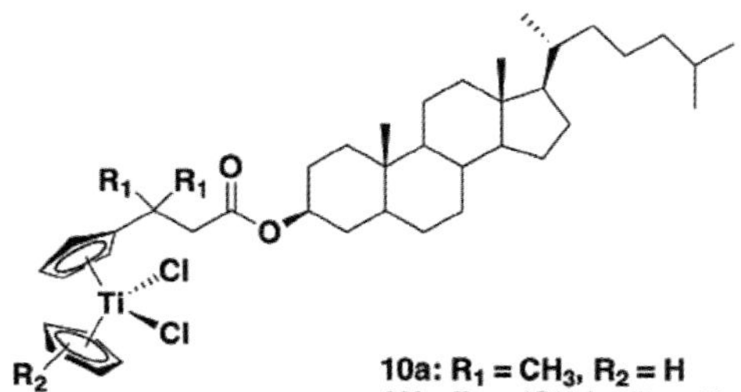

10a: $R_1 = CH_3$, $R_2 = H$
10b: $R_1 = (CH_2)_5$, $R_2 = H$
10c: $R_1 = CH_3$, $R_2 = CH_3$

Figure 4.10 Molecular structures of gelators **9** and **10**.

1D developed fibers. Very recently, a titanocene-based cholesterol gelator has been synthesized [45]. It forms a gel in organic solvents with different polarity. In the gel state, it can form twisted helical fibers confirmed by AFM and CD studies.

4.2.2
Pyrene-Based Chiral Gelators

Pyrene, one of the most commonly used fluorophore probes for molecular aggregation, has been utilized to devise pyrene-containing gelators. Maitra *et al.* [46] reported both achiral and chiral pyrene-based gelators **11a** and **11b**. Both derivatives form a gel with alcohols, hexane, cyclohexane, decane, and dodecane. Helical aggregates are formed during the gelation of **11b** evidenced by CD studies (Figure 4.11b). Later, pyrene-containing oligo(glutamic acid) gelator (**12**) [47] was reported. In the gel state, the gelator molecules are self-assembled through intermolecular hydrogen bonding between peptide residues and $\pi-\pi$ interaction between the pyrene moieties (Figure 4.11).

4.2.3
Diaminoycyclohexane-Based Chiral Gelators

Another interesting building block for chiral gelators is 1,2Diaminocyclohexane that can be used to form alkyl amide derivatives (Figure 4.12). Only the *trans*-isomer with

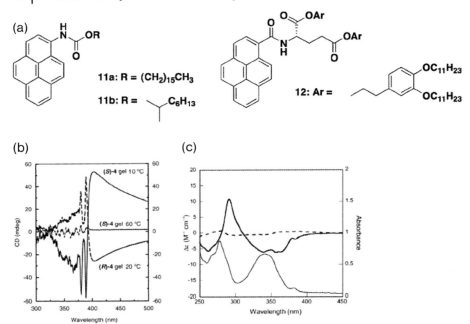

Figure 4.11 (a) Chemical structures of gelators **11** and **12**, (b) CD spectra of gels of (R)-**11b** and (S)-**11b** in cyclohexane at the indicated temperature, (c) UV-vis and CD spectra of gelator **12** of concentration 4.0×10^{-2} mol. (*Thin solid line*: UV-vis spectra at room temperature in the gel phase, *thick broken line*: CD spectra at room temperature in the sol phase and *thick solid line*: CD spectra at room temperature in the gel phase), and (d) schematic illustration of gelator in the gel.

long alkyl chains forms a gel with a variety of organic solvents, including alcohols, ethers, esters, ketones, hydrocarbons, polar aprotic solvents, edible oils, and mineral oils. It is important to note that the *cis*-isomer as well as the *trans*-isomer with the alkyl chain bearing less than six methylene units do not form gels in the used

13: *trans* (1R, 2R)
14: *trans* (1S, 2S)

Figure 4.12 Chemical structure of diaminocyclohexane-based chiral gelators.

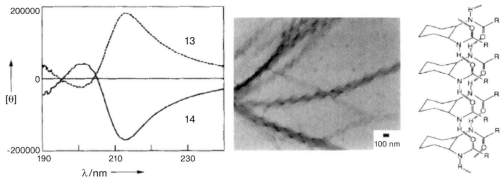

Figure 4.13 (a) CD spectra of gelators **13** and **14** with same concentration (1 mM) in acetonitrile at 20 °C. (b) TEM image of gel of **13** in acetonitrile stained by OsO_4. (c) Schematic representation of molecular packing in the gel fibers.

solvents [48, 49]. Intermolecular hydrogen bonding between the amide groups of adjacent molecules, van der Waals interaction of long alkyl chains and the projection angle between the alkyl chains of the cyclohexane platform are the responsible driving force for packing in 1D aggregates.

CD investigations of the gels from **13** and **14** have revealed that the molecular chirality in the helical stacking and the large amplification of the CD signal originates from the chiral aggregates of the LMWG molecules in the gel state. Subsequent TEM observation of the gel negatively stained by OsO_4 has confirmed the presence of helicity in the gel fibers. These experimental data have nicely explained the helix-formation mechanism with the help of molecular modeling (Figure 4.13). According to these lines of information, two equatorial amide-NH and amide-CO are antiparallel to each other and perpendicular to the cyclohexyl ring. Therefore, it is possible to form an extended molecular tape stabilized by two intermolecular hydrogen bonds between neighboring molecules.

4.2.4
OPV-Based Chiral Gelators

A serendipitous observation for gelation of oligo(*p*-phenylenevinylene)s (OPVs) derivatives, consisting of a short linear π-conjugated system, has been reported by Ajayaghosh *et al.* [50–53]. **OPV1** having two hydroxylmethyl groups and six dodecyloxy side chains forms gels with nonpolar organic solvents. The introduction of long hydrocarbon chains in OPV facilitates the van der Waals interactions and simultaneously inhibits the crystallization, leading to formation of a gel. The hydroxymethyl groups favor the formation of the linear hydrogen-bonded polymeric structures that construct the arrays of superstructures from OPV gelators (Figure 4.14). X-ray diffraction studies overwhelmingly support the multilayer lamellar assemblies through hydrogen bonding and π-stacking, in which each OPV is arranged perpendicularly to the growth axis of the fiber.

(a)

(b)

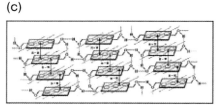

(c)

Figure 4.14 (a) Chemical structure of **OPV1**, (b) SEM image of dried gel of **OPV1** from toluene. (c) Probable self-assembly of **OPV1**, showing the hydrogen bond and π-stack-induced three-dimensional network formation.

OPV2, a chiral analogue of **OPV1**, also forms a gel in organic solvents though the gelation ability is significantly reduced with the insertion of chiral side chains. However, the introduction of chiral side chains has expressed the self-assembly in a helical sense depending upon the chirality of asymmetric carbon. CD measurements of **OPV2** have shown the strong bisignate Cotton effect corresponding to π–π* transition (400 nm). The concentration and temperature variant CD spectroscopic studies of **OPV2** have suggested the involvement of two different kind of helical aggregates that are also confirmed by morphological observations through field emission scanning electron microscopy (FE-SEM) as well as by AFM. A schematic representation of self-assembly of the chiral **OPV2** to a left-handed double helical rope-like structure is shown in Figure 4.15.

Feringa and van Esch et al. have shown in a wonderful piece of work that a reversible, photoresponsive self-assembling molecular system can be successfully designed in which the molecular and supramolecular chirality communicate [54]. A

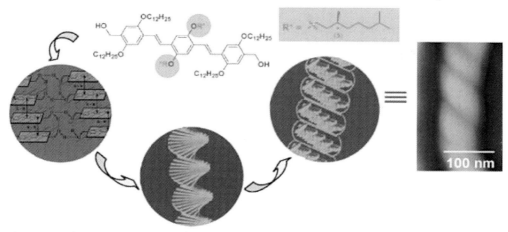

Figure 4.15 Schematic representation of self-assembly of **OPV2** into helical coiled-coil superstructures. A magnified AFM image of the superstructure is shown on the right.

dithienylethene photochromic unit functionalized with (R)-1-phenylethylamine-derived amides (Figure 4.16) was synthesized, which exists as two antiparallel, interconvertible open forms **15** with P- and M-helicity and cyclizes in a reversible manner upon irradiation with ultraviolet (UV) light to two diastereomers of the closed product **16**. The light-induced switching between **15** and **16** was followed by changes in electronic properties and conformational flexibility of the molecules. The amide groups incorporated in **15** induce gel formation in organic solvents at room temperature. Interestingly, the chirality present in **15** was expressed in a supramolecular aggregated system.

The open switch **15** in solution was found to be CD silent, whereas upon gel formation a strong CD absorption was observed due to the locking of the M- or P-helical conformation of the open form **15** in the gel state. It was found that the photoactive supramolecular system comprised of two different aggregation states α and β that could include either the open **15** or the closed **16** form, leading to a total of four different states. The aggregation and switching processes by which these four states can be addressed is summarized in Figure 4.17. A solution of open form **15**

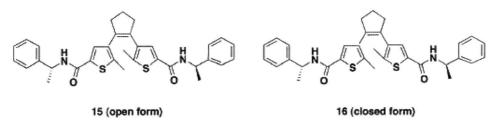

Figure 4.16 Chemical structure of dithienylethene-based organogelator.

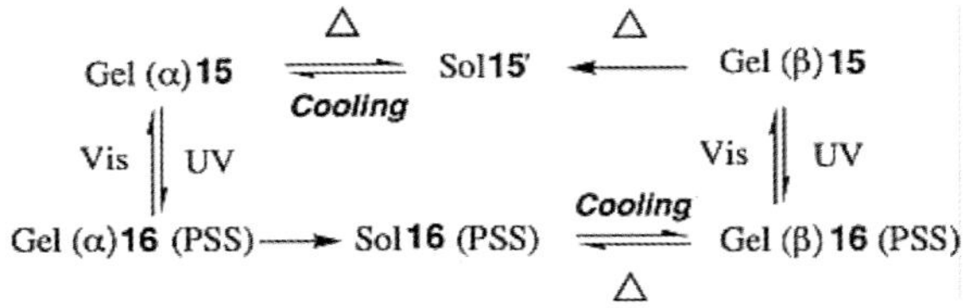

```
                        Δ              Δ
    Gel (α) 15    ⇌    Sol 15'   ←    Gel (β) 15
                Cooling
    Vis ↕ UV                           Vis ↕ UV
                        Cooling
    Gel (α) 16 (PSS) → Sol 16 (PSS) ⇌  Gel (β) 16 (PSS)
                          Δ
```

Figure 4.17 Aggregation and switching processes of **15** and **16** by thermal and light stimuli.

gives a stable gel (α) **15** (*P*-helicity) upon cooling. Photocyclization gives a metastable gel (α) **16** (PSS) (*P*-helicity) [PSS; photostationary state] with high diastereoselectivity (96%), which is fully reversible. Heating the gel (α) **16** (PSS) leads to a solution of **16** that gives a stable gel (β) **16** (PSS) (M-helicity). Irradiation of gel (β) **16** (PSS) with visible light results in a metastable gel (β) **15**, which in turn can be reconverted to the stable gel (β) **16** (PSS) by UV irradiation. Lastly, a heating and cooling cycle results in the transformation of the metastable gel (β) **15** to the original stable gel (α) **15** via the solution of **15**. The outstanding ability of this system to control chirality at different hierarchical levels would attract huge attention from advanced technology requirements such as molecular memory systems and smart materials.

4.3
Chiral Hydrogels

4.3.1
Chiral Fatty Acids

After the discovery of helical assemblies of 12-hydroxystearic acid in the gel phase by Tachibana *et al.*, Fuhrhop *et al.* [55] extended the similar kind of observation on serine dodecylamide (D- and L-) in water on the basis of cryo-TEM studies (Figure 4.18). It is interesting to note that in low-polarity solvents like toluene, L-serine dodecylamide produces multilayer tubules. The aggregation in water depends on pH and the ribbons and tubules are seen only in the pH range between 4.9 and 6.4, where amide and $COO^{(-)}$–COOH hydrogen bonds are cooperatively operating in water. Hydrogen-bonded two chains hydrated by large volume of water stabilized the aggregates to form ribbons and tubules in water and the greater extent of hydration induces the curvature. In organic solvents, the polar head groups are assembled in a reversed micelle fashion. The hydrogen-bonded chains are not solvated, whereas only the region of alkyl chains swells and intercalates to grow up multilayered tubules are grown. The racemate gives only platelets. The helical fibers are stable only if the chiral fatty acids are pure enantiomers and the racemate crystallizes as platelets. This phenomenon is known as "*chiral bilayer effect*" [56] that has been extensively investigated by Fuhrhop *et al.* with potassium tartaric dodecylamide [57] and *N*-octyl and *N*-dodecyl glyconamide [58, 59]. In all cases, bilayer helices are formed in water with a thickness of 4.0 nm.

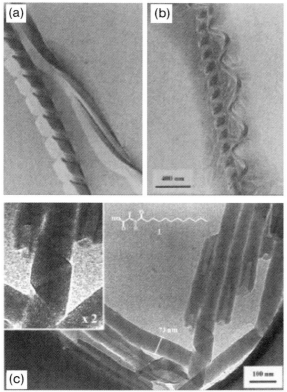

Figure 4.18 TEM pictures of (a) D-serine dodecylamide, (b) L-serine dodecylamide in water and (c) L-serine dodecylamide in toluene.

4.3.2
Chiral Sugar-Based Gelators

Shinkai et al. [60] devised and synthesized the sugar-based aqueous gelators containing azobenzene moieties linked with saccharides through amide linkage (Figure 4.19). Gelator **17** behaves as a super gelator as it forms a gel as low as 0.05 wt%.

A UV-vis study shows that in the gel state the absorption maxima shifted to lower wavelength with respect to these in the solution sate. This spectral change indicates the aggregation of azobenzene moieties in an H-type fashion. CD spectroscopy

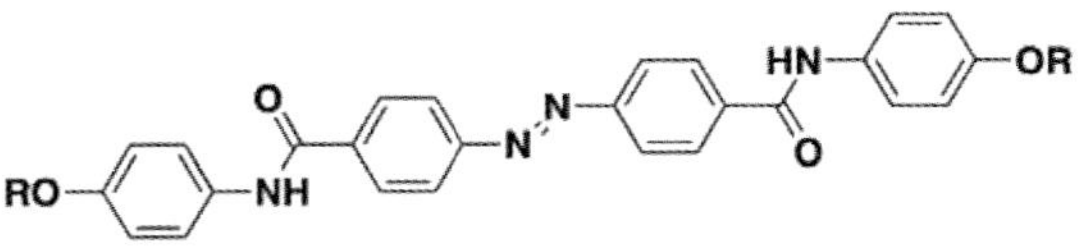

17: R = β-D-glucopyranoside
18: R = α-D-glucopyranoside
19: R = α-D-galactopyranoside
20: R = α-D-mannopyranoside

Figure 4.19 Chemical structures of sugar-based gelators.

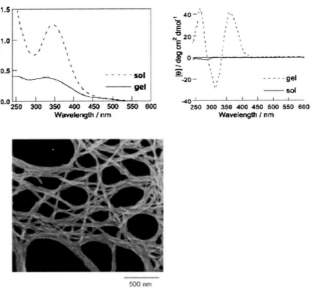

Figure 4.20 (left) UV-vis spectra, (middle) CD spectra of the gel sate (0.08 wt% of **17** in water) and in the DMSO solution. (right) SEM picture of xerogel prepared from the same concentration.

reveals the strong positive Cotton effect and the negative Cotton effect. This behavior implies that azobenzene moieties are stacked in a clockwise direction, that is, with right-handed helicity. This microscopic helicity is expressed in the macroscopic helicity as observed in SEM investigations (Figure 4.20). Of the four compounds shown in Figure 4.19, only gelator **17** (β-isomer) forms hydrogel and others (**18**:α-isomer), **19**, and **20** fail to form gels in water. The energy-minimized structure of gelator **17** is almost planar, which allows the $\pi-\pi$ stacking of azobenzene moieties as well as hydrogen bonding of OH of saccharides. In contrast, others have L-shaped sugar moieties attached with azobenzene, causing the hindrance to cooperative $\pi-\pi$ stacking and hydrogen-bonding interactions.

Other types of chiral sugar-based gelators containing aldopyranoside group, an aminophenyl moiety and a long alkyl chain has been reported to form hydrogels [61]. The gelation ability is enhanced with the double aminophenyl aldopyranoside moieties compared to single aminophenyl aldopyranoside moiety and the difference is attributed to the increase of hydrophilicity in water.

4.3.3
Miscellaneous Chiral Hydrogelators

Xu and coworkers [62] reported that pyrene-modified vancomycin is a novel type of hydrogelator (Figure 4.21). It forms a gel at 0.36 wt % in water. A CD spectrum of the hydrogel produces a relatively strong band at around 460 nm that is absent in the

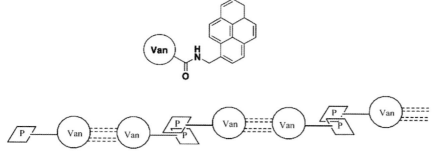

Figure 4.21 Schematic representation of stacking pyrene-based vancomycin gelators.

sol state. This is attributed to the presence of π–π stacking of pyrene moieties in the gel.

4.3.3.1 The Future of Chiral Gels in Nanoscience and Nanotechnology

In summary, we have seen throughout this chapter that a phenomenal number of gelator molecules can be designed by applying various novel strategies, especially with focusing chirality. Some of these molecules have the additional properties of responsiveness towards stimulus. This adds tremendous utility to the gelation chemistry as a whole, since these molecules can be further exploited in the field of sensors, memories, new materials, and so on. The field of organogel chemistry has grown rapidly in the past few years [63]. Due to their well-defined structure, coexistence of highly ordered fibers with a liquid phase, large interfacial area and ability to entrap solutes within the network pores, organogels are highly promising candidates for the design of sensors, drug delivery, catalysis, membrane and separation technology, or as templates for other materials [64]. It is extremely likely that with time and ever-increasing interests of researchers around the globe, organogels will find a wide and new range of applications, not only in nanotechnology but also in biotechnology, which might be beyond our present-day understanding and imagination.

References

1 Weiss, R.W. and Terech, P. (eds) (2006) *Molecular Gels Materials with Self-Assembled Fibrillar Networks*, Springer, The Netherlands.
2 Hanabusa, K., Yamada, M., Kimura, M. and Shirai, H. (1996) *Angewandte Chemie-International Edition*, **286**, 1540.
3 van Esch, J.H. and Feringa, B.L. (2000) *Angewandte Chemie-International Edition*, **39**, 2263.
4 Estroff, L.A. and Hamilton, A.D. (2004) *Chemical Reviews*, **104**, 1201.
5 Terech, P. and Weiss, R.G. (1997) *Chemical Reviews*, **97**, 3133.
6 Grownwald, O. and Shinkai, S. (2001) *Chemistry – A European Journal*, **7**, 4329.
7 Sangeetha, N.M. and Maitra, U. (2005) *Chemical Society Reviews*, **34**, 821.
8 Smith, D.K. (ed.) (2007) Special issue on Low Molecular Weight Organic Gelators. *Tetrahedron*, **63**, 7271–7494.

9 Fages, F. (ed) (2005) *Low Molecular Mass Gelators, in Topics in Current Chemistry*, Vol. 256, Spinger-Verlag, Berlin, Heidelberg.
10 Lioyd, D.J. (1926) *Colloid Chemistry*, Vol. 1 (ed. Alexander, J.), The Chemical Catalog Co, pp. 767–782.
11 Ferry, J.D. (1961) *Viscoelastic Properties of Polymers*, 3rd edn, Wiley, NY, p. 391.
12 Flory, P.J. (1974) *Faraday Discussions of the Chemical Society*, **97**, 7.
13 Derossi, D., Kajiwara, K., Osada, Y. and Yamauchi, A.(eds) (1991) *Polymer Gels, Fundamental and Biomedical Applications*, Plenum Press, New York.
14 Guenet, J.-M. (1992) *Thermoreversible Gelation of Polymers and Biopolymers*, Academic Press, London.
15 Dickinson, E.(ed.) (1991) *Food, Polymers, Gels and Colloids*, Royal Society of Chemistry, Cambridge, England.
16 Daniel, C., Dammer, C. and Guenet, J.-M. (1994) *Polymer*, **35**, 4243; Malik, S., Rochas, C. and Guenet, J.-M. (2005) *Macromolecules*, **38**, 4888.
17 Jana, T. and Nandi, A.K. (2000) *Langmuir*, **16**, 3141.
18 Malik, S. Jana, T. and Nandi, A.K. (2001) *Macromolecules*, **34**, 275;Malik, S. and Nandi, A.K. (2004) *The Journal of Physical Chemistry. B*, **108**, 597.
19 van Bommel, K.J.C., Friggeri, A. and Shinkai, S. (2003) *Angewandte Chemie-International Edition*, **42**, 980.
20 Sugiyasu, K., Fujita, N., Takeuchi, M., Yamada, S. and Shinkai, S. (2003) *Organic and Biomolecular Chemistry*, **1**, 895; Ajayaghosh, A., George, S.J. and Praveen, V.K. (2003) *Angewandte Chemie-International Edition*, **42**, 332; Mukhopadhayay, P., Iwashita, Y., Shirakawa, M., Kawano, S.-i., Fujita, N. and Shinkai, S. (2006) *Angewandte Chemie-International Edition*, **45**, 1592.
21 Kawano, S.-i., Fujita, N., van Bommel, K.J.C. and Shinkai, S. (2003) *Chemistry Letters*, **32**, 12;Malik, S., Fujita, N., Kawano, S.-i. and Shinkai, S. (2007) *Tetrahedron*, **63**, 7326.
22 Tachibana, T. and Kambara, H. (1969) *Bulletin of the Chemical Society of Japan*, **42**, 3422.
23 Tachibana, T., Yoshizumi, T. and Hori, K. (1979) *Bulletin of the Chemical Society of Japan*, **52**, 34.
24 Klyne, W. (1960) *The Chemistry of Steroids*, Wiley, New York.
25 Lin, Y.C. and Weiss, R.G. (1987) *Macromolecules*, **20**, 414.
26 Lin, Y.C., Kachar, B. and Weiss, R.G. (1989) *Journal of the American Chemical Society*, **111**, 5542.
27 Mukkamale, R. and Weiss, R.G. (1995) *Journal of the Chemical Society. Chemical Communications*, 375.
28 Terech, P., Furman, I. and Weiss, R.G. (1995) *The Journal of Physical Chemistry*, **99**, 9558.
29 Ostuni, E., Kamaras, P. and Weiss, R.G. (1996) *Angewandte Chemie (International Edition in English)*, **35**, 1324.
30 Terech, P., Ostuni, E. and Weiss, R.G. (1996) *The Journal of Physical Chemistry*, **100**, 3759.
31 George, M. and Weiss, R.G. (2006) *Accounts of Chemical Research*, **39**, 489.
32 Murata, K., Aoki, M., Nishi, T., Ikeda, A. and Shinkai, S. (1991) *Journal of the Chemical Society. Chemical Communications*, 1715.
33 Murata, K., Aoki, M., Suzuki, T., Harada, T., Kawabata, H., Komori, T., Ohseto, F., Ueda, K. and Shinkai, S. (1994) *Journal of the American Chemical Society*, **116**, 6664.
34 Sakurai, K., Ono, Y., Jung, J.H., Okamato, S., Sakurai, S. and Shinkai, S. (2001) *Perkin Transactions*, **2**, 108.
35 Snip, E., Shinkai, S. and Reinhoudt, D.N. (2001) *Tetrahedron Letters*, **42**, 2153.
36 Kawano, S.-i., Fujita, N., van Bommel, K.J.C. and Shinkai, S. (2003) *Chemistry Letters*, **32**, 12.
37 Kawano, S.-i., Fujita, N. and Shinkai, S. (2005) *Chemistry – A European Journal*, **11**, 4735.
38 Geiger, C., Stanescu, M., Chen, L. and Whitten, D.G. (1999) *Langmuir*, **15**, 2241.

39 Duncan, D.C. and Whitten, D.G. (2000) *Langmuir*, **16**, 6445.
40 Wang, R., Geiger, C., Chen, L., Swanson, B. and Whitten, D.G. (2000) *Journal of the American Chemical Society*, **122**, 2399.
41 Tian, H.J., Inoue, K., Yoza, K., Ishi-i, T. and Shinkai, S. (1998) *Chemistry Letters*, 871.
42 Ishi-i, T., Jung, J.H., Iguchi, R. and Shinkai, S. (2000) *Journal of Materials Chemistry*, **10**, 2238.
43 Ishi-i, T., Iguchi, R., Snip, E., Ikeda, M. and Shinkai, S. (2001) *Langmuir*, **17**, 5825.
44 Ishi-i, T., Ono, Y. and Shinkai, S. (2000) *Chemistry Letters*, **29**, 808.
45 Klowonn, T., Gansauer, A., Winkler, I., Lauterbach, T., Franke, D., Nolte, R.J.M., Feiters, M.C., Borner, H., Hentschel, J. and Dotz, K.Z. (2007) *Chemical Communications*, 1894.
46 Maitra, U., Polturi, V.K., Sangeetha, N.M., Babu, P. and Raju, A.R. (2001) *Tetrahedron Asymmetry*, **12**, 477.
47 Kamikawa, Y. and Kato, T. (2007) *Langmuir*, **23**, 274.
48 Hanabusa, K., Yamada, M., Kimura, M. and Shirai, H. (1996) *Angewandte Chemie (International Edition in English)*, **35**, 1949.
49 Hanabusa, K., Shimura, K., Hirose, K., Kimura, M. and Shirai, H (1996) *Chemistry Letters*, **25**, 885.
50 Ajayaghosh, A. and George, S.J. (2001) *Journal of the American Chemical Society*, **123**, 5148–5149.
51 Ajayaghosh, A. and George, S.J. (2005) *Chemistry – A European Journal*, **11**, 3217–3227.
52 George, S.J., Ajayaghosh, A., Jonkheijm, P., Schenning, A.P.H.J. and Meijer, E.W. (2004) *Angewandte Chemie-International Edition*, **43**, 3422–3425.
53 Ajayaghosh, A. and Praveen, V.K. (2007) *Accounts of Chemical Research*, **40**, 644–656.
54 de. Jong, J.J.D., Lucas, L.N., Kellogg, R.M., van Esch, J.H. and Feringa, B.L. (2004) *Science*, **304**, 278.
55 Boettcher, C., Shade, B. and Fuhrhop, J.-H. (2001) *Langmuir*, **17**, 873.
56 Fuhrhop, J.-H., Schnieder, P., Rosenberg, J. and Boekema, E. (1987) *Journal of the American Chemical Society*, **109**, 3387.
57 Fuhrhop, J.-H., Demoulin, C., Rosenberg, J. and Boettcher, C. (1990) *Journal of the American Chemical Society*, **112**, 2827.
58 Koening, J., Boettcher, C., Winkler, H., Zeitler, E., Talmon, Y. and Fuhrhop, J.-H. (1993) *Journal of the American Chemical Society*, **115**, 693.
59 Svenson, S., Koening, J. and Fuhrhop, J.-H. (1994) *The Journal of Physical Chemistry*, **98**, 1022.
60 Kobayashi, H., Friggeri, A., Koumoto, K., Amaike, M., Shinkai, S. and Reinhoudt, D.N. (2002) *Organic Letters*, **4**, 1423.
61 Jung, J.H., Shinkai, S. and Shimizu, T. (2002) *Chemistry – A European Journal*, **8**, 2648.
62 Xing, B., Yu, C.-W., Chow, K.-H., Ho, P.L., Fu, D. and Xu, B. (2002) *Journal of the American Chemical Society*, **124**, 14846.63.
63 For recent examples see (a) Yagai, S., Ishii, M., Karatsu, T. and Kitamura, A. (2007) *Angewandte Chemie-International Edition*, **46**, 8005; (b) Hirst, A.R., Huang, B., Castelletto, V., Hamley, I.W. and Smith, D.K. (2007) *Chemistry – A European Journal*, **13**, 2180; (c) Miravet, J.F. and Escuder, B. (2007) *Tetrahedron*, **63**, 7321; (d) Meziane, R.A., Brehmer, M., Maschke, U. and Zentel, R. (2008) *Soft Materials*, **4**, 1237; (e) Brizard, A., Stuart, M., van Bommel, K., Friggeri, A., de Jong, M. and van Esch, J. (2008) *Angewandte Chemie-International Edition*, **47**, 2063; (f) Piepenbrock, M.-O.M., Lloyd, G.O., Clarke, N. and Steed, J.W. (2008) *Chemical Communications*, 2644; (g) Sugiyasu, K., Kawano, S.-i., Fujita, N. and Shinkai, S. **(2008)** *Chemistry of Materials*, **20**, 2863; (h) Seo, J., Chung, J.W., Jo, E.-H. and Park, S.Y. (2008) *Chemical Communications*, 2794.
64 (a) Jung, J.H., Ono, Y. and Shinkai, S. (2000) *Angewandte Chemie-International Edition*, **39**, 1862; (b) Kobayashi, S., Hamasaki, N., Suzuki, M., Kimura, M., Shirai, H. and Hanabusa, K. (2002) *Journal of the American Chemical Society*, **124**, 6550;

(c) Sone, E.D., Zubarev, E.R. and Stupp, S.I. (2002) *Angewandte Chemie-International Edition*, **41**, 1705; (d) Jung, J.H., Lee, S.H., Yoo, J.S., Yoshida, K., Shimizu, T. and Shinkai, S., (2003) *Chemistry – A European Journal*, **9**, 5307; (e) Marx, S. and Avnir, D. (2007) *Accounts of Chemical Research*, **40**, 768; (f) Llusar, M. and Sanchez, C., (2008) *Chemistry of Materials*, **20**, 782.

5
Expression of Chirality in Polymers

Teresa Sierra

5.1
Historical Perspective on Chiral Polymers

Looking back for the presence of chiral polymers in nanoscale structures, it is apparent that they have always been there. Indeed, proteins, DNA and polysaccharides are biological macromolecules that are composed of chiral monomers, L-amino acids, nucleotides and monosaccharides, which adopt chiral conformations and chiral superstructures responsible for the origins of life. These chiral polymers, because of the specificity inherent in their chiral composition, are the basis of numerous biostructures that develop crucial biological roles. For example, proteins (in the form of enzymes) are catalysts that accelerate chemical reactions in a very specific way. In the form of antibodies, proteins develop highly specific recognition events. The most efficient information storage device consists of a chiral polymer, DNA, with a double-helical structure, which forms the chromosomes in which the genetic code is stored. From a chemical point of view, it is worth emphasizing the importance of the chemical structures of these biopolymers in their function, not only as far as covalent bonds between monomers are concerned, but also because of the possibility of noncovalent intra- and intermacromolecular interactions, which permit the development of their specific functions. The double helix of DNA [1], the triple helix of collagen [2, 3] or the complex structure of the tobacco mosaic virus, consisting of a helical RNA chain surrounded by the helical stacking of peptidic building blocks [4], are illustrative of the importance of chiral self-assembly processes in properties and function. Furthermore, these structures are dynamic and can be affected by the environment and external conditions, which can modify the implementation of specific functions.

The great performance of chiral biomacromolecules is the perfect model for synthetic chemists devoted to functional chemical structures. As a result, it is not strange that researchers have envisioned great potential in synthetic chiral polymers as a means to emulate and control, from the laboratory, processes as specific as those controlled by natural chiral polymers. Moreover, these processes could be effective not only in biological events but also in other properties such as electrical, optical, and mechanical systems.

Chirality at the Nanoscale: Nanoparticles, Surfaces, Materials and more. Edited by David B. Amabilino
Copyright © 2009 WILEY-VCH Verlag GmbH & Co. KGaA, Weinheim
ISBN: 978-3-527-32013-4

For the reasons outlined above, it is clear that research into chiral polymers must deal with the synthesis of the macromolecules as well as with the possibilities of achieving well-defined chiral architectures with the help of noncovalent interactions, especially H-bonding and π-interactions. Beyond the preparation of these chiral polymeric systems, a third important aspect is their final chirality and how it can be expressed, manipulated and amplified in a controlled way.

A survey of the literature makes it clear that the greatest advance concerns the synthesis of chiral macromolecules. For example, chiral polymers with defined helicity in their main chain were possible after the discovery of Ziegler–Natta catalysts and the possibility of achieving the first highly isotactic polymer from a chiral vinyl monomer, which had a helical structure in the solid state [5], thus giving rise to the development of chiral polyolefins [6]. For fifty years the development of polymerization strategies that allow a variety of chiral macromolecules with prevailing helicity [7, 8] has been mainly focused on three strategies [9]: (i) asymmetric polymerization, which creates configurationally chiral polymers, (ii) helix-sense-selective polymerization, using chiral monomers, chiral catalysts or chiral auxiliaries, and (iii) enantiomer-selective procedures, which enable the incorporation into a helical macromolecule of only one of the enantiomers from a racemic monomer. The combination of these polymerization procedures with unique phenomena, such as cooperative effects responsible for chiral amplification [10–12] (including sergeant and soldiers [13] or majority rule [14] effects), has led to the control of secondary helical structures, as the most abundant and simplest higher structural order, in a variety of chiral polymeric systems. Thus, for example, chiral polymers with either fluxional helical conformations with low-energy helix-inversion barriers, for example, polyisocyanates [15], poly(cis-acetylenes) [16–18] or polysilanes [19, 20], or stable helical structures with high-energy helix-inversion barriers, for example, polyisocyanides [21, 22], or polymethacrylates with bulky side groups [6], are typical examples of helical macromolecules that have given rise to interesting phenomena related to the control and tuning of their handedness. In addition to the helical conformation of the polymeric backbone, chiral expression in polymers includes chiral crosslinked materials, such as polymer-stabilized chiral mesophases [23–25], as well as chiral aggregates frequently formed by interchain interactions in π-conjugated systems such as polythiophene [26, 27] or polyphenylenevinylidene [28, 29].

More recently, a great impulse has been given to the development of the second aspect mentioned above, which concerns the accomplishment of well-defined chiral architectures derived from chiral polymers [30, 31]. This has been possible thanks to the increasing knowledge on supramolecular synthetic strategies as well as to the incorporation of powerful characterization techniques that allow molecular architectures to be studied at the nanoscale.

It is interesting that both levels of the synthetic/assembly process are complemented with the possibility of tuning at will the chirality of the final polymeric system. This is a key issue that relies on the design of active systems that can respond to certain external physical, chemical, electric or electromagnetic stimuli, whose information is transferred to the final architecture. This procedure gives rise to dynamic phenomena that involve controlling, inducing and switching chirality [32, 33]. These possibilities represent the basis for the establishment of chiral polymers as smart functional materials for applications in biological processes as well as in materials science and nanotechnology. On account of this role of chiral polymers as responsive materials, the spectrum of their applications is very broad and the search for new perspectives continues. Thus, in addition to the well-known possibilities for the preparation of chiral stationary phases for HPLC [34, 35], the dynamism of chiral polymers makes them suitable for chiral sensors, chiro-optical switches, memory elements for information storage, highly effective chiral catalysts and conductive materials, amongst other applications.

This chapter is intended to present an overview of the research carried out on chiral polymers and their implication in nanoscale phenomena. Accordingly, illustrative examples, mostly corresponding to the first years of the current millennium, have been selected in an attempt to show the latest significant advances in the control of chiral-helical architectures based on polymers and the properties and functions derived from these systems. For this reason, synthetic aspects have not been directly considered but those aspects of polymer design that provide chirality control (Section 5.2), induction (Section 5.3) or switching (Section 5.4) of the final chiral architectures are covered. Likewise, the control of chiral polymeric architectures restricted within nanodomains makes it interesting to discuss the potential of chiral block-copolymers and nanosegregation phenomena (Section 5.5). Finally, the capability of chiral polymers to use their helical architecture to template the formation of chiral molecular assemblies, as well as to promote the organization of functional molecules that favors the optimum performance of their properties, is treated at the end of this chapter (Section 5.6).

5.2
Chiral Architecture Control in Polymer Synthesis

In this section an overview is given of the design and synthetic strategies through which the expression of the chirality of the polymeric structure at the nanometer scale can be controlled. The section has been divided into three parts that, in a simple way, deal with the moment in which the chiral expression of the polymer architecture is accomplished. Thus, the covalent fixation of chiral supramolecular organizations into polymer structures has led to controlled chiral organizations in polymers and so polymerization procedures based on chiral assembly are considered in the first part. The second issue concerns the control during the polymerization process and includes those factors in the reaction process that can affect the final chiral architecture, that is, solvent, chiral templates, external stimuli. The existence of a type of reinforcement upon polymerization, in which polymers prepared by usual polymerization techniques can be further organized into chiral architectures by additional noncovalent interactions such as internal hydrogen bonding or by self-aggregation procedures, is considered in the last part of this section.

5.2.1
Polymerization of Chiral Assemblies

The covalent fixation of preorganized chiral helical supramolecular assemblies is a useful approach in the synthesis of chiral polymers with a controlled helical nanostructure. For example, H-bonded supramolecular polymeric structures, helical superstructures built upon π-stacking interactions or chiral aggregated systems based on liquid-crystalline interactions have been employed.

5.2.1.1 Chiral Organization Through H-Bonding Interactions

The participation of H-bonding interactions to organize monomers into polymerizable helical superstructures has been reported as a useful strategy. Covalent fixation of chiral columnar assemblies in apolar solvents has been performed in such a way that chiral polymers are prepared, which retain the helical organization of the previously self-assembled columnar architecture. Polymerizable columnar assemblies have been reported that consist of disc-like molecules that organize into columnar aggregates stabilized by H-bonding interactions. Benzene tricarboxamides (**1**) [36] organize into columnar assemblies that are made chiral through the sergeant and soldiers effect using a tiny amount of a chiral component. In this way, a nonreactive chiral benzene tricarboxamide derivative (**2**) was described as acting as a structure-directing agent that drives helicity towards a given handedness, thus promoting chiral amplification due to the helical superstructure. Upon light-induced polymerization the final polysorbate (**3**) backbone retains the induced chirality even after the chiral agent is removed. Furthermore, a memory effect is observed for the polymer, which undergoes unfolding–refolding cycles that retain the helical sense determined by the chiral molecule.

Using a similar strategy, triazine triamides **4** [37] were also reported to self-assemble into helical columnar aggregates through the addition of a tiny amount of a reactive chiral triazine derivative, **5**. Polymerization of the resulting assembly by ring-closing metathesis polymerization in THF allowed the chiral-helical architecture to be fixed (Figure 5.1). AFM and TEM observations showed a fibrous structure with a diameter of 2–4 nm, which corresponds to the size of the triazine monomer.

H-bonding interactions have also helped in the organization of diacetylenic monomers and this has enabled the preparation of chiral polydiacetylenes with a well-defined 1D helical structure controlled at the nanometer scale [38]. The process is based on UV-induced topochemical polymerization of the previously assembled diacetylene monomers. The polymerization is possible whenever the diacetylene

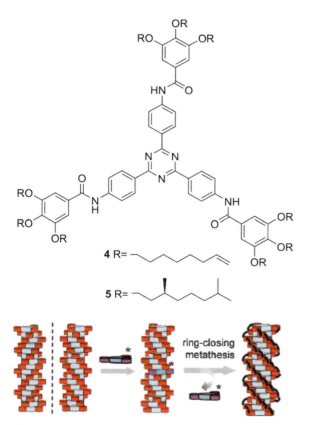

Figure 5.1 Cartoon representation of chiral amplification for the one-dimensional stacking of achiral **4** in the presence of tiny amounts of chiral **5**, and subsequent fixation of the chiral structure by ring-closing metathesis. Reproduced with permission from [37]. Copyright Wiley-VCH Verlag GmbH & Co. KGaA 2006.

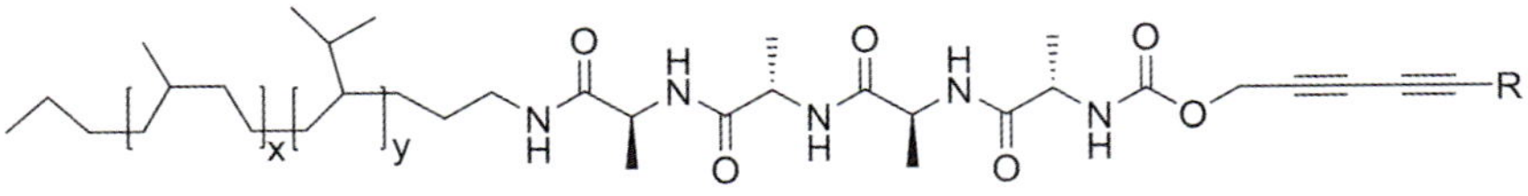

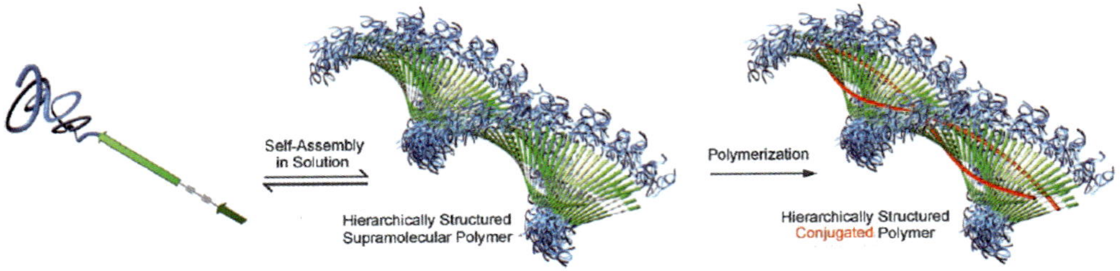

Figure 5.2 Schematic representation of the "self-assemble into a hierarchical structure, then polymerize" approach described by Jahnke et al. Reproduced with permission from [42]. Copyright Wiley-VCH Verlag GmbH & Co. KGaA 2008.

monomers are placed at an appropriate distance and with an appropriate packing angle. The strategy takes advantage of the possibility of tuning the hierarchical supramolecular organization of diacetylenic monomers [39–41] through a biomimetic approach that uses the interaction between oligopeptide fragments to give β-sheet peptidic assemblies. Thus, macromonomers (**6**) have been designed that consist of three fragments: the diacetylenic group, an oligopeptide segment and a hydrophobic polymeric block. The supramolecular structures formed by this type of supramolecular synthons range from tape-like to helical bundles depending on the molecular structure of the chiral monomer [42]. As a result, previously formed supramolecular polymers that are susceptible to degradation but with well-defined hierarchical nanoscale organization can be converted into conjugated macromolecules with a stable, covalently fixed, chiral-helical architecture (Figure 5.2).

5.2.1.2 Chiral Organization Through π-Stacking Interactions

Following this strategy based on performing polymerization upon aggregation, Aida et al. recently reported the first example of a conductive nanocoil with defined handedness. In previous publications, they described the formation of graphite-like nanotubes that consist of amphiphilic hexabenzocoronenes (HBC) [43] and can adopt a nanocoiled structure with uniform diameter and pitch [44]. The incorporation of pendant norbornene units in the amphiphilic molecule **7** allows ring-opening metathesis polymerization, which gives rise to a polymeric helical superstructure consisting of π-stacks of HBC (Figure 5.3a). Interestingly, these stacks exhibit conductivity upon doping. Incorporation of a chiral monomer (**8**) permits, through a sergeant-and-soldiers mechanism, the selective formation of

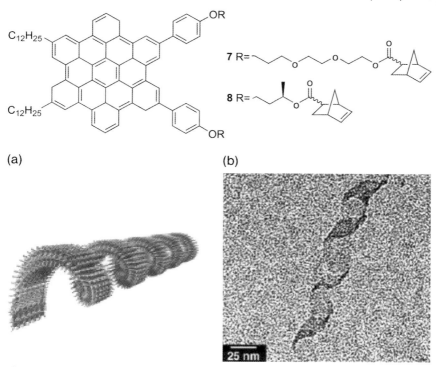

Figure 5.3 (a) Representation of a one-handed nanocoil formed through a sergeants and soldiers effect in the coassembly of **7** with **8**. (b) TEM image of the sample prepared with 20 mol% of **8**. Reproduced with permission from [45]. Copyright Wiley-VCH Verlag GmbH & Co. KGaA 2008.

nanocoils with one-handed helical chirality, which could be visualized by TEM (Figure 5.3b) [45].

5.2.1.3 Chiral Organization Through Mesogenic Driving Forces

Liquid-crystalline interactions offer a particular control of the supramolecular organization of molecules that makes them useful to control the chiral supramolecular organization of monomers susceptible to polymerization. This approach allows a chiral organization to be covalently fixed within a polymeric material.

Covalent capture of a chiral liquid-crystalline organization was developed to prepare chiral polymeric networks from cholesteric mesophases. Thus, *in-situ* photopolymerization of liquid crystals, described by Broer *et al.* [46, 47], was applied to the preparation of polymeric cholesteric materials, which are of practical interest for optical components of liquid-crystal displays. A cholesteric mesophase consisting of a mixture of reactive liquid-crystalline monomers (**9**) and a chiral dopant (**10**) can be photopolymerized with UV light in the presence of a photoinitiator. This process

5 Expression of Chirality in Polymers

CH₂=CHCOO~~~~~O—⌬—C(=O)—O—⌬—O—C(=O)—⌬—O~~~~~

9, Nematic monoacrylate Cr 100°C N 175°C I

CH₂=CHCOO~~~~~O—⌬—C(=O)—O—⌬—O—C(=O)—⌬—O~~~~OCOCH=CH₂

10, Chiral liquid crystal diacrylate Cr 70°C Ch 92°C I

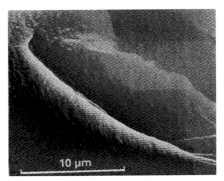

Figure 5.4 SEM image of the fracture surface of a pitch-gradient cholesteric network. Reprinted by permission from MacMillan Publishers Ltd: Nature [23], copyright 1995.

gave rise to a chiral polymeric network with a stable helical superstructure. Furthermore, it has been reported that it is possible to tune carefully the helical pitch of the final crosslinked chiral structure by controlling the molecular structure of the chiral dopant [48] or the diffusion of monomers [49] during the polymerization process. Thus, films with a pitch gradient along the thickness (Figure 5.4) – of interest for broad-band reflective polarizers [23] – or photopatterned cholesteric layers for reflective color filter arrays [50] have been prepared.

Liquid-crystalline elastomers that keep the helical organization of a cholesteric phase were described by Finkelmann *et al.* [24] To obtain cholesteric elastomeric networks with a monodomain structure, a two-step procedure was performed. In the first step a partial addition of polymethylhydrogensiloxane (**11**) with mesogens **12** and **13** and crosslinking component **14** in toluene gave rise to a weakly crosslinked gel. The second and key step was the deswelling process followed by the completion of the reaction. Deswelling was only allowed in one direction by slow evaporation of the solvent under centrifugation. This anisotropic deswelling simultaneous with the phase transition to the cholesteric mesophase is sufficient to cause a macroscopic orientation of the choles-teric phase structure, which is locked in by the completion of the crosslinking process.

The synthesis of polyisocyanides **15** with long-distance chiral induction from stereogenic centers that are remote from the skeleton has been based in the mesogenic driving forces that lead some compounds to organize into mesophases [51]. Chiral isocyanide monomers with a promesogenic structure made it possible for chiral induction to pass as far as 21 Å from the stereogenic center to the carbon at which polymerization takes place. Parallel behavior between the chiral induction in the helical polymer and in cholesteric mesophases by these monomers was observed and can be explained in terms of steric arguments governing both processes. Indeed, odd–even rules proposed for cholesteric mesophases were observed for the chiral senses induced in the synthetic polymers [52, 53].

5.2.2
Control of Chiral Architecture During Polymerization

Control of chirality during the polymerization process using chiral solvents, chiral templates or an external chiral stimulus, such as light, have been postulated as useful strategies to achieve chiral nanoscale architectures such as films consisting of helical fibers or fibril bundles.

5.2.2.1 Polymerization in Chiral Solvents

Polymerization procedures in which chiral information comes from the solvent and is transferred to the growing polymeric chain represent an approach that can lead to well-controlled helical architectures. Indeed, flexible macromolecules, such as polyacrylamides or polymethacrylamides, can be produced in an isotactic configuration from bulky monomers using (+)- or (−)-menthol as solvent under radical conditions [54]. Likewise, the use of a chiral solvent, either (R)- or (S)-limonene, in helix-sense-selective polymerization processes has recently been reported for the synthesis of polysilanes [55]. However, the use of a solvent that provides chiral information not only at the molecular level but also within its own supramolecular organization has been used to further control the chirality of the polymer, with control not only of the chain chirality but also of the growing fibers. This approach was first described by Akagi *et al.* in 1998 for the synthesis of helical polyacetylene [56]. The use of liquid crystals as polymerization solvents for the synthesis of conjugated polymers had previously been shown as a useful strategy to prepare well-aligned polyacetylene films with high electrical conductivities [57–60]. Consequently, the use of a chiral nematic phase as the polymerization solvent was envisaged as endowing a helical organization on the resulting polyacetylene fibers and hence new electromagnetic and optical properties. A chiral-helical polyacetylene was prepared by a polymerization method using a chiral liquid crystal as the solvent for a homogeneous Ziegler–Natta catalyst $(Ti(O-n-Bu)_4/Et_3Al)$. The liquid-crystal solvent consisted of a nematic liquid crystal doped with a chiral binaphthol derivative. SEM studies on the resulting polyacetylene films showed the formation of helical fibrillar morphologies (Figure 5.5). It was concluded that the polyacetylene chain adopts a helical conformation and that these helical chains are further organized into fibril bundles with the same screw sense as that of the chiral nematic liquid crystal (N^*-LC) used as solvent.

Later, an investigation was carried out into the influence of the chiral dopant responsible for the cholesteric organization of the solvent and the structure of the helical fibers prepared. The use of chiral dopants with catalytic activity, that is, a chiral titanium complex, showed that it is possible to merge both roles – chiral induction and catalytic activity – into the same compound and to obtain the same type of fibrillar morphology for the resulting polyacetylene [61]. Detailed investigations into the influence of the helical twisting power of the chiral dopant on the helical morphology

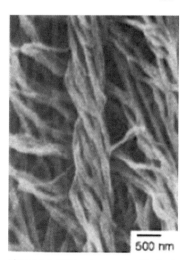

Figure 5.5 SEM micrograph of helical polyacetylene film synthesized in a chiral nematic liquid crystal with a chiral binaphthol derivative as the chiral dopant. From [56]. Reprinted with permission of AAAS.

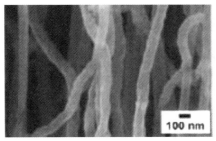

Figure 5.6 SEM micrographs of helical polyacetylene films synthesized in left-handed N*-LCs containing 0.75 mol% of a binaphthol chiral dopant. Helical pitch, 470 nm. Reprinted with permission from [64]. Copyright American Chemical Society 2007.

of polyacetylene films were also performed using crown-ether-derived binaphthyls [62]. Indeed, it was shown that the twisting power of the dopant increased on decreasing the size of the crown-ether ring. On increasing the twisting power it was found that the interdistance between the fibril bundles of the helical polyacetylene decreased, the diameter of a fibril bundle decreased, but the diameter of a fibril did not change. The interdistance between the fibril bundles of the helical polyacetylene was equal to about half of the helical pitch of the N*-LC, and the screw direction of the polyacetylene fibrils was opposite to that of the N*-LC. The influence of temperature on the twisting power of the dopant and hence the helical morphology has also been described [63]. It was found that the helical morphology of polyacetylene could be dictated by controlling the helical pitch of the cholesteric liquid crystals used as solvent for the polymerization. The conducting properties of the polyacetylene and their relationship with the helical structure of the polymer made it of interest to investigate the possibility of attaining single fibrils of the conjugated polymers (Figure 5.6) [64]. The use of chiral dopants with a high twisting power led to a cholesteric solvent with a helical pitch as tight as 270 nm. This solvent depressed the formation of bundles and resulted in a bundle-free fibril morphology consisting of single fibrils [65, 66].

Conducting polymers other than polyacetylene have been the target for the application of this chiral induction strategy using different polymerization methods. For example, chiral polythiophene-phenylenes **16** have been prepared from achiral monomers by a polycondensation reaction in a chiral nematic solvent [67]. It is proposed that the chirality of the polymers comes from the induced asymmetry in the aggregation of the polymeric chains and is stable in solution due to π-interactions between polymeric chains, a situation that is only disrupted by heating [68]. The polymers prepared in this way exhibit an exciton-splitting signal in the CD spectra in the absorbance region of the polymeric backbone and also show circular polarized luminescence.

16

(a)

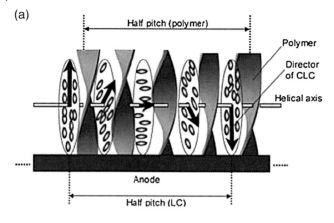

(b)

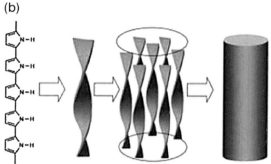

Figure 5.7 Possible polymerization mechanism of pyrrole in chiral nematic liquid crystals. (a) The polymer grows from the anode to the cathode side between the cholesteric organization. (b) The aggregation of polypyrrole forms a fibril structure. Reproduced with permission from [70]. Copyright Wiley-VCH Verlag GmbH & Co. KGaA 2007.

Electrochemical polymerization, which is an effective method to prepare conducting polymers, has also been carried out in chiral nematic liquid-crystalline electrolytes in order to prepare chiral conjugated polymers. It is proposed that the polymer grows from the anode to the cathode side within the chiral nematic matrix giving rise to vertically aligned fibril structures of chiral polypyrrole that replicate the liquid-crystal texture (Figure 5.7) [69, 70]. SEM observation of the resulting films showed periodic distances in the polypyrrole films that correspond to the distance between lines in the fingerprint texture of the cholesteric mesophase, that is, half the pitch of the cholesteric medium (Figure 5.8). This is consistent with the role proposed for the cholesteric liquid crystal as a guide along which fibrils of polypyrrole grow. This strategy, which is called chiral electrochemical polymerization, has also been successfully applied to the preparation of films of chiral PEDOT (**17**) [71],

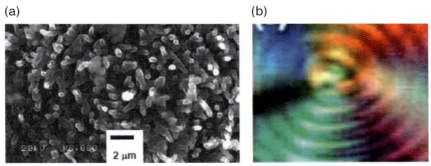

Figure 5.8 (a) SEM images of a polypyrrole film (LC free) synthesized by electrochemical polymerization within a chiral nematic liquid crystal. (b) Polarizing optical microscopy image of the chiral nematic liquid crystal mixture containing the monomer. Reproduced with permission from [69]. Copyright Wiley-VCH Verlag GmbH & Co. KGaA 2006.

poly[1,4-bis(3′,4′-ethylenedioxythienyl)phenylene] (**18**) [72] and polybithiophene (**19**) polymers [73].

17   **18**   **19**

5.2.2.2 Polymerization with Chiral Templates

Other chiral control procedures during polymerization have dealt with the use of chiral templates as described for the preparation of chiral polyaniline nanofibers. The synthesis of polyaniline nanofibers in aqueous solution relies on the use of concentrated camphor sulfonic acid solution as a chiral inductor and the use of aniline oligomers to accelerate the polymerization reaction along with ammonium sulfate as an oxidant. Under these conditions, films of entangled polyaniline nanofibers with helical morphologies were prepared as observed by TEM (Figure 5.9) [74, 75]. In contrast to the chemical method, electrochemical polymerization on ITO substrates using similar templates, camphor sulfonic acid and aniline oligomers, gave rise to polyaniline films with higher optical activity [76].

Water-processable chiral nanofibers that consist of supramolecular complexes of polyaniline chains (**20**) with a one-handed-helical structure intertwined with chondroitin sulfate (**21**) are obtained by chemical polymerization of the corresponding aniline monomer in the presence of **21** as a chiral polymeric template [77]. The supramolecular chiral complex polyaniline-**21** may broaden the applications of this conducting polymer in capillary electrophoresis.

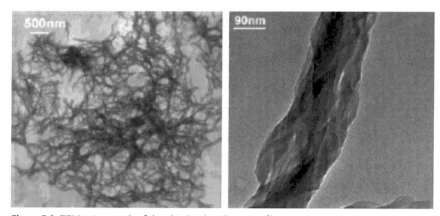

Figure 5.9 TEM micrograph of the chiral polyaniline nanofiber network (left) and magnification of a fiber bundle with helical nanofibers embedded (right). Reprinted with permission from [74]. Copyright American Chemical Society 2004.

5.2.2.3 Polymerization of Chiral Assemblies by Circularly Polarized Radiation

Finally, it is interesting to point out the use of an external chiral agent such as circularly polarized UV radiation to promote the formation of chiral polydiacetylenes (**22**) with stable optical activity. The photopolymerization process is carried out on monomer films prepared by vacuum evaporation and provides optical activity that is stable even after the thermochromic phase transition occurs at high temperatures [78]. Similar types of experiment leading to the preparation of optically active polydiacetylenes with designed chirality have been described for LB films of the monomers photopolymerized using circularly polarized radiation [79].

5.2.3
Chiral Architecture Control upon Polymerization: Noncovalent Interactions

There are a number of chiral macromolecules that can undergo hierarchical helical self-organization, a process that gives rise to chiral superstructures resulting from intra- or intermacromolecular noncovalent interactions. This process can be considered as organization beyond the polymerization process, and allows control over the formation of nanostructures that may exhibit well-defined 1D, 2D or 3D architectures. Such control of helical superstructures can be achieved by different strategies. On the one hand, an appropriate design of the constituent monomers, which incorporate groups susceptible to H-bonding or π-interactions, allows the formation of chiral helical superstructures that are frequently based on mesomorphic approaches. On the other hand, π-conjugated polymeric chains can form chiral supramolecular aggregates under the appropriate solvent and temperature conditions.

5.2.3.1 Control of the Chiral Architecture by H-Bonding Interactions
An extensively used practice to induce the formation of higher-order structures from helical polymers imitating protein features has been attained by using amino-acid-derived monomers to build up chiral polymers, such as polyisocyanides and polyacetylenes.

Nolte *et al.* demonstrated that is possible to further stabilize the 4_1 helical conformation of the polyisocyanide backbone by means of hydrogen-bonding interactions between appropriately chosen side groups [80]. In this way, polymerization of isocyanodipeptides derived from α-amino acids leads to macromolecules (**23**) whose helical backbone is stabilized by H-bonds between amide groups of contiguous dipeptide side groups along the polymer chain. As a consequence, a β-sheet-like arrangement can be formed with defined arrays of hydrogen bonds along the polymeric backbone (Figure 5.10) [81]. This leads to well-defined and rigid helical polymers that can further stabilize their superstructural chirality into lyotropic cholesteric mesophases [82]. When isocyanide monomers bear tripeptide side groups (**24**), the stereochemistry of the α-amino acid residues has a strong influence on the formation of well-defined hydrogen-bonded arrays, since the polymerization process is less sterically favorable than in the case of isocyanodipeptides [83].

Helical polymers of isocyanopeptides **25** derived from β-amino acids have also been synthesized [84]. In contrast to the kinetically stable helical macromolecules that are formed upon polymerization of α-amino acid based isocyanopeptides, a dynamic helical model was proposed for the β-amino-acid-derived polyisocyanopeptides (Figure 5.11).

The potential of these polyisocyanopeptides for the synthesis of new functional helical macromolecules was highlighted in a recent study in which acetylene-functionalized peptide side groups were incorporated in the system (**26**), a characteristic that allows derivatization of the polymer through click chemistry [85].

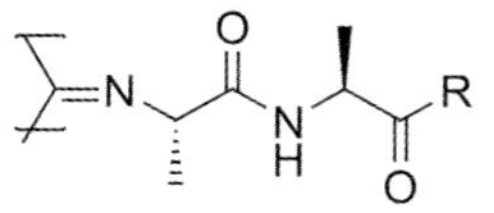

23 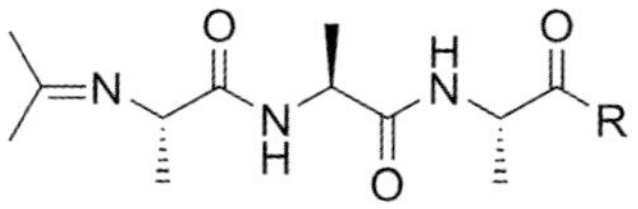   **24**

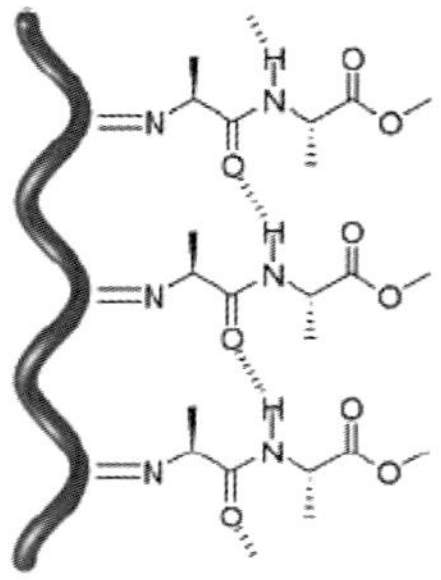

Figure 5.10 Schematic representation of the hydrogen-bonding arrays between the side chains in polyisocyanodipeptide **23**. Reproduced with permission from [81]. Copyright Wiley-VCH Verlag GmbH & Co. KGaA 2006.

25

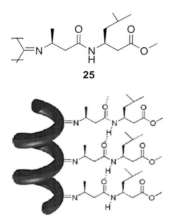

Figure 5.11 Schematic representation of the hydrogen-bonding arrays between the side chains in polyisocyanodipeptide **25**, derived from β-amino acids. Reproduced with permission from [84]. Copyright Wiley-VCH Verlag GmbH & Co. KGaA 2006.

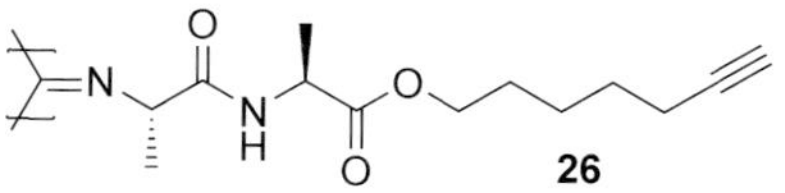

26

In a similar way to the polyisocyanopeptides described above, polyphenylisocyanide **27** with L-alanine pendants represents a class of stiff helical polymers that is

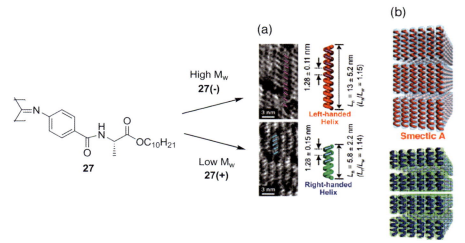

Figure 5.12 (a) AFM images of polymers **27**(−) and **27**(+) on HOPG and schematic representations of the left-handed helical **27**(−) and right-handed helical **27**(+). (b) Schematic representation of the 3D smectic ordering of the one-handed helical polymers **27**-(−) and **27**-(+) in the liquid-crystalline state. Reprinted with permission from [87]. Copyright American Chemical Society 2008.

stabilized by four sets of intramolecular H-bonds, which can impart further organization into mesophases. For example, lyotropic cholesteric behavior has been reported for this polymer when prepared by conventional polymerization techniques [86]. However, polymers prepared by a living polymerization technique, which leads to diastereomeric helical polymers of different molecular weights, **27**(−) and **27**(+), and low molecular weight dispersion, organize themselves into well-defined 2D and 3D smectic supramolecular arrangements on substrates or in the liquid-crystalline state, respectively (Figure 5.12). The higher-order organization was visualized and measured by AFM and X-ray diffraction [87].

The dynamic helical structure of polyacetylenes has also been functionalized with amino-acid-derived side groups (**28**), which endow stability to the helical structure through H-bonds between side groups, with the possibility of tuning the stability by external stimuli such as solvent and pH conditions. Indeed, it is possible to control the morphology of nanofibers formed by evaporation of solvent under different pH conditions (Figure 5.13a). Under neutral conditions helically bundled nanofibers of macromolecular assemblies were detected by AFM (Figure 5.13b). In contrast, under basic conditions, unraveled nanofibers with sizes corresponding to the single macromolecule were observed (Figure 5.13c) [88]. Recently, Yashima *et al.* reported the formation of ordered two-dimensional crystals of one-handed helical polyacetylene **29** chains in parallel arrangement on graphite. AFM direct visualization and X-ray measurements allowed the determination of the nanoscale dimensions of the polyphenylacetylene assembly (Figure 5.14) [89].

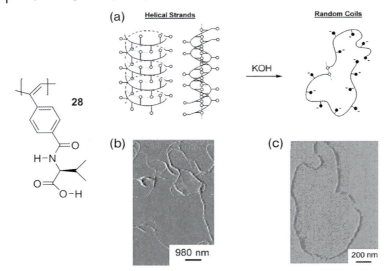

Figure 5.13 (a) Schematic representation of single- and double-stranded helixes of **28** via intra- and interchain hydrogen bonding. Ionization of carboxylic acid in basic conditions destroys the hydrogen bonds leading to the formation of random coils. (b) AFM images of helical nanofibers formed by **28** on mica at room temperature under neutral conditions. (c) AFM images of helical nanofibers formed by **28** on mica at room temperature under basic conditions. Reprinted with permission from [88]. Copyright American Chemical Society 2001.

Other polyphenylacetylenes with amino-acid-derived pendant groups, that is, L-leucine [90], L-lysine [91] or L-valine [92], have been reported that show helical structures stabilized by intra- and interchain hydrogen bonding. The dynamic nature of the polyacetylene backbone makes some of these systems very versatile in terms of tuning helicity in a reversible way [92].

This sort of bioinspiration to synthesize helical polymers with well-defined H-bonded arrays has led to novel polymer structures and to functionality in the resulting

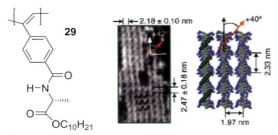

Figure 5.14 Magnified AFM phase image of **29**. A possible model was proposed according to the results of X-ray structural analysis. Reproduced with permission from [89]. Copyright Wiley-VCH Verlag GmbH & Co. KGaA 2006.

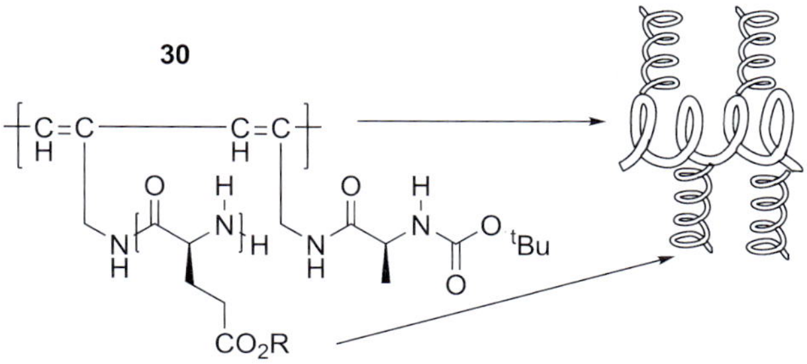

Figure 5.15 Cartoon representation of the helical-helical graft polymer **30** [93].

Figure 5.16 Schematic representation of the double-strand helically arranged hydrogen bonds in **31**. Reproduced with permission from [101]. Copyright Wiley-VCH Verlag GmbH & Co. KGaA 2005.

polymer. Masuda *et al.* prepared a novel polymeric structure, namely a helical-helical graft polymer, formed by a polyacetylene main chain-carrying helical polypeptides as side groups (**30**) (Figure 5.15) [93]. Polyphenylacetylenes with oligopeptides side chains proved effective as organocatalysts for the asymmetric epoxidation of chalcone [94]. The one-handed helical array of the side oligopeptide groups seems to be essential, since the monomers showed almost no enantioselectivity.

The presence of the amido group in polypropargylamides (**31**) has been exploited to achieve polyacetylenes with rigid conformations due to intramolecular hydrogen bonding along the polymer chain [95–99]. The model proposed is a helical conformation of the main chain surrounded by two helically arranged hydrogen-bonded strands (Figure 5.16) [100]. Moreover, these helical polymers aggregate in a poor solvent to afford lyotropic cholesteric mesophases with a helical superstructure, the sign of which depends on the solvent and temperature [101].

5.2.3.2 Control of the Chiral Architecture by π-Stacking and Steric Factors

In 1990, Ringsdorf *et al.* described a strategy to build helical architectures with controlled chirality by exploiting the mesogenic driving forces that govern the formation of columnar mesophases [102]. Indeed, polymer **32** that adopted a helical architecture addressed by π-stacking interactions between disk-like triphenylenes in the main chain was reported. The chiral architecture was biased towards a given screw sense by using chiral tails in the triphenylene pendant groups. Furthermore, considering the electron-donating character of the triphenylene unit, the possibility of controlling the optical activity of the columnar organizations in films by charge-transfer interaction with a chiral electron-acceptor compound was demonstrated

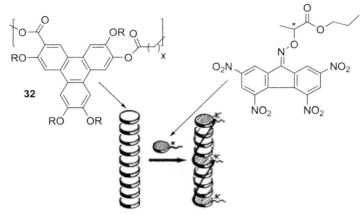

Figure 5.17 Schematic representation of the columnar organization of polymer **32** before and after doping with a chiral derivative of tetranitrofluorenone as electron-acceptor. Reproduced with permission from [102]. Copyright Wiley-VCH Verlag GmbH & Co. KGaA 1990.

(Figure 5.17). Thus, from achiral polymers, circular dichroism was observed in the absorption region of the triphenylene group when a chiral derivative of tetranitrofluorenone was incorporated as an electron-acceptor in the system.

A strategy that uses columnar organizations to build up controlled architectures in dendronized polymers has been widely developed by V. Percec. Dendronized polystyrenes and polymethacrylates have been described that present cylindrical or spherical morphologies depending on the structure of the side dendritic side-group [103]. As far as the polymeric backbone is concerned, the dynamic nature of *cis*-transoidal polyacetylene made this macromolecule suitable for the design of novel cylindrical dendronized systems (e.g., **33**) within which a helical conformation of the polymeric chain is favored (Figure 5.18) [104]. This helical conformation is the origin of chirality in nanoscale architectures, whose dimensions can be controlled by the synthetic strategies and the selection of dendritic side-groups. Detailed structural analysis of these polymers by X-ray diffraction showed that the transfer of chiral information from the dendritic group to the polymer backbone has a strong steric contribution, which influences the dimensions of the columnar organization in the mesophase [105].

5.2.3.3 Chiral Superstructures by π-Interactions: Chiral Aggregates

Polythiophenes are interesting π-conjugated polymers in terms of their properties and potential applications. It is well established that these polymers, when bearing chiral side groups, can present higher-order chiral aggregates even if the polymer does not show optical activity in the single-molecular state. Indeed, CD studies of chiral polythiophenes showed that polymer chains organize into chiral aggregates in poor solvents, at low temperature, or in films [26]. The chiral

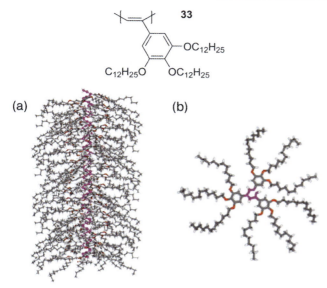

Figure 5.18 (a) Stick-model representation of the structure of the cylindrical macromolecule **33**. (b) Top view of a column stratum. Reprinted with permission from [105]. Copyright American Chemical Society 2006.

aggregates consist of a twisted π-stacking arrangement of the polymeric chains. Moreover, the chiral ordering of polythiophene aggregates can be controlled by a metal-salt-dependent doping-dedoping process [106]. The chirality of the aggregates depends on the doping level and maximum chiral ordering can therefore be achieved by balancing doping by polymer–metal coordination and dedoping by aggregate formation.

Chiral induction through metal coordination has been described for polythiophene **34** bearing an optically active oxazoline side groups [107]. Studies with different metal salts showed that induced CD appeared in most cases along with slight changes in the UV/vis spectra (no color change) when recorded in a good solvent for the polymer. These results indicate that the chirality may not be induced by chiral π-stacked aggregates of the polymer, but by the chirality of the main chain; for example, a predominantly one-handed helical structure induced by intermolecular coordination of the oxazoline groups to metal ions [108].

When copper(II) coordination takes place in a poor solvent, in which π-stacked aggregates are favored, it is possible to turn off the chirality of the system through a doping process (Figure 5.19). The system can revert to the chiral aggregate after an undoping process using EDTA to complex the copper cation [109].

Another example of optically active conjugated polymers that show chiral π-stacked aggregates consists of polycarbazole polymers (**35**). Polycarbozole **35** bearing citronellol-derived chiral side chains, show optical activity in molecularly dispersed

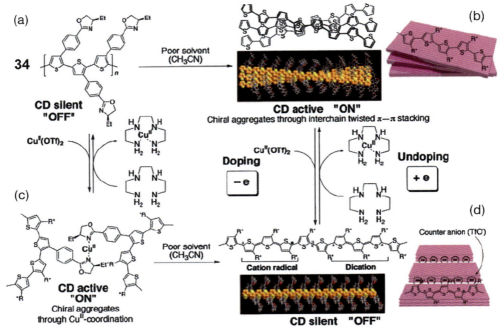

Figure 5.19 Schematic illustrations of chiral supramolecular aggregate formation of **34** (a) in the presence of a poor solvent such as acetonitrile (b) and Cu(II) (c). A possible structure of the doped **34** aggregates (d) and two different switching processes of supramolecular chirality of **34** aggregates are also shown. Reprinted with permission from [109]. Copyright American Chemical Society 2002.

solution and in the aggregated state. In contrast to the usual situation in polythiophenes, theoretical calculations show that chiral polycarbazoles most probably adopt a helical conformation of their main chain, which is maintained upon formation of the chiral aggregates [110].

In a similar way to π-conjugated polythiophenes, σ-conjugated polysilanes can form supramolecular chiral aggregates, which can be controlled by the appropriate choice of the ratio of poor/good solvent [111]. Studies on poly(n-alkylaryl)silanes (**36**) indicate that, depending on the length of the side chains, the aggregates formed by the addition of a nonsolvent to the polymer solution may show CD spectra opposite to that of the polymer in a good solvent, which corresponds to the chirality of the polymer chain. Consequently, the chirality of the aggregate can be tuned by adjusting the size of the side chains and the good/poor solvent ratio [112].

5.3
Asymmetry Induction in Nonchiral Polymers

Given the difficulties that can be encountered in the synthesis of chiral macromolecules with well-defined helical superstructures, an interesting alternative is to establish a procedure that allows the induction of chirality in a previously synthesized achiral polymer. Dynamic helical polymers can adopt a prevailing handedness through noncovalent interaction with chiral molecules, including acid–base interactions, hydrogen bonding, host–guest complexation, and also by the action of an external stimuli such as light.

In this section an overview of experiments carried out to induce chirality into polymeric chains is presented according to the type of external chiral auxiliary used and the way in which it interacts with the polymeric chain.

5.3.1
Induction Through Noncovalent Interaction with Chiral Molecules

The induction of chirality associated with a polymer chain conformation with a preferred helical sense has most commonly been achieved by promoting interactions between the polymer and small discrete chiral molecules. Most of the induction experiments have been based on acid–base interactions. Nevertheless, other noncovalent linkages – such as host–cation and metal-coordination – have proven successful in the induction of chirality into achiral polymers.

5.3.1.1 Chiral Induction by Acid–Base Interactions
Polyacetylenes, polyisocyanides, polyvinylpyridines, polyanilines and polythiophenes, which are polymers capable of expressing chirality in the form of helical, have mainly been the focus for acid–base chiral-induction processes.

The induction of chirality associated with a preferred helical sense using acid–base interactions has been extensively studied in polyacetylenes [32]. In 1995, Yashima et al. reported the first experiments into the induction of optical activity into a random coil polyphenylacetylene formed from achiral carboxyphenyl side groups (**37**) [113]. Acid–base complexation with amines led to the formation of a prevailing helix, the sign of which depended upon the stereochemistry of the chiral amino guest molecule. The effect derived from this induction process has been used to attain different phenomena. For example, stimuli-responsive polyphenylacetylene gels bearing a carboxylate side group have been described and the swelling properties of these materials in DMSO and water can change on modification of their helical chirality through acid–base interactions with chiral amines [114].

37

Stereoregular polyphenylacetylene **38** bearing an N,N-diisopropylaminomethyl group has been reported to take part in acid–base interaction with various chiral acids including aromatic and aliphatic carboxylic acids, phosphoric and sulfonic acids, and amino acids. The complexes exhibited an induced circular dichroism in the UV-vis region of the polymer backbone. The charged polyphenylacetylene was highly sensitive to the chirality of the acids and this enabled the detection of a small enantiomeric imbalance in the chiral acids in water (Figure 5.20) [115]. Indeed, due to the appearance of chirality-amplification effects expressed in the formation of a cholesteric liquid-crystal phase in water when the polymer interacted with chiral small molecules, chirality induction occurred even with a tiny enantiomeric excess of the corresponding chiral molecule [116, 117]. These experiments showed that polyphenylacetylenes are suitable probe materials to determine the chirality of small molecules whose optical activity is too low to be detected by conventional spectroscopic techniques.

Recent experiments have broadened the scope for these polymers as chiral sensors with the possibility of employing chiral biomolecules, such as neutral carbohydrates and positively charged peptides, proteins, amino sugars, and antibiotics. Indeed, polyphenylacetylenes with strongly acidic functional pendants, such as phosphonic acid (or its ethyl ester) and sulfonic acid, have been shown to interact with these biomolecules to give an induced CD in the UV-vis region of the polymer. These results open the door to the rational design of synthetic receptors for chirality sensing of molecules of biological interest without the need for derivatization [118]. The interest in highly organized polymeric architectures such as those presented by dendronized polymers is illustrated by the preparation of supramolecular complexes between polyphenylacetylene **39** bearing phosphonic side groups with dendron-like bulky amines derived from glutamic acid (**40**) [119]. These polymers show a Cotton effect, the magnitude of which increased significantly on increasing the generation number of the dendron-like chiral amine.

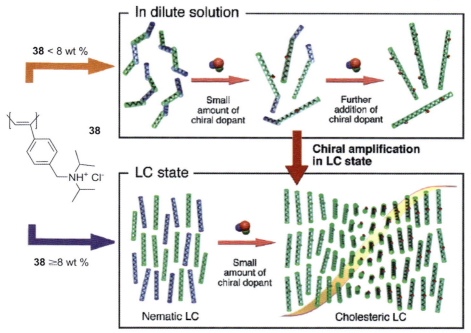

Figure 5.20 Mechanism proposed for the hierarchical amplification process of the macromolecular helicity of **38**, which has interconvertible, right- and left-handed dynamic helical conformations separated by rarely occurring helix reversals in dilute solution. Excess of preferred-sense helical sense (right- or left-handed helix) of **38** is induced with a small amount of chiral dopant, and further addition of the chiral dopant is required for a complete single-handed helix formation in **38** in dilute solution. In the nematic mesophase, the population of helix reversals may be reduced to form the racemic single-handed helices. In the presence of a small amount of chiral dopant, chirality is significantly amplified, resulting in a single-handed helical supramolecular assembly, thus showing a cholesteric mesophase. Reprinted with permission from [117]. Copyright American Chemical Society 2006.

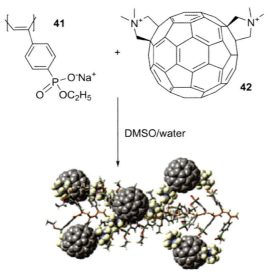

Figure 5.21 Schematic representation of the macromolecular helicity induction on polymer **41** with optically active **42**. Reprinted with permission from [120]. Copyright American Chemical Society 2004.

Furthermore, the phosphonic-acid-derived polyacetylene **41** has also been provided with helical chirality by interaction with strategically designed, geometrically favorable, chiral cationic C_{60} bisadduct **42** in DMSO/water systems [120]. The resulting supramolecular system affords a useful method to align fullerene units within a controlled helical array with a predominant screw-sense (Figure 5.21).

Polyphenylacetylenes **43** and **44**, bearing achiral oligoglycine residues, can adopt in water a one-handed helical conformation through interaction with chiral oligopeptides **45** and **46**, respectively, by means of ionic, H-bonding and hydrophobic interactions [121]. The presence of additional amide residues in the pendant groups makes these induced chiral polymers interesting as candidates for selective recognition and inclusion processes for specific chiral guests to produce more sophisticated supramolecular helical assemblies with achiral dyes (see Section 5.6).

In contrast to the well-known fluxionality of polyphenylacetylenes and the possibility of modifying their helical conformation, it was less clear how the conformational rigidity of polyphenylisocyanides would allow external control of the chirality of its polymeric chain. In 2002, Yashima *et al.* demonstrated that polyphenylisocyanides with small substituents cannot maintain a 4_1 helical conformation [122]. Thus, the polymeric backbone of polyphenylisocyanide **47** bearing carboxy groups can show a dynamic, one-handed helical conformation through acid–base interactions in DMSO with chiral amines and aminoalcohols, in a similar way to that observed for polyphenylacetylene **37**. The conformational dynamics of polyphenylisocyanides were demonstrated using NMR and CD spectroscopies and molecular-dynamics simulations.

Polyvinyl polymers with pyridine pendant groups can interact with small chiral molecules, which induce helical structures with a prevailing sense into the polymeric chain [123]. Poly(4-vinylpyridine) **48** has been shown to interact with small biomolecules such as amino acids, which play a double role. On the one hand the amino acid determines the chirality of the helical secondary structure of the polymer by means of acid–base interactions in water under acidic conditions. On the other hand, they make the complexes biocompatible and, therefore, potential biofunctional materials with a chiral superstructure as found in biomacromolecules [124].

Acid–base interactions with small chiral molecules have also been used to induce optical activity in polyanilines. Results were described for polyaniline films cast from a *m*-cresol solution that contained camphorsulfonic acid as a chiral inductor [125]. Induction experiments in solutions of polyaniline in N-methylpyrrolidinone and dimethylformamide were carried out using phosphoric, sulfonic and carboxylic chiral acids as inductors [126]. Phosphoric acids used for the experiments did not have a hydrogen-bonding acceptor site, indicating that hydrogen bonding between the acid and polyaniline is not a prerequisite for the polymer to adopt an optically active form.

Recent studies on acid–base interactions in solution that promote the formation of chiral films of polyanilines were described for aqueous mixtures of poly(2-methoxyaniline-5-sulfonic acid), **49**, with chiral amines or aminoalcohols [127]. The chiral induction is believed to be initiated by acid–base interactions involving "free" sulfonic acid groups on the polymer chains and the amino group of the chiral molecule. This gives rise to the formation of chiral aggregates in which assemblies of polymer **49** chains adopt a predominantly one-handed helical arrangement.

In a similar way to polyanilines, the dynamic axial chirality of the carboxybiphenol units integrated within an alternating copolymer, poly[(phenyleneethynylene)-alt-(carboxybiphenyleneethynylene)], **50**, could respond to the chirality of optically active diamines, thus showing an ICD due to an excess of a single-handed, axially twisted conformation, in the wavelength region of the π-conjugated polymeric backbone [128].

Polythiophenes have also been the subject of a similar type of chiral induction experiment. The electrostatic interaction in water between ammonium side groups of achiral polythiophene **51** and small chiral bioanions (ATP, **52**) led to the induction of a chiral helical conformation in the conjugated polymeric backbone, which was reinforced by the possibility of π–stacking between nucleobases [129]. The polymer showed a split ICD signal in the region of the π-π* transition wavelength. Moreover, the induced chirality of the polymer underwent inversion with changes in the concentration of ATP. This observation was explained in terms of the interplay between the two types of interaction, electrostatic and π-stacking, that could lead to small energy differences between resulting diastereoisomeric forms of the polythiophene/ATP complex.

The interaction of the polymer with the chiral molecule occurs along the polymer chain in the experiments described above. In contrast, achiral polypeptides (**53**) have been shown to adopt a one-sense helical conformation by interaction with a chiral carboxylic acid through an amino group suitably located at the N-terminus of the peptide. The chirality of the acid is amplified along the peptidic chain according to a so-called noncovalent domino effect [130, 131].

53

5.3.1.2 Chiral Induction by Host–Cation Interactions

Crown-ether side groups have been incorporated into polyphenylacetylene **54** as cation-binding hosts so that interactions with chiral amino acids, amines or aminoalcohols promote the induction of helical chirality into the polymeric backbone [132–134]. These polyacetylenes could also be of interest as chiral sensors since they are highly sensitive to a small enantiomeric excess of a host chiral amino acid due to a significant cooperative interaction [135]. Thermally driven on-off switching has also been described as a potential application of this type of system. Indeed, temperature can be used as an external stimulus to control the complexation with the chiral guest and hence the helicity of the polymer [136]. On the other hand, the use of chiral bis(amino acids) (**55**) gave rise to the formation of gels in certain solvents by means of noncovalent intermolecular crosslinking (Figure 5.22). The chiral bis(amino acid) induces a one-handed helical structure into the polyphenylacetylene (**54**), which seems to be essential for the gelation process [137].

The strategy for the induction of chirality through complexation of chiral amino acids and crown-ether side groups was found to be a suitable approach to obtain fullerene-based controlled arrangements [138]. A helical array of C_{60} units around the helical backbone of a polyphenylacetylene, **56**, with pendant fullerenes and 18-aza-crown-ether side groups was obtained with a preferred screw sense (Figure 5.23). These studies open up new possibilities for fullerene-containing materials for electronic and optoelectronic applications.

The possibility of using the complexation of chiral amines with crown-ether side groups to induce chirality into polymers has also been successfully employed with polyisocyanates **57** [139] and **58** (Figure 5.24) [140].

5.3.1.3 Chiral Induction by Metal Coordination

Even though there is only one example, it is interesting to note the possibility of inducing helical chirality along the backbone of a polyaniline-derived polymer by metal complexation. Complexation of the emeraldine base of poly(o-toluidine), **59**, with a chiral palladium(II) complex, **60**, bearing one labile coordination site led to the formation of the chiral conjugated polymer complex, **61**, which exhibited induced circular dichroism based on the induction of chirality into the π-conjugated backbone [141]. The chirality of the podand ligand was considered to regulate a propeller twist of the π-conjugated backbone. This strategy provided an efficient and feasible route to chiral d,π-conjugated complexes, in which the metals introduced are believed

144 | 5 Expression of Chirality in Polymers

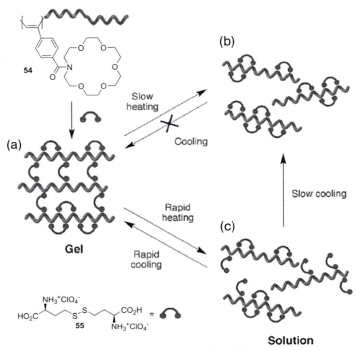

Figure 5.22 Schematic illustrations of the one-handed helicity induction and gelation of **54** upon complexation with **55** (A) and the mechanism of irreversible (A and B) and reversible (A and C) gel–sol transition upon slow heating and cooling and rapid heating and cooling cycles, respectively. Reprinted with permission from [137]. Copyright American Chemical Society 2005.

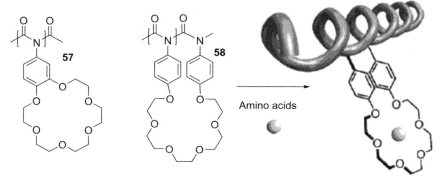

Figure 5.24 (a) Schematic representation of one-handed helicity induction of polyisocyanate **58** upon complexation with L-aminoacids. Reprinted with permission from [140]. Copyright American Chemical Society 2004.

to play an important role as metallic dopants. These chiral conjugated complexes could be promising functional materials for asymmetric redox catalysis.

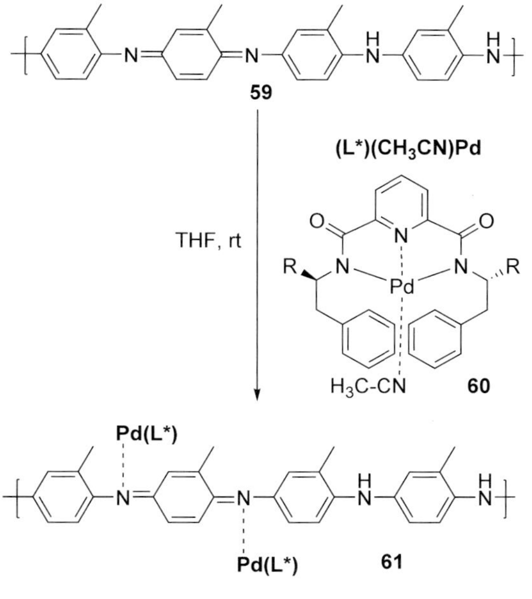

Figure 5.23 Schematic representation of macromolecular helicity induction in copolymer **56** upon complexation with L-alanine. The fullerene groups and crown-ether rings arrange in a helical way around the helical polymeric backbone. Reproduced by permission of The Royal Society of Chemistry from [138].

5.3.2
Induction Through Noncovalent Interaction with Chiral Polymers

In addition to small chiral molecules, chiral-helical polymers have also been used to induce a preferred helicity in achiral polymers by means of intermacromolecular noncovalent interactions.

The chirality induced into an anionic polyelectrolyte, the sodium salt of poly(4-carboxyphenylisocyanide) – **62** – through the use of a chiral amino alcohol (**63**) and subsequent removal of the chiral inductor, was successfully replicated in water in an oppositely charged polyelectrolyte, the hydrochloride of polyphenylacetylene **38** bearing an N,N-diisopropylaminomethyl groups (Figure 5.25). The process involves the formation of an interpolymer helical assembly with controlled helicity [142]. These helical polyelectrolytes are held together by the simple attraction of opposite charges. Nevertheless, the interpolymer complexation was more difficult to control compared to the complexation between a polymer and small molecules since it was highly influenced by external conditions such as the pH and salt concentration.

Experiments into induction by chiral polymers have also been carried out on films of polysilanes. Transfer of chiral information to polymer solid films was achieved using a chiral polymeric inductor – poly[n-decyl-(S)-2-methylbutylsilane], **64** – immobilized on a quartz substrate, either by grafting or spin coating, which acts as a command surface (Figure 5.26) [143]. It was reported that chirality transfer and, moreover, chiral amplification into achiral polysilanes was achieved depending on the rigidity of the optically inactive polymer, semirigid poly(n-decyl-3-methylbutylsilane), **65**, and after thermal annealing. A tiny amount of immobilized optically active polysilane could induce and amplify the optical activity in the optically inactive polysilane layer by thermal treatment. It is presumed that weak van der Waals

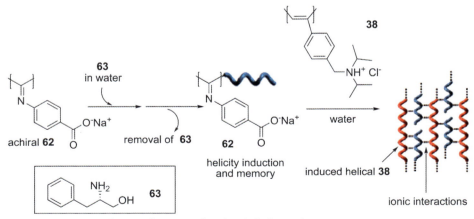

Figure 5.25 Schematic illustration of one-handed helicity induction and memory in **62** and the replication of the macromolecular helicity in optically inactive **38** in water. Reprinted with permission from [142]. Copyright American Chemical Society 2004.

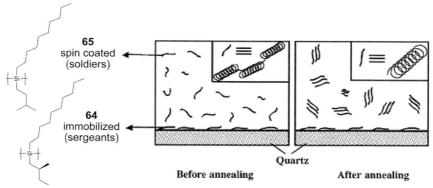

Figure 5.26 Schematic presentation of thermodriven chirality transfer and amplification in "soldier" optically inactive polysilane (**65**) from "sergeant" optically active helical polysilane **64** on quartz. Reprinted with permission from [143]. Copyright American Chemical Society 2004.

interactions at the surface between these two polymers may result in the transfer and amplification of optical activity into the achiral polymer.

The natural helical structure of biomolecules can be a useful source of chiral polymeric templates for the induction of chirality into polymers. For example, the formation of a chiral supermolecule consisting of a negatively charged conjugated polythiophene derivative (**66**) and a synthetic peptide with positively charged lysine groups (Figure 5.27) has been reported [144]. It is proposed that acid–base complexation between the two polymers gave rise to the induction of one-handed helicity into the main chain of the polythiophene, which shows a strong induced circular dichroism band in the π–π^* transition region in aqueous solutions. Likewise, the complexation of the polythiophene and the synthetic peptide induces the transition from random-coil to helical structure in the synthetic peptide.

5.3.3
Induction Through the Formation of Inclusion Complexes

The inner cavity of a helical polymer can be used to trap small molecules, which, if chiral, can induce a prevailing handedness into the polymeric chain. The formation of this type of inclusion complex is driven by weak interactions, solvophobic or H-bonding, and this has proven to be a useful strategy to build chiral helical structures and, furthermore, to modify the helix sense through structural changes in the chiral guest molecules.

The possibility of inducing helical chirality into achiral polymers by the formation of inclusion complexes was reported by Moore *et al.* for foldamer-type *m*-phenylene ethynylene oligomers (**67**). These compounds could include within their hydrophobic helical cavity (Figure 5.28) small chiral guests such as monoterpenes (e.g., **68**) [145] and rod-like chiral molecules (e.g., **69**) [146]. The stoichiometry of these

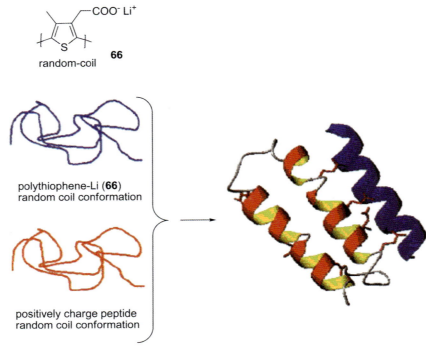

Figure 5.27 Schematic representation of the formation of a chiral supramolecular structure between a random-coil synthetic peptide and random-coil polythiophene **66** that induces helicity in both polymers. Reproduced with permission from [144]. Copyright (2004) National Academy of Sciences, USA.

solvophobically driven, reversible complexes is strictly 1 : 1 and in both cases there is preferential binding of the chiral guest to one of the oligomer's enantiomeric helical conformations. The incorporation of end-capping groups into the rod-like guest does not block the formation of the inclusion complex or the induction of a given sense for

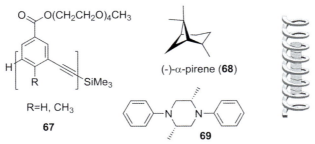

Figure 5.28 Schematic representation of the inclusion complex consisting of foldamer **67** including a chiral guest, **69**, within its helical cavity. Reprinted with permission from [146]. Copyright American Chemical Society 2001.

the helical conformation of the polymer. This means that the formation of the complex involves at least some deformation of the folded helix [147].

In contrast to the hydrophobically driven formation of the inclusion complex, Inouye et al. have successfully demonstrated the possibility of using H-bonding interactions that allow the use of saccharides as chiral-guest inductors of the prevailing helical sense into foldamers derived from m-ethylenepyridine polymers (**70, 71**). The N acceptor atom in these systems remains in the internal cavity of the helix formed by solvophobic π-stacking in polar solvents. This situation allows the incorporation of a saccharide in the inner cavity, which in turn affords an inclusion complex in which the helicity of the foldamer is biased towards a given handedness dictated by the chirality of the saccharide-guest. Consequently, an induced circular dichroism signal is observed in the absorbance band of the polymeric main chain. This system was presented as a useful approach for the recognition of saccharides [148]. Recently, the same authors have shown that it is possible to tune the handedness of the foldamer by mutarotation of the glucose guest (Figure 5.29). This finding is presented as a possible way to monitor the mutarotation of saccharides by recording the ICD signal of helical polymers [149].

Conversely, achiral polythiophene chains functionalized with ammonium groups, **51**, in DMSO/water systems wrapped within the helical cavity of a natural polysaccharide such as schizophyllan (**72**) showed ICD in the absorption region of the π–π* band [150]. The absence of a CD signal at longer wavelength, which is characteristic of a π-stacked arrangement of polythiophene chains, reveals that the chiral induction took place along the chain of the polymer and not because of chiral assemblies between polythiophene chains (Figure 5.30). This represents a suitable approach to prepare stable supramolecular insulated molecular wires with one-handed helical structures.

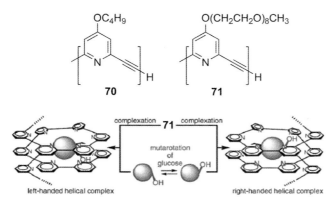

Figure 5.29 Helix inversion of the complex formed between poly (m-ethynylpyridine) **71** and d-glucose induced by mutarotation of d-glucose. Reproduced with permission from [149]. Copyright Wiley-VCH Verlag GmbH & Co. KGaA 2007.

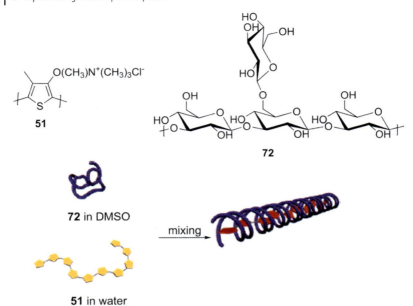

Figure 5.30 Schematic representation of the induction of a helical main-chain conformation into polythiophene **51** by wrapping with schizophyllan (**72**). Reprinted with permission from [150]. Copyright American Chemical Society 2005.

5.3.4
Induction by a Chiral External Stimulus

A racemic mixture can be biased towards an excess of one of the constituent enantiomers by the action of an external chiral agent. In synthetic polymers that adopt dynamic helical conformations, it has been shown that it is possible to displace the equilibrium between the two enantiomeric helixes by the presence or action of chiral solvents or irradiation with chiral light.

5.3.4.1 Solvent-Induced Chirality

In 1993, Green *et al.* [151] demonstrated the possibility of inducing chirality into a stiff polyisocyanate using chiral solvents such as (*S*)-chloromethylbutane. It was concluded that the optical activity induced in the polymer depends on the concentration of the chiral solvent in the polymer domain, which is related to the preferential attraction of the different solvents by the polymer. Hence, it was possible to tune the final induced optical activity in helical polyisocyanates by chiral solvents by controlling the side groups in the polymer that show different levels of attraction for the different solvents. Measurement of the optical activities of these polymers in different solvent mixtures with a chiral component has been presented as a useful method for studying the preferential solvation of this rod-like polymer in different types of solvent [152].

Experiments on chiral solvation to induce optical activity into achiral silicon-containing polymers have also been reported. For example, polyalkylphenylsilanes (**73**) and polyalkylalkylsilanes (**74**), which exist as random coils whose segments can form either plus or minus helixes, adopt a preferred helix sense when dissolved in synthetic (S)-(−)-2-methyl-1-propoxybutane and (S)-(−)-(2-methylbutoxymethyl) benzene [153]. Similar chiral solvation effects, which are responsible for the transformation of the random-coil conformation into the helical one, were described for chiral poly(*m*-phenylenedisilanylene), **75**. Thus, depending on the polarity of the solvent, chiral σ-π conjugated polymers, composed of an alternating arrangement of organosilicon and π-electron units, can show optical activity derived from a biased twist sense of the helical conformation of its polymeric backbone [154].

5.3.4.2 Light-Induced Chirality

Light can be used as a nondestructive external agent to induce chirality into polymeric systems. The three parameters, that is, intensity, wavelength and polarization, make light a versatile stimulus that can give rise to controlled internal changes in photochromic materials.

Most of the studies reported to date have been carried out on azobenzene-containing polymers – particularly side-chain polymers with azobenzene side groups [155, 156]. Thus, elliptically polarized light (EPL) can induce optical activity in films of amorphous polyesters (e.g., **76**), as reported by Nikolova *et al.* and Kim *et al.* [157, 158] Fukuda *et al.* have reported the incorporation of highly birefringent units in the azobenzene polymer so that large optical rotary power is achieved in the films [159]. When azopolymers present mesomorphic organizations (e.g., **77**), circularly polarized light (CPL) can be used to induce chirality [160–162]. The mechanism by which the chirality of the light is transferred to the material is still a matter of debate. The photoenantiomerization of *cis*-isomers of azobenzene and/or transfer of the angular moment from circularly polarized light to the chromophore have been proposed as possible mechanisms. Nevertheless, internal changes in the polymeric structure, which seem to be related to the formation of helical architectures in the bulk, have been proposed to occur. Furthermore, it seems plausible that the transfer of the chirality of the light to mesomorphic systems (nematic **78** or smectic **79**) could be related with phase transitions to chiral mesophases with superstructural helicity such as cholesteric [163–165] and TGB*-like phases, respectively [166].

A different strategy was developed by Takezoe et al., who reported a doped polymeric system (Figure 5.31) consisting of a nematic main-chain polymer (**80**) and the W-shaped azobenzene-containing dopant **81** [167]. CD measurements indicated that optical activity was not only induced into the dopant but also into the main chain polymer. The mechanism proposed is based on the transfer of the angular momentum from light to the medium, which gives rise to photoresolution of the chiral conformation of the W-molecule and the induction of an enantiomeric excess (Figure 5.31b). The W-shaped dopant acts as a chiral trigger, the molecular chirality of which is subsequently transferred to the host polymer. However, the model proposed is free from macroscopic helical structures, the formation of which is hindered in this type of rigid system.

Recently, the induction of optical activity in a solution resulting from the photoresolution of a racemic polyisocyanate was discovered by Feringa et al. [168]. The poly(n-hexylisocyanate) **82** incorporates a photochromic group in a terminal position, and this acts as a chiral trigger when stimulated with unpolarized 365 nm radiation. This group, as a light-driven molecular motor, undergoes unidirectional rotation in a four-step cycle through a combination of two photochemically induced cis–trans-isomerizations, each followed by an irreversible thermal isomerization. The cis-isomers promote a preference for a given handedness in the polymeric backbone (Figure 5.32). Accordingly, it is possible to induce chirality into the system in addition

5.3 Asymmetry Induction in Nonchiral Polymers

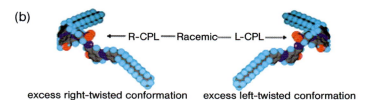

Figure 5.31 (a) Structure and composition of the doped photoresponsive polymeric system. (b) The most stable conformation calculated by MOPAC of the W-shaped azobenzene molecule **81** is the twisted conformation. Reprinted with permission from [167]. Copyright (2006) by the American Physical Society.

to having fully reversible control of its helicity. Polisocyanate **83**, which incorporates a structurally modified bistable chiroptical switch, allows the molecular chirality to be transmitted to the supramolecular organization of a cholesteric mesophase via macromolecular chirality induction. Furthermore, the magnitude and sign of the cholesteric helical pitch is fully controllable by light using two different wavelengths

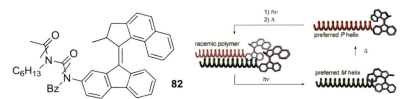

Figure 5.32 Schematic illustration of the reversible induction and inversion of helicity into polyisocyanate **82**. A thermal isomerization of the rotor inverts the preferred helicity. Racemization occurs upon a subsequent photochemical and thermal isomerization step. Reproduced with permission from [168]. Copyright Wiley-VCH Verlag GmbH & Co. KGaA 2007.

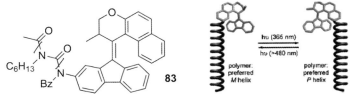

Figure 5.33 Schematic illustration of the reversible inversion of helicity into polyisocyanate **83** fully controlled by light. Reprinted with permission from [169]. Copyright American Chemical Society 2008.

(Figure 5.33). This allows avoiding thermal steps that would affect the chiral supramolecular mesomorphic organization [169].

5.4
Chiral Memory Effects. Tuning Helicity

Chiral memory effects rely on the performance of a two-state on-off function based on chirality-responsive polymers. Thus, it is important to tune the helicity of the polymer through an external stimulus and also to achieve permanence of the resulting helical handedness once the stimulus is removed. This possibility of switching between stable chiral states controlled by an external agent (temperature, solvent, light, etc.) should make these systems suitable for information storage.

In this section, an overview is given on experiments carried out to achieve memory effects in chiral polymers. The discussion includes studies on the inversion or racemization of chiral helicity in these systems that should provide the basis for switchable memory based on chirality. Experiments have been carried out on both chiral and achiral polymers.

5.4.1
Memory Effects from Chiral Polymers

As a first step to design chiral memory devices, the way to achieve helix inversion in dynamic chiral polymeric systems has been thoroughly studied. The action of different external agents, such as temperature, solvents or light, has been used to trigger chirality switching by the inversion of helicity in different types of helical polymer.

5.4.1.1 Temperature- and/or Solvent-Driven Memory Effects
Investigations into thermally driven helix-sense-inversion processes were described for polyisocyanates [11] both in solution and in the solid state. In solution, it was found that the inversion temperature could be tuned by controlling the composition of different-competing chiral comonomers according to the majority rule concept [170]. In the solid state, the helical reversal was affected by the transition temperature between the melt and the glass state. More interestingly, the effect was

amplified when the chirality of polymers such as **84** and **85** was transferred to the supramolecular organization of their respective cholesteric phases.

Likewise, the helix sense of polypropiolic esters **86** can undergo inversion through changes in temperature and solvent due to the small energetic barrier for helix reversal [171]. The helix sense determined by the chirality in the side chain, in some cases, can be elegantly inverted to the opposite sense by the action of an external stimulus such as temperature or solvent [16]. The action of these agents is strongly related to the structure of the side chain. Therefore, if the length of the alkyl side chain is appropriately controlled, the helix inversion can be driven by a change in solvent even at ambient temperature.

Thermally driven helix inversion has also been described for poly(N-propargylamides) bearing chiral pendant groups. These polymers, which show a helical conformation stabilized by hydrogen-bonding interactions between amide groups, may undergo helix inversion in solution promoted by a change in temperature [101, 172, 173]. A promising type of behavior was found for propargylamide polymer **87** with menthol-derived chiral side groups, which undergoes helicity inversion when heated to 70 °C. On heating the polymer for more than 30 min, the inversion is not thermally reversible and the induced helix sense is maintained even on cooling the sample to room temperature. In order to make polymer **87** suitable for a memory device, it is necessary to make possible the recovery of the original helix conformation. Indeed, it is possible to revert to the original helix sense by an acid-catalyzed helix deformation and reprecipitation from toluene, which affords the initial handedness (Figure 5.34) [174].

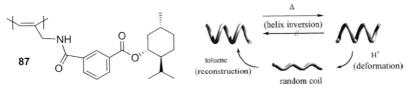

Figure 5.34 Scheme that represents the cycle of inverting and restoring the original helix sense of polymer **87**. Reprinted with permission from [174]. Copyright American Chemical Society 2003.

A combination of temperature with host–guest interactions in the β-cyclodextrin (β-CyD)-derived side chains of helical polyphenylacetylene **88** was reported as a new system with tunable optical activity that also involves color changes associated with modifications in the helical conformation of the chiral polymer [175]. The novelty of the system lies in the changes brought about by complexation. The helical conformation of the polymer is adjustable not only in terms of sense but also in the pitch, and this strongly affects the color in solution. This behavior may provide a new approach for sensors. However, a memory effect was not found in these systems on removal of the cofactor (guest molecules) that affected their chirality.

The versatility of polyphenylacetylenes for chiral-inversion processes resulting from their fluxional nature was assessed in recent experiments on 2D chiral surfaces (Figure 5.35). The inversion of macromolecular helicity of the dynamic chiral polyphenylacetylene **29** on graphite was visualized directly by AFM. This system provides a 2D switchable chiral surface that experiences inversion of helicity upon exposure to the vapor of a specific organic solvent [176].

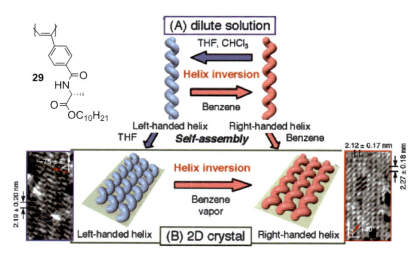

Figure 5.35 Schematic illustration of the macromolecular helicity inversion in dilute solution and 2D crystal state. (a) The helix sense of **29** in benzene inverts to the opposite helix-sense in THF and chloroform. (b) Rod-like helical **29** self-assembles into 2D helix bundles with the controlled helicity upon exposure of each organic solvent vapor. The one-handed 2D helix bundles of **29** further invert to the opposite handedness by exposure to benzene vapor. AFM images of both enantiomeric helical conformations are shown. Reprinted with permission from [176]. Copyright American Chemical Society 2006.

Stiff helical polysilanes consisting of stretchable, rotating silicon–silicon single bonds in a helical backbone bearing chiral pendant groups are also good candidates for memory processes based on helix-sense inversion induced by temperature changes. Certain polysilanes (e.g., **89**) were reported to show switchable helicity with a dynamic memory in isooctane solution at a given temperature [177, 178]. The casting of the polymer solution from the pre- and post-transition temperatures led to films that gave positive and negative Cotton circular dichroism (CD) signals, respectively. This result suggested that the helical sense of the polymer below and above transition temperature in the homogeneous solution was memorized in their corresponding cast films [179]. Furthermore, these polymers were proposed for chiro-optical memory systems with rewritable (RW) and write-once read many (WORM) modes in solid films. Plus and minus helix transitions are thermally reversible but strongly dependent on molecular weight [180].

A different approach for memory devices is based on the aggregating properties of polysilanes. Depending on the solvent and temperature, polysilanes can form chiral aggregates with amplified optical activity with respect to the dissolved polymer. The chiro-optical characteristics of microaggregates of poly(alkyl-alkoxyphenylsilane)s could be tuned by the appropriate choice of good/poor solvent ratio (solvent polarity), solvent addition order and temperature [111]. In order to memorize the chirality of the aggregates, a strategy was designed that was based on the preparation of microcapsules of polyureaurethane within which chiral aggregates of polydialkylsylanes (**90** and **91**) maintain the chiro-optical properties determined by the preparation temperature. The optical activity was only erased by destroying the aggregates [181].

5.4.1.2 Light-Driven Memory Effects

A promising strategy to achieve chiral memory systems is the possibility of using light as an external stimulus that allows switching of chirality between two stable states by means of photochemical processes. As already seen in the induction of chirality into achiral polymers, this requires the use of photoresponsive groups that can undergo some structural change upon irradiation, and this change can be transferred to the material properties.

Figure 5.36 Schematic representation of the transition from M- to P-helical polyisocyanates induced by a photochemical *trans–cis*-isomerization of the azo chromophore. Differences in the interactions between the asymmetric centers in the side chains and the polymer backbone lead to a preference for the M- or P-helical conformation. Reprinted with permission from [182]. Copyright American Chemical Society 1995.

The photochemically induced *cis–trans*-isomerization process of azobenzene has been extensively used to promote changes in properties. The use of azobenzene side chains in helical polymers has provided the possibility of controlling the chirality of the macromolecules upon UV irradiation, which promotes *trans*-to-*cis*-isomerization of the azo group, and subsequent reversal to the *trans*-configuration by heating. Early works on these switchable systems were undertaken by Zentel on THF solutions of chiral polyisocyanates bearing chiral azobenzene side groups (Figure 5.36) [182]. Later, these experiments were extended to thin films of azobenzene-containing chiral polyisocyanates in a PMMA matrix [183]. The chiro-optical properties of the film could be reversibly switched. Below the T_g, the photochemically modified helix conformation was stable, despite thermal relaxation of the azo chromophores.

Angiolini *et al.* reported the possibility of photomodulating the chirality of azobenzene chiral polymer films using circularly polarized radiation to induce photochemical isomerization of the azobenzene chromophores [184–187]. Methacrylic polymers **92** bearing a chiral spacer and azobenzene side groups have been investigated for reversible optical storage. These chiral polymers show reproducible writing–erasing cycles upon linearly and circularly polarized radiation, which promotes inversion of their original chirality and recovering by irradiation with the opposite radiation.

5.4.2
Memory Effects from Achiral Polymers

Remarkable memory phenomena involving induced chirality in achiral-helical polymers have mostly been reported for polyacetylenes and polyisocyanides. In general, experiments consisted in the induction of a prevailing helical sense into an achiral polyacetylene by the incorporation of optically active amines. The induced helicity is memorized when the chiral amine is replaced by an achiral one [188]. This memory effect was reported for a variety of pendant groups in the polymeric chain, that is, carboxyphenyl [189] and phosphonic phenyl in DMSO [190]. The presence of TsOH acid was important in the latter case [191]. A further step in the induction of chiral memory in polyacetylenes was recently described for polymer **93**, which showed a so-called dual-memory phenomenon (Figure 5.37) [192]. In this case, the helicity induced in the polymer by complexation to an enantiomerically pure amine (**94**) can be switched to the opposite helical conformation by heating at 65 °C. Once induced, both enantiomeric helixes can be memorized when the chiral amine is removed and replaced by an achiral one (**95**). The helicity stored in the polymeric chain remains stable at room temperature within the corresponding complex with the achiral amine.

The achiral sodium salt of poly(4-carboxyphenylisocyanide) **62** folds into a one-handed helix through configurational isomerization (*syn/anti*-isomerization) around the C=N backbone through noncovalent interaction with optically active amines, optically active quaternary ammonium salts or L-amino acids in organic solvent–water mixtures [193, 194]. The induced helicity is retained when the optically active compounds are completely removed, and even when the material is further modified with achiral amines through amido linkages. However, the helicity gradually decreases with temperature and completely disappears at 80 °C. Approaches described to stabilize the chiral memory of these systems rely on the formation of supramolecular architectures such as hydrogels or liquid crystals. Hydrogels were obtained by

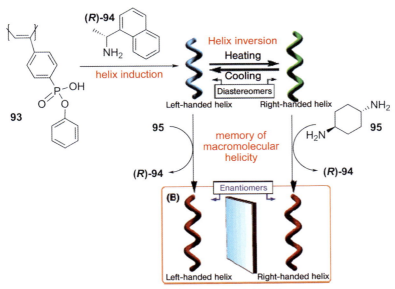

Figure 5.37 Schematic illustration of induced one-handed helicity in optically inactive **93**, helix inversion with temperature, and subsequent memory of the diastereomeric macromolecular helicity by replacing the chiral amine **94** by an achiral one **95**. Reprinted with permission from [192]. Copyright American Chemical Society 2005.

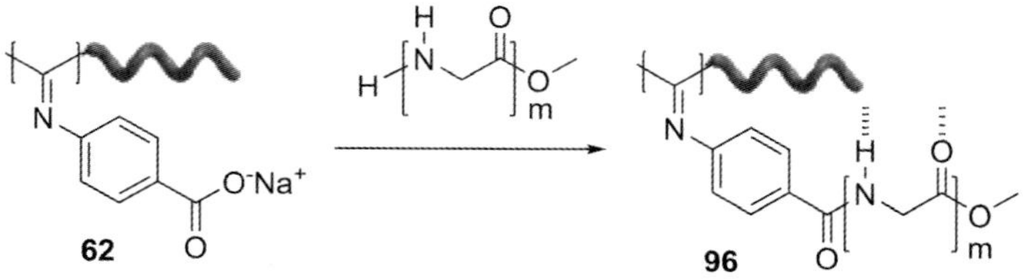

Figure 5.38 Schematic illustration of the modification of the side groups of the polymer with memory of previously induced macromolecular helicity. The side groups of helical **62** can be modified with optically inactive oligoglicines without loss of chiral memory. Polymer **96** gives rise to mesomorphic cholesteric organizations stabilized by hydrogen bonding. Reproduced with permission from [196]. Copyright Wiley-VCH Verlag GmbH & Co. KGaA 2007.

crosslinking poly(4-carboxyphenylisocyanide) using chiral diamines. The hydrogels obtained in this way showed chiral memory when the chiral diamines were replaced by achiral diamines. Moreover, the hydrogel was stable on heating and the induced chirality was maintained even at 90 °C, the temperature at which the helical polymer without crosslinking loses its memorized chirality [195]. Cholesteric lyotropic liquid-crystalline behavior has been found for helical poly(4-carboxyphenylisocyanide) **96**, which is prepared by modification of **62** with oligoglicines through amido linkages. The stability of the chiral helical mesophase is favored by well-defined arrays of intramolecular hydrogen bonds (Figure 5.38). These polyisocyanides with achiral amide residues present chiral memory that in some cases is stable even at 100 °C in N,N-dimethylformamide, thus providing a robust heat-resistant helical scaffold to which a variety of functional groups can be introduced [196].

An interesting chiral-memory effect, stabilized by the formation of gel architectures, was recently revealed in syndiotactic polymethylmethacrylate, st-PMMA, **97** [197]. This polymer was known to adopt a helical disposition (74/4 helix with an inner cavity of about 1 nm) in aromatic solvents to give thermoreversible gels [198]. It has been found that when the solvent is chiral, for example, 1-phenylethanol (**98**), the polymer adopts a prevailing helical sense, which is maintained after removing the optically active solvent. On the other hand, the diameter of the helical cavity allows the alignment of fullerene units given the size of the C_{60} molecule. Thus, st-PMMA/C_{60} complexes were prepared in which 1D fullerene arrangements were wrapped by a helical polymer (**97**), the helicity of which could be tuned according to the chirality of the gelling medium and memorized when the chiral solvent **98** was replaced by an achiral one (Figure 5.39). The gel exhibited a vibrational circular dichroism in the polymer IR region and this is indicative of a prevailing helix sense, even in the absence of a chiral solvent. Moreover, an induced CD was observed in the absorption region of the encapsulated C_{60}.

In addition to the induction phenomena already commented in Section 5.2, the possibility of switching chirality induced into achiral polymeric films by means of the photoisomerization process of azobenzene groups has also been investigated [199]. Experiments mainly consisted in the reversible modulation of the optical activity by irradiation with circularly polarized light of opposite handedness. Light-driven chiro-

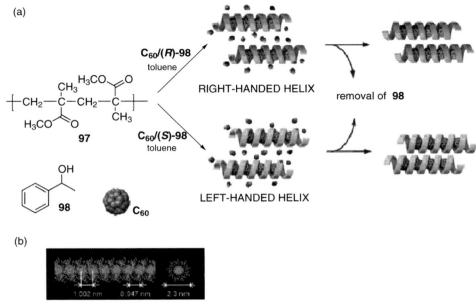

Figure 5.39 (a) Schematic illustration of right- and left-handed helicity induction in the C_{60}-encapsulated in st-PMMA (**97**) in the presence of (**R**)-**98** or (**S**)-**98** and chiral memory upon removing the chiral solvent. (b) Calculated structure for the C_{60}/st-PMMA. Reproduced with permission from [197]. Copyright Wiley-VCH Verlag GmbH & Co. KGaA 2008.

optical switches can thus be designed based on amorphous polymers as well as on liquid-crystal systems.

Recently, Fukuda et al. reported the possibility of using the rotation of polarization to promote rewritable, multilevel optical data storage in amorphous azobenzene polymeric systems [200]. The system works as a bistable switch in which state "1" correspond to the chiral structure induced into the polymer by elliptically polarized light, which gives rise to rotation of the polarization plane of the reading light. The state "0" is achieved after erasing the photoinduced chirality with circularly polarized radiation.

Experiments in liquid-crystal systems may have further interest for applications since modulation or switching of chirality affects the helical superstructure, and hence optical properties, inherent to the cholesteric organization, which amplifies the chiro-optical effect.

5.5
Chiral Block-Copolymers and Nanoscale Segregation

Block-copolymers have been defined as soft materials consisting of two or more polymer fragments (blocks) with different chemical natures and linked through covalent bonds [201]. This characteristic enables them to present nanoscale segregation, which gives rise to the formation of ordered spherical, cylindrical and lamellar phases that have regularly shaped and uniformly spaced nanodomains in the solid state as well as in a selective solvent for one of the blocks [202, 203].

Chiral block-copolymers have been investigated to achieve morphologies in bulk materials, in the mesophase when one of the blocks has mesomorphic properties and in the presence of solvents when an amphiphilic system is used in the macromolecule. This section covers examples in which segregated structures are obtained in these three systems.

5.5.1
Chiral Block-Copolymers: Nanoscale Segregation in the Bulk

Block-copolymers can organize in the bulk to give well-organized one-, two- or three-dimensional nanostructures [204]. Nanohelical superstructures within block-copolymers were described early on for ternary block-copolymers consisting of polystyrene-block-polybutadiene-block-PMMA. However, chirality was not reported in the material since the formation of the helical domain occurred in a racemic form (plus and minus) [205].

Chirality-driven morphologies in block-copolymers were recently reported in diblock-copolymers **99** consisting of left-handed helical blocks of L-polylactic acid (PLLA) within a polystyrene (PS) matrix. When the composition of the block-copolymers was rich in PS fractions [206], the morphology of the bulk was a hexagonal packing of helical nanodomains, which was attributed to the effect of chirality interacting with the immiscibility of constituent blocks. The dynamic nature of this system was later reported. Thus, the morphology of the self-assembled helices could be tuned in a reversible manner by external stimuli (crystallization and shearing), which promoted interconversion of the helix into cylinders depending on melting and annealing processes (Figure 5.40) [207]. In another example, a PS-PLLA block-copolymer rich in PLLA fractions forms a bulk core-shell cylindrical microstructure. A scrolling mechanism due to the chiral effect of the PLLA block was proposed to account for the formation of this morphology (Figure 5.41). An initial bilayer microstructure is twisted and bent, resulting in a helical curvature at the interface. This process occurs in different solvents with distinct selectivity and evaporation rates. The scrolling of the helical domains gives rise to a core-shell cylinder PS nanostructure [208], which upon degradation of PLLA turned into a tubular nanostructure in a polystyrene matrix.

5.5.2
Chiral Block-Copolymers: Nanoscale Segregation in the Mesophase

It is interesting to note the expression of chirality in block-copolymers when one of the phases presents mesomorphic behavior. This allows the modulation of mesophase-derived properties within a nanoscopic segregated structure in which not only

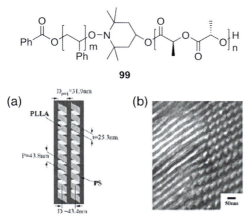

99

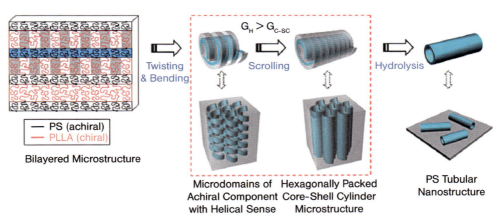

Figure 5.40 (a) Schematic representation of the nanohelical morphology of **99** [206]. (b) TEM micrograph for phase transformation from helices to cylinders during crystallization of **99** at 110 °C. Reproduced with permission from [207]. Copyright Wiley-VCH Verlag GmbH & Co. KGaA 2005.

Figure 5.41 Cartoon representation of the formation of PS tubular nanostructures from the core–shell cylinder microstructure of **99** rich in PLLA. Reproduced with permission from [208]. Copyright Wiley-VCH Verlag GmbH & Co. KGaA 2006.

is the morphology important because of phase separation of the linked blocks but also the liquid-crystalline organization of the mesogenic block. Some examples have been reported in which the liquid-crystal block is chiral and shows a SmC* mesophase, which is capable of having ferroelectric properties derived from its noncentrosymmetric organization. One approach to achieve fast-switching ferroelectric polymers consisting in the preparation of block-copolymers **100** consisting of a liquid-crystalline system and an amorphous block, poly(isobutyl vinyl ether), by living-sequential polymerization [209, 210]. It was reported that the properties of ferroelectric liquid crystals – such as birefringence and monostable switching characteristics – were improved in these materials, making this a good approach for active matrix addressed displays.

100

Recently, triblock-copolymers that show segregation of chiral SmC domains were reported [211]. The triblock polymer **101**, prepared by ring-opening metathesis polymerization, consisted of an outer block of methyltetracyclodecene and an inner block containing a chiral mesogenic monomer. The interest in triblock-copolymers lies in the possibility of achieving thermoplastic elastomeric materials, which in the case of a liquid-crystalline block promotes the formation of highly oriented domains using thermoplastic processing methods. This approach could be of interest for SmC* domains to unwind their helical superstructure, making the system suitable for electro-optical applications. The results of the study indicate that, depending on the composition of the blocks, either monolayer SmC* domains or bilayer SmC* domains are formed at room temperature. Dynamomechanical analysis of the triblock-copolymers revealed an elastic plateau above the T_g of the mesogenic block, indicating that these systems exhibit elastomeric behavior at elevated temperature and could form the basis of interesting materials for liquid-crystal shape-memory elastomers.

101

Other block-copolymer systems containing a chiral mesogenic block have been reported and are mentioned below, but a manifestation of chirality in the segregated structures was not described for these materials.

One of these systems consisted of a chiral liquid-crystal block and polystyrene as an amorphous block. These reports mainly focused on the structural study of the morphology of the segregated phases, one of which showed liquid-crystalline behavior [212, 213]. More specifically, the system PS-SCLCP **102** was studied in the bulk as well as in thin films. The material had a hexagonal-packed cylinder nanostructure, comprising either glassy polystyrene cylinders in a liquid-crystal matrix or LC cylinders in a polystyrene matrix depending on the volume fraction of each block [214].

5.5 Chiral Block-Copolymers and Nanoscale Segregation

[Structure 102]

A supramolecular strategy was described to induce liquid-crystalline behavior into one of the blocks of the diblock-copolymer **103** with photoactive groups. Complexation of a chiral acid with the azopyridine side group through H-bonding interactions provided nematic and smectic behavior in an LC block. Nevertheless, helical morphology was not reported for this nanostructure [215].

[Structure 103]

5.5.3
Chiral Block-Copolymers: Nanoscale Segregation in Solvents. Amphiphilic Block-Copolymers

Block segregation can also occur in the presence of a selective solvent due to the different nature of the constituent blocks.

An intense investigation was carried out by Nolte *et al.* on the development of different morphologies based on amphiphilic block-copolymers **104** with one rigid block derived from polyisocyanodipeptides and a flexible block consisting of PS. The incompatibility between these two segments is very marked due to the rigidity of the polyisocyanodipeptide, which shows a high persistence length due to the formation of a β-sheet structure between side groups along the helical conformation of the polyisocyanide backbone (Figure 5.42) [216]. This type of chiral block-copolymer allows control of the formation of either micelles, vesicles or bilayer aggregates by the appropriate choice of solvent. Thus, illustrating the expression of the chirality of the

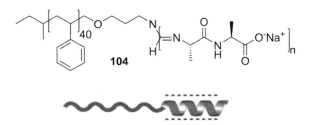

Figure 5.42 Schematic representation of diblock polymer **104** with a flexible polystyrene block and the rigid polyisocyanodipeptide. From [216]. Reprinted with permission of AAAS.

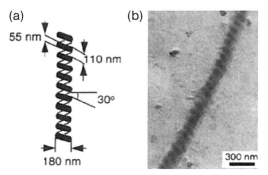

Figure 5.43 (a) Schematic representation of the helix of **104**. (b) TEM image of the superhelix formed by **104** in a sodium acetate buffer of pH 5.6. From [216]. Reprinted with permission of AAAS.

polyisocyanide block at the nanoscale, the aggregation behavior of charged diblock-copolymers in aqueous systems leads to the formation of superhelixes with handedness opposite to that of the helical polyisocyanide-derived block in a situation comparable to natural systems (Figure 5.43) [216, 217]. Furthermore, the aggregation architecture of these systems can be controlled by both external and intrinsic factors: solvent, temperature, pH, size ratio between blocks, and interactions between the polar-head group and different anions.

A compositional modification of the peptide moiety of the polyisocyanide block with the incorporation of thiophene groups led to amphiphilic block-copolymers (**105**) that present highly organized stacks of thiophene groups along the helical chain

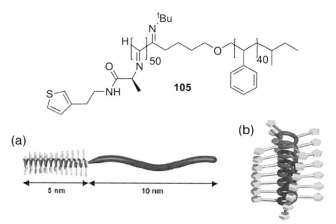

Figure 5.44 (a) Schematic representation of the amphiphilic block-copolymer **105** with a helical polyisocyanide block and a flexible polystyrene block. (b) Helical stacks of thiophene groups along the helical polyisocyanide chain. Reproduced with permission from [218]. Copyright Wiley-VCH Verlag GmbH & Co. KGaA 2003.

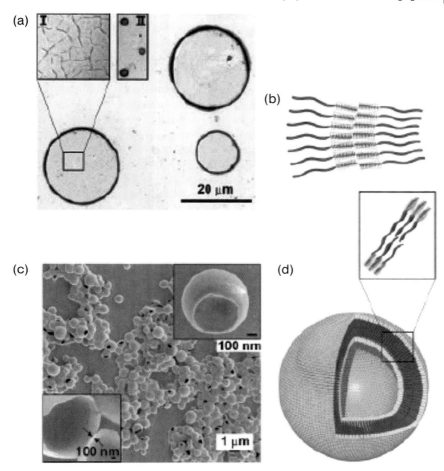

Figure 5.45 (a) TEM image of **105** vesicles formed in CHCl₃. (b) Schematic representation of the vesicle wall in CHCl₃. (c) SEM image of **105** vesicles formed in THF/water. (d) Schematic representation of a vesicle formed and the membrane showing the proposed bilayer structure. Reproduced with permission from [218]. Copyright Wiley-VCH Verlag GmbH & Co. KGaA 2003.

of the polar block. (Figure 5.44). These block-copolymers formed vesicles in organic (Figures 5.45a and b) and in aqueous solvents (Figures 5.45c and d), with diameters varying between 2 and 22 µm and an average membrane thickness, as determined from the TEM and SEM images, of 27 ±5 nm [218].

Alternatively, the presence of highly organized stacks of thiophene groups (Figure 5.44b) allowed crosslinking within the vesicle membranes. Hence, the vesicle can be covalently fixed into the desired morphology. The polymerization of thiophene groups was carried out by electrochemical oxidation in vesicles formed in THF/water (1 : 5 v/v) as solvent. Furthermore, the vesicles are capable of including

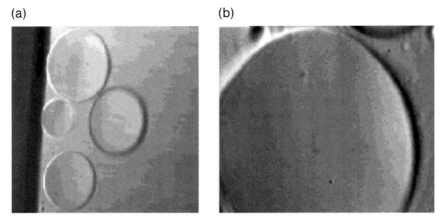

Figure 5.46 (a) Optical micrograph of electroformed vesicles of **105** (Width of the image: about 60 μm). (b) Optical micrograph of two vesicles fused into one vesicle (width of the image: about 80 μm). Reprinted with permission from [220]. Copyright American Chemical Society 2004.

enzymes, as was demonstrated by fluorescence microscopy experiments, thus resulting in potential microreactors.

The use of the electroformation method, which is employed for the preparation of giant vesicles from lipids [219], showed that this type of amphiphilic diblock polymeric system, **105**, formed giant vesicles with diameters from 1 to 100 μm (Figure 5.46) [220]. The giant vesicles prepared in this way consist of membranes that are still fluid and allow microinjection in spite of their polymeric nature. The vesicles also show the ability to encapsulate a biopolymer. In addition, the presence of thiophene groups makes these systems polymerizable, a characteristic that should allow the rigidity of the membrane to be regulated.

The amphiphilic block-copolymer **106** consisting of an optically pure and highly isotactic polycarbosilane and a polyethyleneglycol block was found to form micelles in THF/water mixtures [221]. A strong CD intensity, due to the interaction between naphthyl residues, was observed for concentrations higher than the critical micellar concentration. This observation suggested the existence of a higher aggregation order of the block-copolymers in these micelles that, nevertheless, was strongly dependent on the temperature and hydrophobicity of the solvent.

106

The interest in highly ordered structures of π-conjugated materials led to an investigation into the possibility of preparing chiral main-chain block-copolymers (**107**) containing a sexithiophene block and a chiral oligooxyethylene fragment. Even though sexithiophene fragments showed aggregation behavior in dioxane, the aggregates did not present a chiral organization, in contrast to the situation found for dispersed polymers or well-defined oligomers. The strong aggregation in the block-copolymers, which makes the system less dynamic than the other two systems, seems to hamper the bias of the aggregates towards a chiral architecture [222].

107

5.6
Templates for Chiral Objects

The chiral helical structure of some polymers has proven to have a useful role as chiral templates for the induction of chirality into different systems, which, furthermore, maintain the induced chirality even when the polymer is removed. This section gathers recent examples that use chiral-helical polymers to induce chirality into supramolecular aggregates of dye molecules and into cavities within a polymer matrix. The final part is devoted to comments on a recent investigation on the use of the helical conformation of chiral polymers to attain alignment of functional molecules, thus promoting their increased functionality.

5.6.1
Templates for Chiral Supramolecular Aggregates

The use of chiral-helical polymers to template chiral supramolecular aggregates has been mainly focused on the achievement of well-organized helical architectures based on dye molecules. The helical conformation of natural as well as synthetic helical polymers has been employed as chiral template.

5.6.1.1 Templating with Natural Helical Polymers
Beyond their biological purpose, chiral biopolymers, that is, DNA, polypeptides and polysaccharides, have an interesting application as templates for chiral aggregates such as porphyrins and cyanine dyes.

Supramolecular organizations based on porphyrins have been the target of intense research due to the high level of interest in the photochemical and redox properties of this building block. Indeed, porphyrins and their aggregates in water play a crucial role in biological processes such as photosynthesis. Chiral aggregates of porphyrin derivatives formed by interaction with helical biopolymers such as single- and double-stranded DNA or polypeptides have been described [223–227].

Of particular interest is the interaction of porphyrin derivatives with polyglutamic acid. This polymer adopts an α-helical conformation when partially protonated, pH below 5.0, and a random-coil conformation under basic conditions. A preformed binary complex [228] between the α-helical poly L-glutamic acid and the copper complex of the tetracationic meso-tetrakis(N-methylpyridinium-4-yl)porphine (CuT$_4$, **108**) can interact with the protonated derivative of the tetraanionic porphyrin (H$_4$TPPS, **109**). The ternary complex thus formed exhibits significant changes in CD spectra, which indicates that both porphyrin derivatives are aggregated onto the helical polymeric chain in a chiral fashion [229]. A study of the aggregation behavior

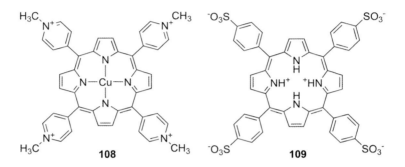

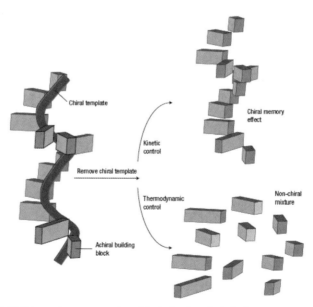

Figure 5.47 Schematic representation of the induction of a helical supramolecular architecture in porphyrin aggregates. A chiral-memory effect determined by kinetic control after removing the DNA chain that acts as chiral template. Reprinted by permission from MacMillan Publishers Ltd: Nature Materials [232], copyright 2003.

of this type of porphyrin system confirmed that net electrostatic interactions between porphyrin monomers and partially protonated polyglutamic acid are the driving force for the self-assembly process that leads to the formation of inert assemblies [230]. Moreover, the central copper atom plays a crucial role in the aggregation dimensions. It must be remarked that these supramolecular porphyrin aggregates are very stable and retain the chirality induced by interaction with the chiral polymeric matrix. Thus, if an excess of poly D-glutamic acid is added to a ternary complex containing poly L-glutamic acid, inversion of chirality was not observed by CD spectroscopy in the Soret region. Moreover, disrupting the helical conformation of the polymeric template by pH-induced conformational transition from the α-helix to a random-coil conformation does not lead to loss of the chirality associated with the supramolecular aggregates or porphyrin derivatives unless other factors, such as temperature or salt concentration, affect their stability (Figure 5.47) [231]. Hence, once templated, these assemblies retain the chiral architecture according to a memory process determined by kinetic inertia [232].

A chiral templating function was also described for DNA on cyanine dyes (**110**). These dyes are addressed to assemble into a helical supramolecular organization through the formation of cofacial dimers within the minor grove of the double helix of DNA (Figure 5.48). One dimer facilitates the aggregation of subsequent dimers directly adjacent to the first in a cooperative fashion until the end of the DNA chain is reached [233].

Chiral aggregation of cyanine dyes was also accomplished using a polysaccharide, carboxymethylamylose (**111**), as the polymeric template [234]. The interaction between **111** and the cyanine dye **112** gives rise to a cooperative interplay through which an extraordinarily large induced circular dichroism is observed from J-aggregates of the achiral **112** in association with a random coil **111**, suggesting that the polysaccharide is transformed into a helix. The CD intensity increases on increasing the degree of carboxyl substitution in the amylose and the pH up to neutral. At neutral conditions, maximum J-aggregation occurs. Furthermore, the intensity of the CD follows the same trend as the fluorescence intensity of the aggregates. These observations led to the conclusion that binding of the J-aggregates onto the template **111** is sterically controlled by the asymmetric environment of

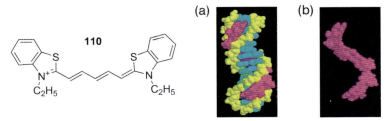

Figure 5.48 (a) Molecular model of an aggregate of three dimers of **110** aligned end to end in the minor grove of a DNA template. (b) The DNA is removed to emphasize the helical architecture of the dye aggregate. Reprinted with permission from [233]. Copyright American Chemical Society 2004.

glucose residues. An increase in carboxyl substitution in **111** gives rise to a rigid (highly fluorescent) cyanine dye superhelix. The CD of the J-aggregates is stable over several months in water at room temperature, suggesting that the cyanine J-aggregates superhelix is strongly resistant to unfolding/dissociation by the solvent.

5.6.1.2 Templating with Synthetic Helical Polymers

Besides helical natural polymers, synthetic polymers have also been reported as useful templates for the formation of chiral aggregates of porphyrins. Yashima *et al.* reported a beautiful strategy (Figure 5.49) in which a one-handed helical conformation is induced into the hydrochloride salt of poly[4-(N,N-diisopropylaminomethyl)phenylacetylene] **38** by an enantiomerically pure binaphthol **113**, which interacts hydrophobically with the main chain of the polymer. This gives rise to a helical disposition of the charged amino groups of the polymer, which interact electrostatically with a tetraanionic porphyrin (**109**). This ionic complexation promotes the formation of J-aggregates in a helical fashion biased to a given handedness, as demonstrated by the appearance of an ICD signal. The induced supramolecular chiral organization is memorized by porphyrins and remains stable even if the helicity of the polymer inverts by interaction with the opposite enantiomer of binaphthol [235].

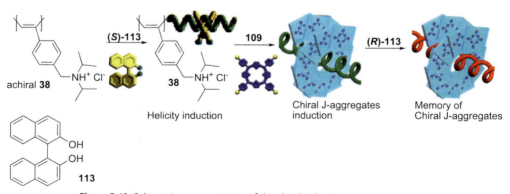

Figure 5.49 Schematic representation of the chiral induction into **38** by complexation with **(S)-113**, formation of supramolecular chiral aggregates of achiral **109**, and memory of the supramolecular chirality of **109** after inversion of the helicity of **38**. Reproduced with permission from [235]. Copyright Wiley-VCH Verlag GmbH & Co. KGaA 2006.

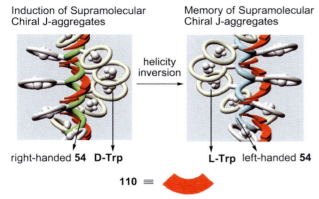

Figure 5.50 Schematic illustration of the formation of supramolecular helical aggregates of cyanine dye **110** by polyacetylene **54** whose chirality is induced by complexation with D-Trp and memory of the supramolecular chirality after helicity inversion of **54** by excess of L-Trp. Reprinted with permission from [236]. Copyright American Chemical Society 2007.

In a similar way, the helical arrangement of cyanine dyes templated by polyphenylacetylene **54** was recently reported [236]. The helical structure induced in water-soluble polyphenylacetylene **54** with crown-ether side groups by a chiral amino acid (D-Trp.HClO$_4$) can trap the achiral benzoxazole cyanine dye **110** within its hydrophobic helical cavity inside the polymer. The resulting complex shows the formation of chiral supramolecular aggregates of the cyanine dyes, which exhibit an induced circular dichroism in the cyanine chromophore region. The supramolecular chiral arrangement of the cyanine dyes was memorized when the template helical polymer inverted into the opposite helicity by addition of the enantiomeric amino acid (Figure 5.50). Thereafter, thermal racemization of the helical aggregates slowly took place.

Selective formation of enantiopure crystals from a racemic mixture has also been achieved using chiral polymers as interactive templates during the crystallization process. Polymethacrylates with side chains derived from D- or L-lysine result in the successful formation of enantiopure crystals from D,L-methionine.HCl [237]. Spherical microparticles of chiral polymers, poly(N-acryl-L-aminoacid)s [238] or poly(N-vinyl-L-aminoacid)s [239] were also shown to be active in similar chiral discrimination processes during amino-acid crystallization occurring on microparticle surfaces. Block-copolymers have also been presented to be useful for this purpose [240]. Hydrophilic chiral block-copolymers **114**, consisting of PEO-b-branched poly(ethyleneimine) with a variety of optically active groups, were employed as additive in the crystallization of calcium tartrate tetrahydrate. The chiral polymer systems, which have yet to be optimized, were demonstrated to slow down the formation of both the racemic crystal and one of the enantiomeric crystals. Moreover, the chiral polymers influences the morphology of the formed chiral crystal creating unusual morphologies.

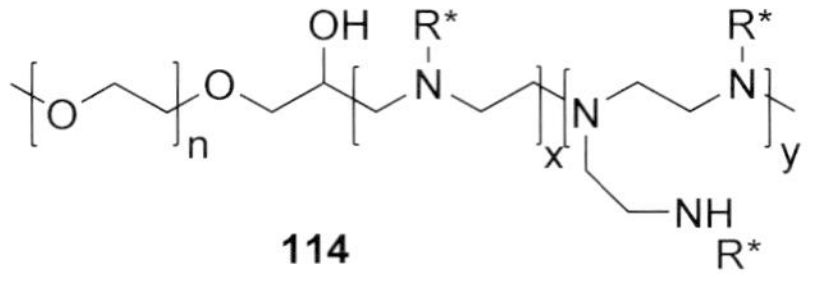

5.6.2
Molecular Imprinting with Helical Polymers

Molecular imprinting is a technique by which highly selective recognition sites can be generated in a synthetic polymer. The intense research activity in this area is due to the broad range of applications described for these systems, such as separation, analytical systems, sensors, synthesis and catalysis [241, 242]. Among these applications, the research on chiral imprinted polymers by virtue of their application in enantiomer separation is particularly well established [243–246]. The strategy is mainly based on the use of chiral molecules as templates to create chiral recognition holes in a polymer matrix.

Recently, a few publications have appeared in the literature concerning the use of chiral-helical polymers to template helical cavities in a polymer matrix, thus broadening the possibilities for imprinted materials.

Single-handed helical poly(methyl methacrylate)s (**115**, **116**, **117**) were used as templates during the formation of gels from different monomers and crosslinking agents [247]. CD experiments in transparent suspensions of the gels confirmed that the helical structure of the template was imprinted in the gel. The recognition ability of the chiral gel was tested by evaluating its capacity to resolved racemates. Gel templated with the polymer represented in the figure showed high chiral recognition ability for small molecules (i.e. *trans*-stilbene oxide, Tröger's base, and flavanone). The study of the influence of different monomers on the chiral recognition ability of the final gel proved that bifunctional monomers such as α-benzyloxymethylacrylic acid gave higher efficiency than a monofunctional vinyl monomer, methacrylic acid [248].

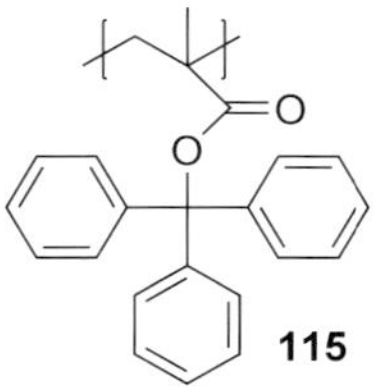

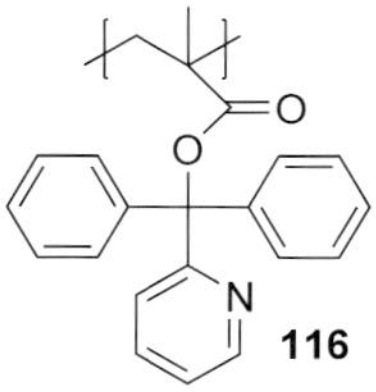

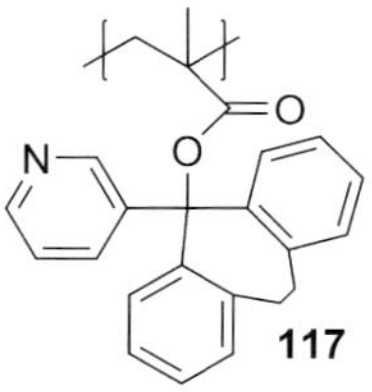

Chiral block-copolymers **118** of polyethyleneoxide and of D-phenylalanine (PEO-b-D-Phe) have also been proposed as efficient templates to imprint helical cavities into mesoporous silica materials. These copolymers perform an efficient role as a surfactant template during the preparation of silica materials with mesoporous structures with hexagonal symmetry, a pore size of 5 nm and a high surface area (about 700 m^2/g). This silica showed enantioselectivity against racemic solutions of valine with an enantioselectivity factor of 2.34 [249].

118

5.6.3
Templating by Wrapping with Helical Polymers

Control of the chirality of molecules or oligomers susceptible to adopting a helical conformation can be achieved by peripheral wrapping with helical polymers that provide a template effect.

This template effect by wrapping has been described for polysaccharides, which were shown to induce helical conformations in oligothiophenes [250] and oligosilanes (Figure 5.51) [251–253]. Thus, the induction of chirality in the form of a twisted conformation occurs through the formation of inclusion complexes. Within these complexes, oligothiophenes or oligosilanes gain optical activity, the sign of which depends on the chirality of the wrapping polysaccharide and was detected by CD signals in the absorption region of the corresponding oligothiophene or oligosilane chromophores. Chiral polymeric templates used for these experiments were carboxymethylamylose (**119**), which adopts a single-stranded helical conformation, and native curdlan (**120**) and schizophylan (**72**), both of which adopt triple-stranded helical conformations.

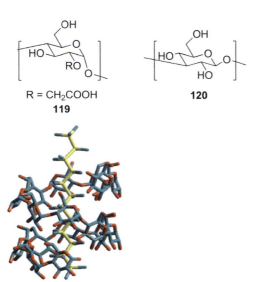

Figure 5.51 Energy-minimized model based on the AMBER force field for an inclusion complex of $Me(SiMe_2)_{12}Me$ with an amylose **119** fragment containing 16 repeating α-1,4-d-glucopyranose residues. Reproduced with permission from [252]. Copyright Wiley-VCH Verlag GmbH & Co. KGaA 2005.

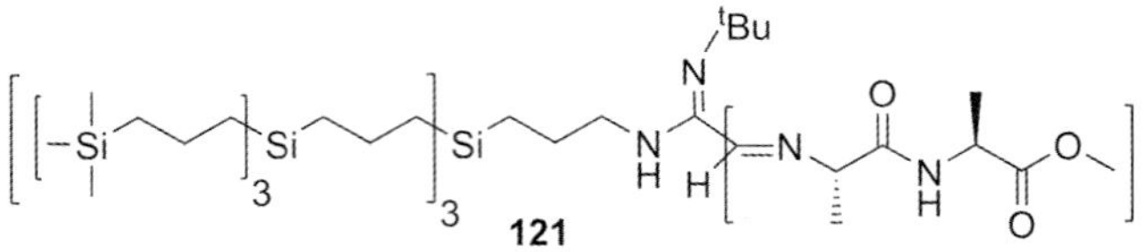

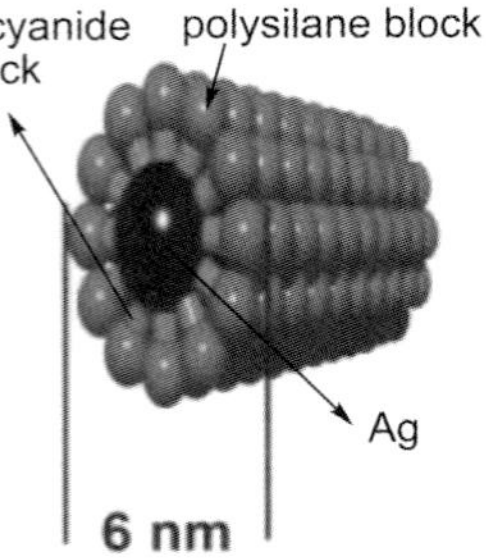

Figure 5.52 Schematic representation of the molecular organization proposed for copolymer **121** in the presence of Ag^+. Reproduced with permission from [254]. Copyright Wiley-VCH Verlag GmbH & Co. KGaA 2002.

Even though there is not transmission of chirality from the polymer to the templated object, it is interesting to comment on the role of micelles formed by block-copolymers (e.g., **121**) consisting of a chiral polyisocyanopeptide block and a dendritic polysilane block, which can template the formation of crystalline silver nanowires. The process consists of the reduction of silver anions coordinated to the peptide groups (Figure 5.52) [254].

5.6.4
Alignment of Functional Groups

Even though it may not be strictly a template effect, the helical structure of some chiral polymers can be exploited to organize active molecules covalently linked to the periphery of the polymeric structure. As a result, the properties associated with the functional molecule are enhanced with respect to the monomeric unit. Experiments carried out with this aim have been mainly based on the stable helical structure of polyisocyanides. Polypeptides, polyacetylenes and foldamers have also been used to align functional molecules in a helical arrangement.

5.6.4.1 Polyisocyanides
A report on this function of helical polymers was presented by Nolte et al., who organized nonlinear optical chromophores as orientationally correlated side groups of a polyisocyanide [255]. In such an organization, each chromophore contributes coherently to the second-order nonlinear response of the polymer structure, which is enhanced with respect to the monomer.

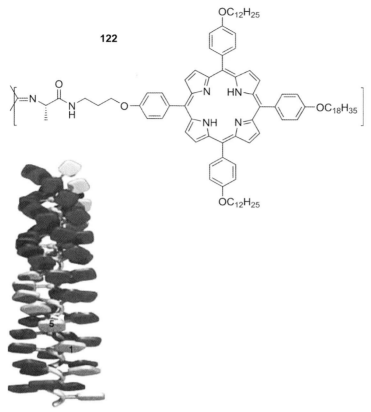

Figure 5.53 Schematic drawing of a molecule of polymer **122**. The porphyrin stack shows that the fifth porphyrin has a slip angle (30°) with respect to the first porphyrin. Reproduced with permission from [256]. Copyright Wiley-VCH Verlag GmbH & Co. KGaA 2003.

Likewise, well-defined arrays of porphyrins attached to a rigid polyisocyanide backbone (**122**) have been reported and these consist of four columns of around 200 stacked porphyrins arranged in a helical fashion along the polymer backbone (Figure 5.53) [256]. The helical polymers are made more rigid by the occurrence of an interside-chain hydrogen-bonding network that gives rise to a rod-like structure with an overall length of 87 nm. CD spectroscopy recorded in solution showed a negative exciton-coupling signal in the Soret band that indicated a left-handed helical disposition of the porphyrin chromophores. Furthermore, photophysical studies showed that at least 25 porphyrins within one column are excitationally coupled.

The same strategy, which combines the helical structure of polyisocyanide with the rigid nature of the polymer backbone promoted by hydrogen-bonding interactions between amino-acid side groups, was undertaken to attain long and well-defined arrays of perylenediimide chromophores (**123**) [257]. Circular dichroism spectroscopy revealed the chiral organization of the chromophore units in the polymer, whereas absorption and emission measurements proved the occurrence of excited-state interactions between those moieties due to the close packing of the chromophore groups.

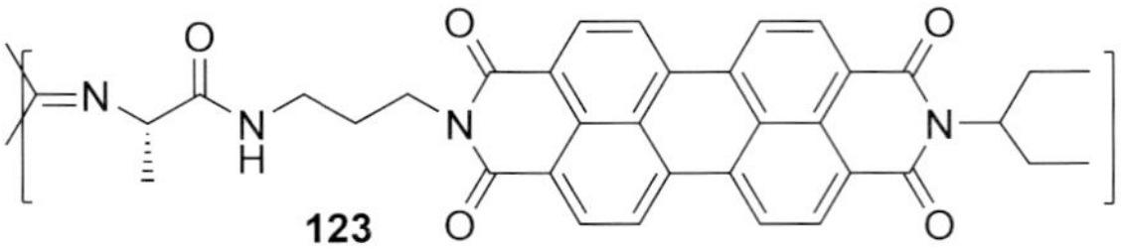

123

Electroactive tetrathiafulvalene (TTF) groups have been organized along the helical polymeric backbone of a polyisocyanide (**124**). Accordingly, polyisocyanides were described that present a multistate chiroptical macromoleular switch based on the redox behavior of TTF) [258]. The system was characterized in its neutral state and oxidized to generate mixed-valence states, which displayed charge mobility in solution. Charge transport along the polymer was evidenced for the mixed-valence state and this contained cation radicals and neutral TTF [259]. Results from these studies confirmed that there is interaction between side functional groups along the stacks promoted by the helical polymer scaffold.

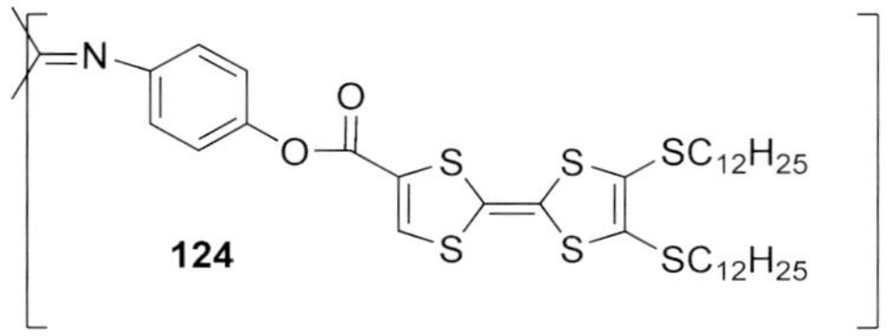

124

5.6.4.2 Polypeptides
The alpha-helix of polypeptides has proven useful to arrange chromophores such as naphthalene [260] or porphyrin [261] (Figure 5.54) to give materials with enhanced electron-transport capability.

5.6.4.3 Polyacetylenes
Polyacetylene polymers also have the capability to arrange chromophores to give enhanced fluorescence with respect to molecule itself. This is the case for polyacetylene **125** with pyrene side groups, which was reported to show large excimer-based fluorescence due to a very stable helical structure that provided a stereoregular arrangement of the chromophores [262].

125

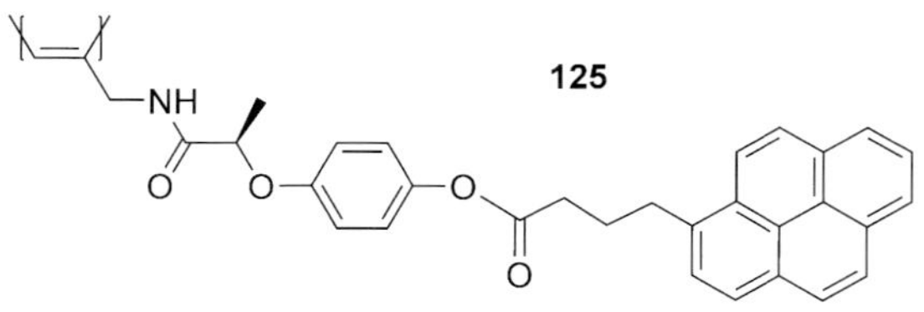

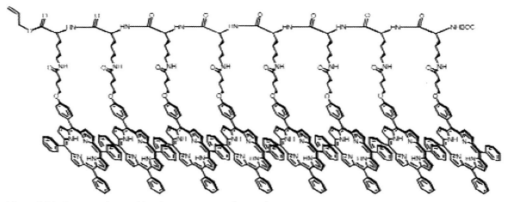

Figure 5.54 Octamer obtained by oligomerization of a porphyrin functionalized derivative of L-lysine. Reproduced with permission from [261]. Copyright Wiley-VCH Verlag GmbH & Co. KGaA 2001.

5.6.4.4 Foldamers

Polymeric chains based on the foldamer concept can be used to arrange functional molecules in a controlled way such that interactions between them are improved. Meijer *et al.* described the poly(ureidophthalimide) **126** with oligophenylenevinylidene chromophores (OPV) as side groups (Figure 5.55). This polymer adopts a stable

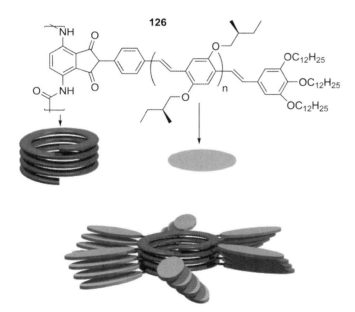

Figure 5.55 Cartoon representation of the chiral arrangement of OPV chromophores along the helical main chain of foldamer **126**. Reprinted with permission from [263]. Copyright American Chemical Society 2006.

helical conformation in heptane, with the previous history of the polymer in solution (either $CHCl_3$ or THF) having a large influence on the structure. It was stated that the intramolecular hydrogen bonding of the phthalimide units is responsible for the initial chiral alignment of the OPV chromophores [263].

5.7
Outlook

The expression of chirality in polymers can be as simple as the helical conformation of their main chain or as complex as the quaternary structure of a protein. Within this interval, chiral polymers offer a variety of possibilities that are possible thanks to a deep knowledge acquired in polymerization procedures and self-assembly processes based on intra- and intermacromolecular chain interactions. In this chapter, referred literature has been selected in order to illustrate the manifestation of chirality at the nanoscale using polymers. It has been shown that it is already possible to understand and control the lab-synthesis of well-organized mono-, two- and three-dimensional chiral architectures based on optically active polymers, and achiral polymers in which chirality is induced by external agents. Nevertheless, it is worth emphasizing that there has been an increasing concern to implement functional polymeric systems on the basis of the wide spectrum of chiral architectures that can be achieved. The first steps to attain this goal have already been taken. Some of these chiral arrangements have demonstrated the possibility of developing functional systems that use chirality as a tunable property, which can be stored and read in memory systems, opening practical advanced applications for these complex nanostructures. Chiral block-copolymers offer advantages in the formation of chiral objects, or cavities, with controlled shape and dimensions thanks to nanosegregation processes between chiral and nonchiral blocks. Finally, the chiral information of optically active organizations based on polymers acting as a guide has been transferred to achiral systems, thus promoting their defined organization and the performance of improved properties.

Through these advanced functional systems, chiral polymeric architectures are able to break into technologically important fields – from chiral recognition, chiral separation, chiral sensing and asymmetric catalysis to applications in electronics, optoelectronics, combination of electro- and magneto- properties, information storage, and so on – on the basis of their particular chemical and physical properties. This leads us to conclude that, nowadays, the motivation for research in this subject is not limited to the understanding and control of the formation of well-defined chiral polymeric nanostructures, but also to achieve their practical application.

Acknowledgements

I sincerely thank José Luis Serrano for his support and helpful discussions. Financial founding from CYCIT project MAT2006-13571-CO2-01, FEDER founding and DGA are acknowledged.

References

1. Watson, J.D. and Crick, F. (1953) *Nature*, **171**, 737–738.
2. Ramachandran, G.N. and Kartha, G. (1954) *Nature*, **174**, 269–270.
3. Ramachandran, G.N. and Kartha, G. (1955) *Nature*, **176**, 593–595.
4. Klug, A. (1983) *Angewandte Chemie-International Edition*, **22**, 565–582.
5. Pino, P. and Lorenzi, G.P. (1960) *Journal of the American Chemical Society*, **82**, 4745.
6. Hopkins, T.E. and Wagener, K.B. (2002) *Advanced Materials*, **14**, 1703–1715.
7. Kawakami, Y. and Tang, H.Z. (2000) *Designed Monomers and Polymers*, **3**, 1–16.
8. Nakano, T. and Okamoto, Y. (2001) *Chemical Reviews*, **101**, 4013–4038.
9. Okamoto, Y. and Nakano, T. (1994) *Chemical Reviews*, **94**, 349–372.
10. Green, M.M., Peterson, N.C., Sato, T., Teramoto, A., Cook, R. and Lifson, S. (1995) *Science*, **268**, 1860–1866.
11. Green, M.M., Park, J.-W., Sato, T., Teramoto, A., Lifson, S., Selinger, R.L.B. and Selinger, J.V. (1999) *Angewandte Chemie-International Edition*, **38**, 3138–3154.
12. Green, M.M., Cheon, K.S., Yang, S.Y., Park, J.W., Swansburg, S. and Liu, W.H. (2001) *Accounts of Chemical Research*, **34**, 672–680.
13. Green, M.M., Reidy, M.P., Johnson, R.D., Darling, G., O'Leary, D.J. and Willson, G. (1989) *Journal of the American Chemical Society*, **111**, 6452–6454.
14. Green, M.M., Garetz, B.A., Munoz, B., Chang, H., Hoke, S. and Cooks, R.G. (1995) *Journal of the American Chemical Society*, **117**, 4181–4182.
15. Mayer, S. and Zentel, R. (2001) *Progress in Polymer Science*, **26**, 1973–2013.
16. Nomura, R., Nakako, H. and Masuda, T. (2002) *Journal of Molecular Catalysis A-Chemical*, **190**, 197–205.
17. Yashima, E., Maeda, K. and Nishimura, T. (2004) *Chemistry – A European Journal*, **10**, 43–51.
18. Lam, J.W.Y. and Tang, B.Z. (2005) *Accounts of Chemical Research*, **38**, 745–754.
19. Fujiki, M. (2001) *Macromolecular Rapid Communications*, **22**, 539–563.
20. Fujlki, M., Koe, J.R., Terao, K., Sato, T., Teramoto, A. and Watanabe, J. (2003) *Polymer Journal*, **35**, 297–344.
21. Millich, F. (1972) *Chemical Reviews*, **72**, 101–113.
22. Nolte, R.J.M. (1994) *Chemical Society Reviews*, **23**, 11–19.
23. Broer, D.J., Lub, J. and Mol, G.N. (1995) *Nature*, **378**, 467–469.
24. Kim, Sung T. and Finkelmann, H. (2001) *Macromolecular Rapid Communications*, **22**, 429–433.
25. Kikuchi, H., Yokota, M., Hisakado, Y., Yang, H. and Kajiyama, T. (2002) *Nature Materials*, **1**, 64–68.
26. Langeveld-Voss, B.M.W., Janssen, R.A.J. and Meijer, E.W. (2000) *Journal of Molecular Structure*, **521**, 285–301.
27. Brustolin, F., Goldoni, F., Meijer, E.W. and Sommerdijk, N. (2002) *Macromolecules*, **35**, 1054–1059.
28. Satrijo, A. and Swager, T.M. (2005) *Macromolecules*, **38**, 4054–4057.
29. Guo, P.Z., Tang, R.P., Cheng, C.X., Xi, F. and Liu, M.H. (2005) *Macromolecules*, **38**, 4874–4879.
30. Cornelissen, J., Rowan, A.E., Nolte, R.J.M. and Sommerdijk, N. (2001) *Chemical Reviews*, **101**, 4039–4070.
31. Elemans, J.A.A.W., Rowan, A.E. and Nolte, R.J.M. (2003) *Journal of Materials Chemistry*, **13**, 2661–2670.
32. Maeda, K. and Yashima, E. (2006) *Topics in Current Chemistry*, **265**, 47–88.
33. Yashima, E. and Maeda, K. (2008) *Macromolecules*, **41**, 3–12.
34. Nakano, T. (2001) *Journal of Chromatography. A*, **906**, 205–225.
35. Yamamoto, C. and Okamoto, Y. (2004) *Bulletin of the Chemical Society of Japan*, **77**, 227–257.
36. Wilson, A.J., Masuda, M., Sijbesma, R.P. and Meijer, E.W. (2005) *Angewandte*

Chemie-International Edition, **44**, 2275–2279.
37 Ishi-i, T., Kuwahara, R., Takata, A., Jeong, Y., Sakurai, K. and Mataka, S. (2006) *Chemistry – A European Journal*, **12**, 763–776.
38 Zhou, W.D., Li, Y.L. and Zhu, D.B. (2007) *Chemistry, an Asian Journal*, **2**, 222–229.
39 Jahnke, E., Lieberwirth, I., Severin, N., Rabe, J.P. and Frauenrath, H. (2006) *Angewandte Chemie-International Edition*, **45**, 5383–5386.
40 Jahnke, E., Millerioux, A.S., Severin, N., Rabe, J.P. and Frauenrath, H. (2007) *Macromolecular Bioscience*, **7**, 136–143.
41 Jahnke, E., Severin, N., Kreutzkamp, P., Rabe, J.P. and Frauenrath, H. (2008) *Advanced Materials*, **20**, 409–414.
42 Frauenrath, H. and Jahnke, E. (2008) *Chemistry – A European Journal*, **14**, 2942–2955.
43 Hill, J.P., Jin, W.S., Kosaka, A., Fukushima, T., Ichihara, H., Shimomura, T., Ito, K., Hashizume, T., Ishii, N. and Aida, T. (2004) *Science*, **304**, 1481–1483.
44 Yamamoto, Y., Fukushima, T., Jin, W., Kosaka, A., Hara, T., Nakamura, T., Saeki, A., Seki, S., Tagawa, S. and Aida, T. (2006) *Advanced Materials*, **18**, 1297–1300.
45 Yamamoto, T., Fukushima, T., Kosaka, A., Jin, W., Yamamoto, Y., Ishii, N. and Aida, T. (2008) *Angewandte Chemie-International Edition*, **47**, 1672–1675.
46 Broer, D.J. (1993) *Polymerisation Mechanisms*, Vol. 3 (eds. Fouassier, J.P and Rabek, J.F.), Elsevier, London, p. 383.
47 Broer, D.J., Boven, J., Mol, G.N. and Challa, G. (1989) *Makromolekulare Chemie-Macromolecular Chemistry and Physics*, **190**, 2255.
48 van de Witte, P., Brehmer, M. and Lub, J. (1999) *Journal of Materials Chemistry*, **9**, 2087–2094.
49 Broer, D.J., Mol, G.N., van Haaren, J. and Lub, J. (1999) *Advanced Materials*, **11**, 573–578.
50 Lub, J., van de Witte, P., Doornkamp, C., Vogels, J.P.A. and Wegh, R.T. (2003) *Advanced Materials*, **15**, 1420–1425.
51 Amabilino, D.B., Serrano, J.L., Sierra, T. and Veciana, J. (2006) *Journal of Polymer Science Part A-Polymer Chemistry*, **44**, 3161–3174.
52 Ramos, E., Bosch, J., Serrano, J.L., Sierra, T. and Veciana, J. (1996) *Journal of the American Chemical Society*, **118**, 4703–4704.
53 Amabilino, D.B., Ramos, E., Serrano, J.L., Sierra, T. and Veciana, J. (1998) *Journal of the American Chemical Society*, **120**, 9126–9134.
54 Hoshikawa, N., Hotta, Y. and Okamoto, Y. (2003) *Journal of the American Chemical Society*, **125**, 12380–12381.
55 Holder, S.J., Achilleos, M. and Jones, R.G. (2006) *Journal of the American Chemical Society*, **128**, 12418–12419.
56 Akagi, K., Piao, G., Kaneko, S., Sakamaki, K., Shirakawa, H. and Kyotani, M. (1998) *Science*, **282**, 1683.
57 Akagi, K., Ito, M., Katayama, S., Shirakawa, H. and Araya, K. (1989) *Molecular Crystals and Liquid Crystals*, **172**, 115–123.
58 Akagi, K., Katayama, S., Ito, M., Shirakawa, H. and Araya, K. (1989) *Synthetic Metals*, **28**, D51–D56.
59 Park, Y.W., Park, C., Lee, Y.S., Yoon, C.O., Shirakawa, H., Suezaki, Y. and Akagi, K. (1988) *Solid State Communications*, **65**, 147–150.
60 Akagi, K., Shirakawa, H., Araya, K., Mukoh, A. and Narahara, T. (1987) *Polymer Journal*, **19**, 185–189.
61 Piao, G.Z., Kawamura, N., Akagi, K., Shirakawa, H. and Kyotani, M. (2000) *Polymers for Advanced Technologies*, **11**, 826–829.
62 Akagi, K., Guo, S., Mori, T., Goh, M., Piao, G. and Kyotani, M. (2005) *Journal of the American Chemical Society*, **127**, 14647–14654.
63 Goh, M., Matsushita, T., Kyotani, M. and Akagi, K. (2007) *Macromolecules*, **40**, 4762–4771.
64 Goh, M., Kyotani, M. and Akagi, K. (2007) *Journal of the American Chemical Society*, **129**, 8519–8527.

65 Goh, M.J., Kyotani, M. and Akagi, K. (2006) *Current Applied Physics*, **6**, 948–951.

66 Jin, Z.X., Wang, Z.Y., Shi, Z.J., Lee, H.J., Park, Y.W. and Akagi, K. (2007) *Current Applied Physics*, **7**, 367–369.

67 Goto, H. and Akagi, K. (2005) *Angewandte Chemie-International Edition*, **44**, 4322–4328.

68 Goto, H. (2007) *Macromolecules*, **40**, 1377–1385.

69 Goto, H. and Akagi, K. (2006) *Journal of Polymer Science Part A-Polymer Chemistry*, **44**, 1042–1047.

70 Goto, H. (2007) *Journal of Polymer Science Part A-Polymer Chemistry*, **45**, 1377–1387.

71 Goto, H. and Akagi, K. (2004) *Macromolecular Rapid Communications*, **25**, 1482–1486.

72 Goto, H., Jeong, Y.S. and Akagi, K. (2005) *Macromolecular Rapid Communications*, **26**, 164–167.

73 Goto, H. and Akagi, K. (2005) *Macromolecules*, **38**, 1091–1098.

74 Li, W. and Wang, H.L. (2004) *Journal of the American Chemical Society*, **126**, 2278–2279.

75 Li, W.G., Bailey, J.A. and Wang, H.L. (2006) *Polymer*, **47**, 3112–3118.

76 Li, W. and Wang, H.L. (2005) *Advanced Functional Materials*, **15**, 1793–1798.

77 Yuan, G.L. and Kuramoto, N. (2004) *Macromolecular Chemistry and Physics*, **205**, 1744–1751.

78 Manaka, T., Kon, H., Ohshima, Y., Zou, G. and Iwamoto, M. (2006) *Chemistry Letters*, **35**, 1028–1029.

79 Zou, G., Kohn, H., Ohshima, Y., Manaka, T. and Iwamoto, M. (2007) *Chemical Physics Letters*, **442**, 97–100.

80 Cornelissen, J., Donners, J., de Gelder, R., Graswinckel, W.S., Metselaar, G.A., Rowan, A.E., Sommerdijk, N. and Nolte, R.J.M. (2001) *Science*, **293**, 676–680.

81 Cornelissen, J., Graswinckel, W.S., Rowan, A.E., Sommerdijk, N. and Nolte, R.J.M. (2003) *Journal of Polymer Science Part A-Polymer Chemistry*, **41**, 1725–1736.

82 Metselaar, G.A., Wezenberg, S.J., Cornelissen, J., Nolte, R.M. and Rowan, A.E. (2007) *Journal of Polymer Science Part A-Polymer Chemistry*, **45**, 981–988.

83 Metselaar, G.A., Adams, P., Nolte, R.J.M., Cornelissen, J. and Rowan, A.E. (2007) *Chemistry – A European Journal*, **13**, 950–960.

84 Wezenberg, S.J., Metselaar, G.A., Rowan, A.E., Cornelissen, J., Seebach, D. and Nolte, R.J.M. (2006) *Chemistry – A European Journal*, **12**, 2778–2786.

85 Schwartz, E., Kitto, H.J., de Gelder, R., Nolte, R.J.M., Rowan, A.E. and Cornelissen, J. (2007) *Journal of Materials Chemistry*, **17**, 1876–1884.

86 Kajitani, T., Okoshi, K., Sakurai, S.I., Kumaki, J. and Yashima, E. (2006) *Journal of the American Chemical Society*, **128**, 708–709.

87 Onouchi, H., Okoshi, K., Kajitani, T., Sakurai, S.i., Nagai, K., Kumaki, J., Onitsuka, K. and Yashima, E. (2008) *Journal of the American Chemical Society*, **130**, 229–236.

88 Li, B.S., Cheuk, K.K.L., Salhi, F., Lam, J.W.Y., Cha, J.A.K., Xiao, X.D., Bai, C.L. and Tang, B.Z. (2001) *Nano Letters*, **1**, 323–328.

89 Sakurai, S.I., Okoshi, K., Kumaki, J. and Yashima, E. (2006) *Angewandte Chemie-International Edition*, **45**, 1245–1248.

90 Cheuk, K.K.L., Lam, J.W.Y., Chen, J.W., Lai, L.M. and Tang, B.Z. (2003) *Macromolecules*, **36**, 5947–5959.

91 Zhao, H.C., Sanda, F. and Masuda, T. (2006) *Macromolecular Chemistry and Physics*, **207**, 1921–1926.

92 Cheuk, K.K.L., Lam, J.W.Y., Lai, L.M., Dong, Y.P. and Tang, B.Z. (2003) *Macromolecules*, **36**, 9752–9762.

93 Sanda, F., Gao, G.Z. and Masuda, T. (2004) *Macromolecular Bioscience*, **4**, 570–574.

94 Maeda, K., Tanaka, K., Morino, K. and Yashima, E. (2007) *Macromolecules*, **40**, 6783–6785.

95 Gao, G.Z., Sanda, F. and Masuda, T. (2003) *Macromolecules*, **36**, 3938–3943.

96 Zhang, Z.G., Deng, J.P., Zhao, W.G., Wang, J.M. and Yang, W.T. (2007) *Journal*

of *Polymer Science Part A-Polymer Chemistry*, **45**, 500–508.
97 Tabei, J., Sanda, F. and Masuda, T. (2006) *Kobunshi Ronbunshu*, **63**, 286–296.
98 Tabei, J., Nomura, R. and Masuda, T. (2002) *Macromolecules*, **35**, 5405–5409.
99 Nomura, R., Nishiura, S., Tabei, J., Sanda, F. and Masuda, T. (2003) *Macromolecules*, **36**, 5076–5080.
100 Nomura, R., Tabei, J., Nishiura, S. and Masuda, T. (2003) *Macromolecules*, **36**, 561–564.
101 Tabei, J., Nomura, R., Shiotsuki, M., Sanda, F. and Masuda, T. (2005) *Macromolecular Chemistry and Physics*, **206**, 323–332.
102 Green, M.M., Ringsdorf, H., Wagner, J. and Wüstefeld, R. (1990) *Angewandte Chemie-International Edition*, **29**, 1478–1481.
103 Percec, V., Ahn, C.H., Ungar, G., Yeardley, D.J.P., Möller, M. and Sheiko, S.S. (1998) *Nature*, **391**, 161.
104 Rudick, J.G. and Percec, V. (2007) *New Journal of Chemistry*, **31**, 1083–1096.
105 Percec, V., Aqad, E., Peterca, M., Rudick, J.G., Lemon, L., Ronda, J.C., De, B.B., Heiney, P.A. and Meijer, E.W. (2006) *Journal of the American Chemical Society*, **128**, 16365–16372.
106 Zhang, Z.B., Fujiki, M., Motonaga, M. and McKenna, C.E. (2003) *Journal of the American Chemical Society*, **125**, 7878–7881.
107 Yashima, E., Goto, H. and Okamoto, Y. (1999) *Macromolecules*, **32**, 7942–7945.
108 Goto, H., Okamoto, Y. and Yashima, E. (2002) *Chemistry – A European Journal*, **8**, 4027–4036.
109 Goto, H. and Yashima, E. (2002) *Journal of the American Chemical Society*, **124**, 7943–7949.
110 Zhang, Z.B., Motonaga, M., Fujiki, M. and McKenna, C.E. (2003) *Macromolecules*, **36**, 6956–6958.
111 Nakashima, H., Fujiki, M., Koe, J.R. and Motonaga, M. (2001) *Journal of the American Chemical Society*, **123**, 1963–1969.
112 Peng, W.Q., Motonaga, M. and Koe, J.R. (2004) *Journal of the American Chemical Society*, **126**, 13822–13826.
113 Yashima, E., Matsushima, T. and Okamoto, Y. (1995) *Journal of the American Chemical Society*, **117**, 11596–11597.
114 Goto, H., Zhang, H.Q. and Yashima, E. (2003) *Journal of the American Chemical Society*, **125**, 2516–2523.
115 Nagai, K., Maeda, K., Takeyama, Y., Sakajiri, K. and Yashima, E. (2005) *Macromolecules*, **38**, 5444–5451.
116 Maeda, K., Takeyama, Y., Sakajiri, K. and Yashima, E. (2004) *Journal of the American Chemical Society*, **126**, 16284–16285.
117 Nagai, K., Sakajiri, K., Maeda, K., Okoshi, K., Sato, T. and Yashima, E. (2006) *Macromolecules*, **39**, 5371–5380.
118 Onouchi, H., Hasegawa, T., Kashiwagi, D., Ishiguro, H., Maeda, K. and Yashima, E. (2006) *Journal of Polymer Science Part A-Polymer Chemistry*, **44**, 5039–5048.
119 Kamikawa, Y., Kato, T., Onouchi, H., Kashiwagi, D., Maeda, K. and Yashima, E. (2004) *Journal of Polymer Science Part A-Polymer Chemistry*, **42**, 4580–4586.
120 Nishimura, T., Tsuchiya, K., Ohsawa, S., Maeda, K., Yashima, E., Nakamura, Y. and Nishimura, J. (2004) *Journal of the American Chemical Society*, **126**, 11711–11717.
121 Maeda, K., Tsukui, H., Matsushita, Y. and Yashima, E. (2007) *Macromolecules*, **40**, 7721–7726.
122 Ishikawa, M., Maeda, K. and Yashima, E. (2002) *Journal of the American Chemical Society*, **124**, 7448–7458.
123 Sannigrahi, A. and Khan, I.M. (2001) *Polymer Preprints*, **42**, 242–243.
124 Sannigrahi, B., McGeady, P. and Khan, I.M. (2004) *Macromolecular Bioscience*, **4**, 999–1007.
125 Havinga, E.E., Boumann, M.M., Meijer, E.W., Pomp, A. and Simenon, M.M.J. (1994) *Synthetic Metals*, **66**, 93–97.
126 Bodner, M. and Espe, M.P. (2003) *Synthetic Metals*, **135**, 403–404.

127 Strounina, E.V., Kane-Maguire, L.A.P. and Wallace, G.G. (2006) *Polymer*, **47**, 8088–8094.

128 Maeda, K., Morioka, K. and Yashima, E. (2007) *Macromolecules*, **40**, 1349–1352.

129 Li, C., Numata, M., Takeuchi, M. and Shinkai, S. (2006) *Chemistry, an Asian Journal*, **1**, 95–101.

130 Inai, Y., Ousaka, N. and Ookouchi, Y. (2006) *Biopolymers*, **82**, 471–481.

131 Inai, Y., Tagawa, K., Takasu, A., Hirabayashi, T., Oshikawa, T. and Yamashita, M. (2000) *Journal of the American Chemical Society*, **122**, 11731–11732.

132 Nonokawa, R., Oobo, M. and Yashima, E. (2003) *Macromolecules*, **36**, 6599–6606.

133 Nonokawa, R. and Yashima, E. (2003) *Journal of Polymer Science Part A-Polymer Chemistry*, **41**, 1004–1013.

134 Kakuchi, R., Sakai, R., Otsuka, I., Satoh, T., Kaga, H. and Kakuchi, T. (2005) *Macromolecules*, **38**, 9441–9447.

135 Nonokawa, R. and Yashima, E. (2003) *Journal of the American Chemical Society*, **125**, 1278–1283.

136 Sakai, R., Otsuka, I., Satoh, T., Kakuchi, R., Kaga, H. and Kakuchi, T. (2006) *Macromolecules*, **39**, 4032–4037.

137 Morino, K., Oobo, M. and Yashima, E. (2005) *Macromolecules*, **38**, 3461–3468.

138 Nishimura, T., Ohsawa, S., Maeda, K. and Yashima, E. (2004) *Chemical Communications*, 646–647.

139 Sakai, R., Satoh, T., Kakuchi, R., Kaga, H. and Kakuchi, T. (2003) *Macromolecules*, **36**, 3709–3713.

140 Sakai, R., Satoh, T., Kakuchi, R., Kaga, H. and Kakuchi, T. (2004) *Macromolecules*, **37**, 3996–4003.

141 Moriuchi, T., Shen, X.L. and Hirao, T. (2006) *Tetrahedron*, **62**, 12237–12246.

142 Maeda, K., Ishikawa, M. and Yashima, E. (2004) *Journal of the American Chemical Society*, **126**, 15161–15166.

143 Saxena, A., Guo, G.Q., Fujiki, M., Yang, Y.G., Ohira, A., Okoshi, K. and Naito, M. (2004) *Macromolecules*, **37**, 3081–3083.

144 Nilsson, K.P.R., Rydberg, J., Baltzer, L. and Inganas, O. (2004) *Proceedings of the National Academy of Sciences of the United States of America*, **101**, 11197–11202.

145 Prince, R.B., Barnes, S.A. and Moore, J.S. (2000) *Journal of the American Chemical Society*, **122**, 2758–2762.

146 Tanatani, A., Mio, M.J. and Moore, J.S. (2001) *Journal of the American Chemical Society*, **123**, 1792–1793.

147 Tanatani, A., Hughes, T.S. and Moore, J.S. (2002) *Angewandte Chemie-International Edition*, **41**, 325–328.

148 Waki, M., Abe, H. and Inouye, M. (2006) *Chemistry – A European Journal*, **12**, 7839–7847.

149 Waki, M., Abe, H. and Inouye, M. (2007) *Angewandte Chemie-International Edition*, **46**, 3059–3061.

150 Li, C., Numata, M., Bae, A.H., Sakurai, K. and Shinkai, S. (2005) *Journal of the American Chemical Society*, **127**, 4548–4549.

151 Green, M.M., Khatri, C. and Peterson, N.C. (1993) *Journal of the American Chemical Society*, **115**, 4941–4942.

152 Khatri, C.A., Pavlova, Y., Green, M.M. and Morawetz, H. (1997) *Journal of the American Chemical Society*, **119**, 6991–6995.

153 Dellaportas, P., Jones, R.G. and Holder, S.J. (2002) *Macromolecular Rapid Communications*, **23**, 99–103.

154 Sanji, T., Sato, Y., Kato, N. and Tanaka, M. (2007) *Macromolecules*, **40**, 4747–4749.

155 Natansohn, A. and Rochon, P. (2002) *Chemical Reviews*, **102**, 4139–4175.

156 Choi, S.W., Kawauchi, S., Ha, N.Y. and Takezoe, H. (2007) *Physical Chemistry Chemical Physics*, **9**, 3671–3681.

157 Kim, M.J., Shin, B.G., Kim, J.J. and Kim, D.Y. (2002) *Journal of the American Chemical Society*, **124**, 3504–3505.

158 Kim, M.J., Kumar, J. and Kim, J.J. (2003) *Advanced Materials*, **15**, 2005.

159 Sumimura, H., Fukuda, T., Kim, J.Y., Barada, D., Itoh, M. and Yatagai, T. (2006)

Japanese Journal of Applied Physics, **45**, 451–455.
160 Nikolova, L., Todorov, T., Ivanov, M., Andruzzi, F., Hvilsted, S. and Ramanujam, P.S. (1997) *Optical Materials*, **8**, 255–258.
161 Nedelchev, L., Nikolova, L., Matharu, A. and Ramanujam, P.S. (2002) *Appl Phys B*, **75**, 671–676.
162 Pages, S., Lagugne-Labarthet, F., Buffeteau, T. and Sourisseau, C. (2002) *Appl Phys B*, **75**, 541–548.
163 Hore, D., Wu, Y., Natansohn, A. and Rochon, P. (2003) *J Appl Phys*, **94**, 2162–2166.
164 Tejedor, R.M., Millaruelo, M., Oriol, L., Serrano, J.L., Alcala, R., Rodriguez, F.J. and Villacampa, B. (2006) *Journal of Materials Chemistry*, **16**, 1674–1680.
165 Tejedor, R.M., Oriol, L., Serrano, J.L., Partal Ureña, F. and López González, J.J. (2007) *Advanced Functional Materials*, **17**, 3486–3492.
166 Iftime, G., Labarthet, F.L., Natansohn, A. and Rochon, P. (2000) *Journal of the American Chemical Society*, **122**, 12646–12650.
167 Choi, S.W., Ha, N.Y., Shiromo, K., Rao, N.V.S., Paul, M.K., Toyooka, T., Nishimura, S., Wu, J.W., Park, B., Takanishi, Y., Ishikawa, K. and Takezoe, H. (2006) *Physical Review E*, **73**, 021702–021706.
168 Pijper, D. and Feringa, B.L. (2007) *Angewandte Chemie-International Edition*, **46**, 3693–3696.
169 Pijper, D., Jongejan, M.G.M., Meetsma, A. and Feringa, B.L. (2008) *Journal of the American Chemical Society*, **130**, 4541–4552.
170 Tang, K., Green, M.M., Cheon, K.S., Selinger, J.V. and Garetz, B.A. (2003) *Journal of the American Chemical Society*, **125**, 7313–7323.
171 Nakako, H., Nomura, R. and Masuda, T. (2001) *Macromolecules*, **34**, 1496–1502.
172 Tabei, J., Nomura, R., Sanda, F. and Masuda, T. (2004) *Macromolecules*, **37**, 1175–1179.
173 Zhao, H.C., Sanda, F. and Masuda, T. (2005) *Polymer*, **46**, 2841–2846.
174 Tabei, J., Nomura, R. and Masuda, T. (2003) *Macromolecules*, **36**, 573–577.
175 Yashima, E., Maeda, K. and Sato, O. (2001) *Journal of the American Chemical Society*, **123**, 8159–8160.
176 Sakurai, S.I., Okoshi, K., Kumaki, J. and Yashima, E. (2006) *Journal of the American Chemical Society*, **128**, 5650–5651.
177 Fujiki, M. (2000) *Journal of the American Chemical Society*, **122**, 3336–3343.
178 Fujiki, M., Koe, J.R., Motonaga, M., Nakashima, H., Terao, K. and Teramoto, A. (2001) *Journal of the American Chemical Society*, **123**, 6253–6261.
179 Ohira, A., Kunitake, M., Fujiki, M., Naito, M. and Saxena, A. (2004) *Chem Mater*, **16**, 3919–3923.
180 Ohira, A., Okoshi, K., Fujiki, M., Kunitake, M., Naito, M. and Hagihara, T. (2004) *Advanced Materials*, **16**, 1645–1650.
181 Terao, K., Mori, Y., Dobashi, T., Sato, T., Teramoto, A. and Fujiki, M. (2004) *Langmuir*, **20**, 306–308.
182 Maxein, G. and Zentel, R. (1995) *Macromolecules*, **28**, 8438–8440.
183 Mayer, S. and Zentel, R. (2000) *Macromolecular Rapid Communications*, **21**, 927–930.
184 Angiolini, L., Bozio, R., Giorgini, L., Pedron, D., Turco, G. and Dauru, A. (2002) *Chemistry – A European Journal*, **8**, 4241–4247.
185 Angiolini, L., Benelli, T., Bozio, R., Dauru, A., Giorgini, L. and Pedron, D. (2003) *Synthetic Metals*, **139**, 743–746.
186 Angiolini, L., Giorgini, L., Bozio, R. and Pedron, D. (2003) *Synthetic Metals*, **138**, 375–379.
187 Angiolini, L., Benelli, T., Giorgini, L., Mauriello, F. and Salatelli, E. (2006) *Macromolecular Chemistry and Physics*, **207**, 1805–1813.
188 Yashima, E., Maeda, K. and Okamoto, Y. (1999) *Nature*, **399**, 449–451.

189 Hasegawa, T., Morino, K., Tanaka, Y., Katagiri, H., Furusho, Y. and Yashima, E. (2006) *Macromolecules*, **39**, 482–488.

190 Onouchi, H., Kashiwagi, D., Hayashi, K., Maeda, K. and Yashima, E. (2004) *Macromolecules*, **37**, 5495–5503.

191 Hasegawa, T., Maeda, K., Ishiguro, H. and Yashima, E. (2006) *Polymer Journal*, **38**, 912–919.

192 Miyagawa, T., Furuko, A., Maeda, K., Katagiri, H., Furusho, Y. and Yashima, E. (2005) *Journal of the American Chemical Society*, **127**, 5018–5019.

193 Ishikawa, M., Maeda, K., Mitsutsuji, Y. and Yashima, E. (2004) *Journal of the American Chemical Society*, **126**, 732–733.

194 Hase, Y., Ishikawa, M., Muraki, R., Maeda, K. and Yashima, E. (2006) *Macromolecules*, **39**, 6003–6008.

195 Ishikawa, M., Taura, D., Maeda, K. and Yashima, E. (2004) *Chemistry Letters*, **33**, 550–551.

196 Hase, Y., Mitsutsuji, Y., Ishikawa, M., Maeda, K., Okoshi, K. and Yashima, E. (2007) *Chemistry, an Asian Journal*, **2**, 755–763.

197 Kawauchi, T., Kumaki, J., Kitaura, A., Okoshi, K., Kusanagi, H., Kobayashi, K., Sugai, T., Shinohara, H. and Yashima, E. (2008) *Angewandte Chemie-International Edition*, **47**, 515–519.

198 Kusuyama, H., takase, M., Higashihara, Y. and Tseng, H.T. (1983) *Polymer*, **24**, 119–122.

199 Natansohn, A. and Rochon, P. (2002) *Chemical Reviews*, **102**, 4139–4176.

200 Barada, D., Fukuda, T., Sumimura, H., Kim, J.Y., Itoh, M. and Yatagai, T. (2007) *Japanese Journal of Applied Physics*, **46**, 3928–3932.

201 Lazzari, M., Liu, G. and Lecommandoux, S. (2006) *Block Coplymers in Nanoscience*, Wiley-VCH, Weinheim.

202 Klock, H.A. and Lecommandoux, S. (2001) *Advanced Materials*, **13**, 1217.

203 Khandpur, A.K., Forster, S., Bates, F.S., Hamley, I.W., Ryan, A.J., Bras, W., Almdal, K. and Mortensen, K. (1995) *Macromolecules*, **28**, 8796.

204 Bates, F.S. and Fredrickson, G.H. (1990) *Annual Review of Physical Chemistry*, **41**, 525.

205 Krappe, U., Stadler, R. and Voigtmartin, I. (1995) *Macromolecules*, **28**, 4558–4561.

206 Ho, R.M., Chiang, Y.W., Tsai, C.C., Lin, C.C., Ko, B.T. and Huang, B.H. (2004) *Journal of the American Chemical Society*, **126**, 2704–2705.

207 Chiang, Y.W., Ho, R.M., Ko, B.T. and Lin, C.C. (2005) *Angewandte Chemie-International Edition*, **44**, 7969–7972.

208 Ho, R.M., Chen, C.K., Chiang, Y.W., Ko, B.T. and Lin, C.C. (2006) *Advanced Materials*, **18**, 2355–2358.

209 Omenat, A., Hikmet, R.A.M., Lub, J. and Sluis, P.v.d. (1996) *Advanced Materials*, **8**, 906–909.

210 Omenat, A., Hikmet, R.A.M., Lub, J. and vanderSluis, P. (1996) *Macromolecules*, **29**, 6730–6736.

211 Gabert, A.J., Verploegen, E., Hammond, P.T. and Schrock, R.R. (2006) *Macromolecules*, **39**, 3993–4000.

212 Zheng, W.Y. and Hammond, P.T. (1998) *Macromolecules*, **31**, 711–721.

213 Zheng, S.J., Li, Z.F., Zhang, S.Y., Cao, S.K., Tang, M.S., Fen, Q.J. and Zhou, Q.F. (1999) *Chinese Journal of Political Science*, **17**, 579–587.

214 Hamley, I.W., Castelletto, V., Lu, Z.B., Imrie, C.T., Itoh, T. and Al-Hussein, M. (2004) *Macromolecules*, **37**, 4798–4807.

215 Cui, L., Dahmane, S., Tong, X., Zhu, L. and Zhao, Y. (2005) *Macromolecules*, **38**, 2076–2084.

216 Cornelissen, J., Fischer, M., Sommerdijk, N. and Nolte, R.J.M. (1998) *Science*, **280**, 1427–1430.

217 Cornelissen, J., Fischer, M., van Waes, R., van Heerbeek, R., Kamer, P.C.J., Reek, J.N.H., Sommerdijk, N. and Nolte, R.J.M. (2004) *Polymer*, **45**, 7417–7430.

218 Vriezema, D.M., Hoogboom, J., Velonia, K., Takazawa, K., Christianen, P.C.M., Maan, J.C., Rowan, A.E. and Nolte, R.J.M.

(2003) *Angewandte Chemie-International Edition*, **42**, 772–776.
219 Menger, F.M. and Angelova, M.I. (1998) *Accounts of Chemical Research*, **31**, 789–797.
220 Vriezema, D.M., Kros, A., de Gelder, R., Cornelissen, J., Rowan, A.E. and Nolte, R.J.M. (2004) *Macromolecules*, **37**, 4736–4739.
221 Park, S.Y. and Kawakami, Y. (2005) *Macromolecular Chemistry and Physics*, **206**, 533–539.
222 Henze, O., Fransen, M., Jonkheijm, P., Meijer, E.W., Feast, W.J. and Schenning, A.P.H.J. (2003) *Journal of Polymer Science Part A-Polymer Chemistry*, **41**, 1737–1743.
223 Pasternack, R.F., Giannetto, A., Pagano, P. and Gibbs, E.J. (1991) *Journal of the American Chemical Society*, **113**, 7799–7800.
224 Purrello, R., Gurrieri, S. and Lauceri, R. (1999) *Coordination Chemistry Reviews*, **192**, 683–706.
225 Lauceri, R., Purrello, R., Shetty, S.J. and Vicente, M.G.H. (2001) *Journal of the American Chemical Society*, **123**, 5835–5836.
226 Lauceri, R., Campagna, T., Raudino, A. and Purrello, R. (2001) *Inorganica Chimica Acta*, **317**, 282–289.
227 Robert, F.P. (2003) *Chirality*, **15**, 329–332.
228 Purrello, R., Monsu' Scolaro, L., Bellacchio, E., Gurrieri, S. and Romeo, A. (1998) *Inorganic Chemistry Communications*, **37**, 3647–3648.
229 Bellacchio, E., Lauceri, R., Gurrieri, S., Scolaro, L.M., Romeo, A. and Purrello, R. (1998) *Journal of the American Chemical Society*, **120**, 12353–12354.
230 Mammana, A., De Napoli, M., Lauceri, R. and Purrello, R. (2005) *Bioorganic and Medicinal Chemistry*, **13**, 5159–5163.
231 Purrello, R., Raudino, A., Scolaro, L.M., Loisi, A., Bellacchio, E. and Lauceri, R. (2001) *The Journal of Physical Chemistry. B*, **105**, 2474–2474.
232 Purrello, R. (2003) *Nature Materials*, **2**, 216–217.
233 Hannah, K.C. and Armitage, B.A. (2004) *Accounts of Chemical Research*, **37**, 845–853.
234 Kim, O.K., Je, J., Jernigan, G., Buckley, L. and Whitten, D. (2006) *Journal of the American Chemical Society*, **128**, 510–516.
235 Onouchi, H., Miyagawa, T., Morino, K. and Yashima, E. (2006) *Angewandte Chemie-International Edition*, **45**, 2381–2384.
236 Miyagawa, T., Yamamoto, M., Muraki, R., Onouchi, H. and Yashima, E. (2007) *Journal of the American Chemical Society*, **129**, 3676–3682.
237 Zbaida, D., Lahav, M., Drauz, K., Knaup, G. and Kottenhahn, M. (2000) *Tetrahedron*, **56**, 6645–6649.
238 Menaham, T. and Mastai, Y. (2006) *Journal of Polymer Science Part A-Polymer Chemistry*, **44**, 3009–3017.
239 Medina, D.D., Goldshtein, J., Margel, S. and Mastai, Y. (2007) *Advanced Functional Materials*, **17**, 944–950.
240 Mastai, Y., Sedlak, M., Colfen, H. and Antonietti, M. (2002) *Chemistry – A European Journal*, **8**, 2430–2437.
241 Whitcombe, M.J. and Vulfson, E.N. (2001) *Advanced Materials*, **13**, 467–478.
242 Alexander, C., Davidson, L. and Hayes, W. (2003) *Tetrahedron*, **59**, 2025–2057.
243 Maier, N.M. and Lindner, W. (2007) *Analytical and Bioanalytical Chemistry*, **389**, 377–397.
244 Ansell, R.J. (2005) *Advanced Drug Delivery Reviews*, **57**, 1809–1835.
245 Turiel, E. and Martin-Esteban, A. (2004) *Analytical and Bioanalytical Chemistry*, **378**, 1876–1886.
246 Whitcombe, M.J., Alexander, C. and Vulfson, E.N. (2000) *Synlett*, **6**, 911–923.
247 Nakano, T., Satoh, Y. and Okamoto, Y. (2001) *Macromolecules*, **34**, 2405–2407.
248 Habaue, S., Satonaka, T., Nakano, T. and Okamoto, Y. (2004) *Polymer*, **45**, 5095–5100.
249 Gabashvili, A., Medina, D.D., Gedanken, A. and Mastai, Y. (2007) *The Journal of Physical Chemistry. B*, **111**, 11105–11110.

250 Sanji, T., Kato, N. and Tanaka, M. (2006) *Organic Letters*, **8**, 235–238.
251 Sanji, T., Kato, N. and Tanaka, M. (2005) *Chemistry Letters*, **34**, 1144–1145.
252 Sanji, T., Kato, N., Kato, M. and Tanaka, M. (2005) *Angewandte Chemie-International Edition*, **44**, 7301–7304.
253 Ikeda, M., Hasegawa, T., Numata, M., Sugikawa, K., Sakurai, K., Fujiki, M. and Shinkai, S. (2007) *Journal of the American Chemical Society*, **129**, 3979–3988.
254 Cornelissen, J., van Heerbeek, R., Kamer, P.C.J., Reek, J.N.H., Sommerdijk, N. and Nolte, R.J.M. (2002) *Advanced Materials*, **14**, 489–492.
255 Kauranen, M., Verbiest, T., Boutton, C., Teerenstra, M.N., Clays, K., Schouten, A.J., Nolte, R.J.M. and Persoons, A. (1995) *Science*, **270**, 966–969.
256 de Witte, P.A.J., Castriciano, M., Cornelissen, J., Scolaro, L.M., Nolte, R.J.M. and Rowan, A.E. (2003) *Chemistry – A European Journal*, **9**, 1775–1781.
257 de Witte, P.A.J., Hernando, J., Neuteboom, E.E., van Dijk, E., Meskers, S.C.J., Janssen, R.A.J., van Hulst, N.F., Nolte, R.J.M., Garcia-Parajo, M.F. and Rowan, A.E. (2006) *The Journal of Physical Chemistry. B*, **110**, 7803–7812.
258 Gomar-Nadal, E., Veciana, J., Rovira, C. and Amabilino, D.B. (2005) *Advanced Materials*, **17**, 2095–.
259 Gomar-Nadal, E., Mugica, L., Vidal-Gancedo, J., Casado, J., Navarrete, J.T.L., Veciana, J., Rovira, C. and Amabilino, D.B. (2007) *Macromolecules*, **40**, 7521–7531.
260 Yanagisawa, K., Morita, T. and Kimura, S. (2004) *Journal of the American Chemical Society*, **126**, 12780–12781.
261 Solladie, N., Hamel, A. and Gross, M. (2001) *Chirality*, **13**, 736–738.
262 Zhao, H.C., Sanda, F. and Masuda, T. (2006) *Polymer*, **47**, 1584–1589.
263 Sinkeldam, R.W., Hoeben, F.J.M., Pouderoijen, M.J., De Cat, I., Zhang, J., Furukawa, S., De Feyter, S., Vekemans, J. and Meijer, E.W. (2006) *Journal of the American Chemical Society*, **128**, 16113–16121.

6
Nanoscale Exploration of Molecular and Supramolecular Chirality at Metal Surfaces under Ultrahigh-Vacuum Conditions

Rasmita Raval

6.1
Introduction

Supramolecular chiral structures, formed via noncovalent interactions of functional molecular building blocks, have made prominent and invaluable insights into the creation, transfer and amplification of chirality [1–4]. At surfaces, molecular self-assembly is driven by the subtle balances between molecule–surface and molecule–molecule interactions. These forces can be mediated by appropriate choices of surface and molecule and can lead to a special category of assembly, namely chiral surfaces, which possess no mirror or inversion symmetry elements. Such chiral surfaces possess interesting technological potential for heterogeneous enantioselective catalysis, chiral separations, molecular sensing and recognition, nonlinear optics and as command layers for 3D growth.

Generally, the creation of chirality requires sufficient complexity in a system so that all inverse symmetry elements are destroyed. At regular and symmetric low Miller index metal surfaces this complexity is most easily implanted by adsorption of complex organic molecules. In fact, surface chirality can be induced both by the adsorption of molecules that are inherently chiral and also by molecules that are achiral. For the latter, the reduced symmetry at a surface and the loss of rotational and translational freedoms as the molecule is confined at the surface plane render the interface particularly prone to creating asymmetry. Both these influences lead to the removal of molecular and surface reflection symmetry elements, thus creating chirality where none existed before.

Chirality at surfaces is often also expressed in a hierarchical manner and the subsequent self-assembly of adsorbed chiral entities may, in turn, lead to chiral surface organizations that possess no mirror symmetry. The resultant surface chirality may be expressed in a number of ways, dependent, to a large extent, on the relative strengths of the intermolecular and metal–molecule interactions [5, 6]. This review highlights some of these ways, focusing on molecules adsorbed on highly defined single-crystal surfaces metal surfaces in ultrahigh vacuum (UHV). These

Chirality at the Nanoscale: Nanoparticles, Surfaces, Materials and more. Edited by David B. Amabilino
Copyright © 2009 WILEY-VCH Verlag GmbH & Co. KGaA, Weinheim
ISBN: 978-3-527-32013-4

conditions are excellent for the investigation for a number of reasons, including the fact that coverage and temperature can be varied independently in a controlled way, thus making the hierarchical transfer of chirality observable. Furthermore, a range of surface-science techniques including: reflection absorption infrared spectroscopy (RAIRS) that provides vital information concerning the nature of bonds within molecules and between the molecules and the surface; low-energy electron diffraction (LEED) that reveals surface crystallographic structure, and of course scanning tunnelling microscopy (STM) that gives molecularly resolved images of the packing of molecules on conducting surfaces. A more comprehensive discussion on individual systems can be found in the review by Barlow and Raval [5] where a semihierarchical classification of chiral surfaces is also provided. Other excellent reviews on surface chirality have also been published, including liquid/solid interfaces that are not covered here [6–14]. The review commences with the self-assembly of chains of molecules that exhibit 1D chirality, a mechanism that may, for example, lead to the fabrication of chiral nanowires. Then, systems exhibiting 2D chirality where supramolecular interactions induce structures varying in size from small chiral clusters or domains to entire macroscopic surfaces with globally organized chiral arrays are highlighted. The review ends with selected examples of chiral recognition processes at surfaces.

6.2
The Creation of Surface Chirality in 1D Superstructures

1D surface chirality is exhibited in systems where supramolecular interactions are constrained to a specific direction, leading to the formation of simple structures, such as chains of molecules, that either have an intrinsic lack of mirror symmetry, or/and run in a nonsymmetric direction across the metal surface, thus breaking existing mirror symmetry elements. The supramolecular ordering of 4-[trans-2-(pyrid-4-vinyl)]benzoic acid, or PVBA, adsorbed on Ag(1 1 1) is a case in point [15, 16]. PVBA is a planar, flat molecule with a kink between the two aromatic ring systems caused by the alkene link between them, and is achiral in the gas phase, with a mirror symmetry plane coincident with the molecular plane. On Ag(1 1 1), its two-lobed STM image suggests adsorption with its ring system parallel to the surface, thus leading to the direct loss of the molecular mirror plane and the creation of a chiral adsorption motif at the surface. As PVBA can land on the surface with either prochiral face uppermost, it can create both mirror forms of the chiral adsorption motif with equal facility, designated λ-PVBA and δ-PVBA and illustrated in Figure 6.1. The supramolecular assembly of PVBA chiral units on Ag(1 1 1) into 1D chains is dominated by highly directional intermolecular hydrogen-bonding interactions between the benzoic acid group at one end, acting as a hydrogen-bond donor, and the pyridine function at the other end, behaving as a hydrogen-bond acceptor. STM images of Figure 6.1 show that chains of molecules form at the surface, aligned along the three rotationally equivalent high-symmetry $<11\bar{2}>$ directions [16]. Each chain consists of two rows of PVBA molecules and, importantly, is homochiral, that is, composed of one type of

PVBA/Ag(111)

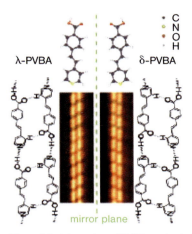

Figure 6.1 Adsorption of PVBA on the Ag(1 1 1) surface. The two chiral adsorption motifs, designated λ-PVBA and δ-PVBA, are shown with the mirror symmetry reflected by a dashed line. STM images show homochiral twin chains held together by H-bonds as in the corresponding models. (Image sizes 40 Å × 135 Å; adsorption temperature 300 K, measured at 77 K.) Reproduced with permission of The American Physical Society from Figure 1 [16].

chiral motif only, either λ or δ. Within a chain, each row of homochiral molecules is held together by strong and directional head-to-tail intermolecular OH- - -N bonds, and the two rows of molecules are positioned in an antiparallel arrangement and held together with weaker lateral H-bonds between a carbonyl group in one row and a H-atom of the pyridine ring in the adjacent row, as shown in Figure 6.1. Given PVBAs facility of creating both the λ or δ motif, the overall adsorption system consists of two types of chains on the surface, each the mirror image of the other. Interestingly, molecular dynamics simulations indicate that chains repel each other so, at low coverages, supramolecular ordering is limited to this 1D organization.

An interesting example of 1D chirality in a system dominated by strong molecule–metal interactions is that provided by tartaric acid on Ni(1 1 0) [17]. Following adsorption of (R,R)-tartaric acid at room temperature, double dehydrogenation of the molecule occurs converting the two carboxylic acid groups into COO- carboxylate functionalities, which bond strongly to the metal. STM data, Figure 6.2, show that the adsorbed bitartrate molecules grow in short strings with a highly preferred growth direction parallel to the high-symmetry [1 $\bar{1}$ 0] crystallographic direction. When the opposite enantiomer (S,S)-tartaric acid is adsorbed, 1D growth is seen along the same [1 $\bar{1}$ 0] crystallographic direction, as shown in the STM image of Figure 6.2. Given that the molecules retain the chiral centers located at the two central carbon atom positions, these aligned short strings represent 1D chirality, *per se*, even though their growth is along a surface symmetry direction. In fact, detailed studies by RAIRS, LEED, STM and density functional theory (DFT) calculations suggest that the

Tartaric acid/Ni(110)

(R,R)-bitartrate   (S,S)-bitartrate

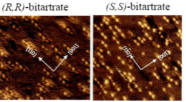

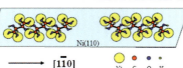

$[1\bar{1}0]$ Ni C O H

Figure 6.2 Adsorption of (R,R)- and (S,S)-tartaric acid on the Ni (1 1 0) surface [17]. STM images of the (R,R)- and (S,S)-bitartrate species at room temperature show a similar growth direction along the $[1\bar{1}0]$ direction. (Image sizes 200 Å × 200 Å.) The schematic depicts the distortion of the bitartrate skeletons and reconstruction of the bonding nickel atoms to give chiral bitartrate-Ni_4 units.

expression of chirality at the surface is inducted at an even deeper level [17]. Specifically, adsorption leads to a highly strained bitartrate–Ni_4 complex at the surface where a strong distortion of the bitartrate skeleton is created alongside a concomitant reconstruction of the bonding nickel atoms, which are pulled away from their symmetric bulk-truncation positions and twisted to give an oblique unit mesh where all the mirror planes are destroyed locally, as shown in Figure 6.2. Thus, the adsorption system possesses arrangements of chiral bitartrate–Ni_4 units in 1D lines, where chirality transfer to the surface is mediated by the strong metal–molecule interaction that leads to reorganization of the underlying metal atoms into chiral arrangements. DFT calculations show that the four O−Ni bonds formed also have a chiral character, with one diagonal pair being equivalent and having a bond length of 2.04 Å, while the opposite diagonal pair possesses a shorter bond length of 1.94 Å, suggesting that chirality may well be communicated into the electronic structure of the adsorption site [18]. Turning to the mirror enantiomer, (S,S)-bitartrate, the adsorption footprint is now in the mirror configuration, with the backbone of the bitartrate molecule distorted in the opposite directions and the underlying metal possessing the mirror chiral reconstruction as depicted schematically in Figure 6.2. We note that for the same molecule on Ni(1 1 1) [19] and Cu(1 1 0), [20–22] the balance of intermolecular and metal–molecule interactions is such that 2D chiral arrays are formed. However, much further work is required to pinpoint the causes that give rise to these critical divergences in organizational behavior.

As the last example of chirality in 1D systems on surfaces, we consider the assembly of the system comprised of nucleic base adenine on Cu(1 1 0). STM images (Figure 6.3) show that when a low coverage of adenine is adsorbed at room temperature and then annealed to 370 K, short chains, made up of adenine dimers,

Adenine/Cu(110)

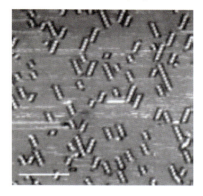

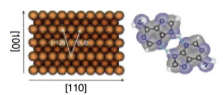

Figure 6.3 Adsorption of adenine on the Cu(1 1 0) surface. STM image of a submonolayer coverage of adenine at room temperature showing dimer chain growth ±19.5° from the [0 0 1] direction. Reproduced with permission of Nature Publishing Group from Figure 1 [23].

form in one of two directions, ±19.5° with respect to the main [0 0 1] symmetry direction of the underlying Cu(1 1 0) surface [23]. These chain growth directions break the reflectional symmetry elements of the surface and, therefore, the 1D structures are essentially chiral. Furthermore, like PVBA, adenine is a prochiral molecule that creates a chiral adsorption motif when the molecular mirror plane is destroyed by adsorbing it flat at a surface. Both mirror forms of the chiral adsorption motif can be created, dictated by which face of the ring system is nearest to the surface. Chen and Richardson [23] argue that the two distinct chiral directions adopted by the adenine chains is an indication of the chirality of the constituent adenine units, that is, each chain is homochiral, consisting of adenine units with the same chiral adsorption motif, which assemble into a specific nonsymmetry direction. Similarly, mirror chiral adsorption motifs assemble in the reflectionally opposite direction. DFT calculations [24] using the general adsorption model adopted in [23] conclude that the self-assembly of adenine on Cu(1 1 0) is directed by mutual polarization and Coulomb attraction. One note of caution needs to be interjected here – adenine is known to possess eight tautomers and a full experimental and theoretical analysis exploring all these possibilities remains to be done. Specifically, RAIRS vibrational data obtained on the adenine/Cu(1 1 0) system by McNutt and coworkers [25] conclude a very different adsorption mode for the molecule.

6.3
The Creation of 2D Surface Chirality

Two-dimensional chirality can arise at a surface as a direct result of the self-organization of the adsorbed molecules into larger structures. These structures may simply be small nanoscale assemblies of molecules whose organization possesses no mirror symmetry elements. Alternatively, the molecules may organize into large, macroscopic domains that do not retain any mirror symmetry elements. Examples of 2D chirality across this lengthscale are given below.

6.3.1
2D Supramolecular Chiral Clusters at Surfaces

One of the first examples of 2D chiral clusters was reported by Böhringer et al. for the adsorption of 1-nitronaphthalene (NN) onto the reconstructed Au(1 1 1) surface [26, 27]. NN adsorbs with the naphthalene ring system parallel to the gold surface and this geometry again transforms a prochiral gas-phase molecule into a chiral adsorbed entity. Individual molecules can, therefore, adopt either mirror chiral adsorption motif, denoted as the l- or r- enantiomer according to whether the nitro group is attached to the left or right carbon ring, respectively. STM images obtained after adsorbing a low coverage of NN at room temperature and subsequent slow cooling to 70 K are shown in Figure 6.4. Here, it can be seen that the NN molecules form small clusters on the fcc areas of the Au(1 1 1) surface, with approximately 85% of the molecules incorporated into decamers. High-resolution STM images show that these decamers are composed of 10 molecules arranged in a modified pinwheel structure, which can take up either one of two chiral organizations, the L- or R-decamer, as shown in Figure 6.4. The decamers are believed to be held together by

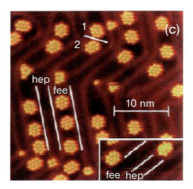
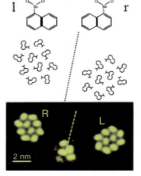

Figure 6.4 Adsorption of NN on a reconstructed Au(1 1 1) surface. The schematic shows the two chiral adsorption motifs and the chirally organized decamer clusters, designated as the l- or r-enantiomers and L- or R-decamers, respectively. The larger STM image shows the formation of the decamers on the fcc sites for a coverage of 0.1 ML, imaged at 50 K with the inset imaged at 20 K. The smaller STM image shows the detail of the chiral decamers imaged at 50 K. Reproduced with permission of The American Physical Society from Figure 1c [27] and of Wiley-VCH GmbH from Figure 2a [26].

intermolecular hydrogen-bonding interactions between the ring CH and nitro O groups and calculations suggest that each decamer is formed from both l- and r- monomers, with a 2 : 8 or 4 : 6 l : r ratio for the L-decamers and a 8 : 2 or 6 : 4 ratio for the R-decamers. Thus, chirality is present at two levels in this system: first, at the local adsorption motif level and, second, at the organization level, driven by intermolecular hydrogen bonding, which produces chiral pinwheels. For this system, the adsorption of the NN to the metal is relatively weak and the clusters can be manipulated by the STM tip and separated in terms of chirality at the surface, that is, the nanoscale equivalent of the famous Pasteur experiment [28]!

Chiral clustering has also been observed for organometallic complexes adsorbed at surfaces. Messina *et al.* [29, 30] reported that the codeposition of iron atoms and 1,3,5-tricarboxylic benzoic acid (trimesic acid, TMA) on a Cu(1 0 0) surface held at 100 K and subsequently annealed the system at room temperature, leads to the formation of chiral complexes stabilized by metal–ligand interactions. The STM images in Figure 6.5 show a central Fe atom surrounded by four TMA molecules. The model depicted in the inset of Figure 6.5 shows that the individual structures of the clusters are chiral with the Fe atoms coordinated to the carboxylate group of each TMA molecule in either a clockwise or anticlockwise fashion. Thus, two types of clusters exist that are mirror images of each other. The chiral "propeller" arrangement around the Fe atom is driven by the strong, unidentate O–Fe bonding that effectively creates a kink in the Fe–O_1–O_2 arrangement, and intermolecular interactions (e.g., steric repulsions and attractive carboxyl-phenyl H-bonds) between neighboring ligands that force all ligands to coordinate with the same chirality. Closer inspection reveals yet another aspect of chirality in this system in that the metal complexes take up one of two orientations with respect to the copper surface, with their principal axes (a line drawn from the center of one TMA molecule to the center of the opposing TMA molecule in Figure 6.5) being either + or −75° from the [0 1 1] direction that is, the clusters are chirally organized with respect to the surface.

One of most beautiful examples of chiral cluster creation comes from the hierarchical assembly of rubrene on Au(1 1 1) [31], where an unprecedented level of complexity and fidelity of chiral transfer over three generations of assembly was

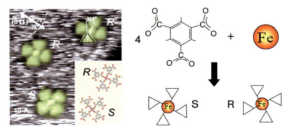

Figure 6.5 Adsorption of Fe atoms and TMA on the Cu(1 0 0) surface. High-resolution STM image of two chiral Fe(TMA)$_4$ clusters, labelled R and S, representing mirror symmetric species, orientated ±75° with respect to the [0 0 1] direction. The schematic illustrates the arrangement and bonding of the carboxylate groups to the Fe atom that determines the chirality of the species. Reproduced with permission of the American Chemical Society from Figure 2 [29].

revealed by STM data. The twisted tetracene backbone of rubrene leads to the creation of a chiral adsorption motif at the Au(1 1 1) surface, with both mirror enantiomers created with equal facility. STM images clearly reveal the absolute chirality of individual species, and provide a powerful mapping of the chiral transfer that occurs upon assembly. Essentially, single rubrene entities assemble into pentagonal supermolecules that further assemble into complex gearwheels and pentagonal chains, culminating in nested decagons that contain 50 rubrene molecules of the same chirality, as shown in Figure 6.6. At each stage of supramolecular assembly, intermolecular forces ensure that the chirality of the starting nucleus is conserved and transmitted with high fidelity to the most complex architectures created at the highest hierarchical order.

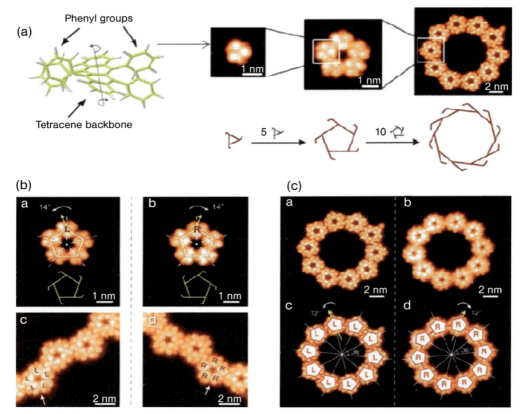

Figure 6.6 (a) STM images of the hierarchical assembly of rubrene on the Au(1 1 1) surface showing how single monomers combine to form pentagonal supermolecules which subsequently assemble to give nested decagons. The chirality of the initial monomer is maintained through the three levels of hierarchy. (b) STM images showing assembly of chiral pentagonal supermolecules to give long chains and gearwheels, with the initial chirality preserved throughout. (c) STM images of decagons assembled from 10 pentagonal supermolecules; again chirality is transferred with high fidelity. Figures adapted from [31], reproduced with permission of Wiley-VCH GmbH.

6.3.2
2D Covalent Chiral Clusters at Surfaces

The power of supramolecular assembly at surfaces to create complex, chiral structures has been amply demonstrated by examples given here. However, there has also been an emerging drive towards creating robust nanostructures that are held together by covalent bonds. Very recently, it has been demonstrated, for the first time, that it is possible to create *covalently connected* chiral nanostructures at surfaces [32] via a thermally induced activation and reaction of tetra(mesityl)porphyrin on Cu(1 1 0) under ultrahigh vacuum conditions. The reaction was followed by STM, whose images – one of which is shown in Figure 6.7 – show that the majority of individual porphyrins are linked up together in lines, angular structures, and grids, with a specific covalent linkage, arising from reaction with the copper, which reduces a methyl group on the mesityl functionality generating a CH_2. radical group, which

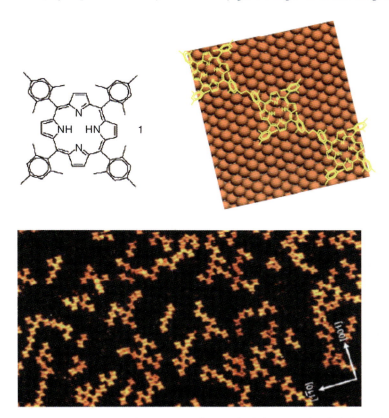

Figure 6.7 STM image (right, $71 \times 38\,nm^2$) showing covalent assembly of tetra(mesityl) porphyrin (**1**) on Cu(1 1 0) produces a combinatorial library of covalently linked porphyrin nanostructures, most of which are chiral due to covalent links being directed along chiral directions (see model, middle), combined with random step-wise condensation processes. STM figures adapted from [32], reproduced with permission of RSC Publishing.

then homocouples the porphyrins. The average STM core-to-core distance between linked porphyrins is approximately 18.5 Å, close to the calculated value for formation of an ethylene linkage (19 Å) between the methyl substituents at the *para* position of the benzene rings relative to the porphyrin ring. The chirality of the nanostructures arises because the alignment of the central pyrrole ring along the high-symmetry axes of the Cu(1 1 0) surface means the connections mediated via the *para* methyl groups all occur along nonsymmetry directions. This combined with a random condensation-type (or stepwise) polymerization, leads to a distribution of sizes and shapes of asymmetric oligomers that contain mainly difunctional units, often three-connected units, and very rarely four-connected ones. This combination of mechanistic and structural detail leads to most of the covalent nanostructures created being chiral, as indicated in the illustration in Figure 6.7.

6.3.3
Large Macroscopic 2-D Chiral Arrays

As the size of two-dimensional assemblies increases at the surface, large chiral arrays or domains can form. Depending on the chirality of the individual adsorption motifs and their overall organization, the handedness of these domains may either be uniquely maintained across the whole surface, which becomes globally homochiral, or coexist with mirror domains giving a surface that is locally chiral but globally achiral.

The phenomenon of a globally homochiral surface created from the supramolecular assembly of molecules was first revealed in STM data of (R,R)-tartaric acid adsorbed on Cu(1 1 0) by Ortega *et al.* [20, 21], as illustrated in Figure 6.8. This

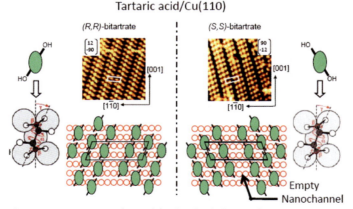

Figure 6.8 STM images obtained for the chiral phases of (R,R)-bitartrate and (S,S)-bitartrate on Cu(1 1 0) together with models of the chiral domains constructed from STM, LEED and RAIRS data. Note the empty nanochannels that are created within the structure. Reproduced with permission of Macmillan Publishers Ltd. from Figure 2 [20].

macroscopic chiral phase consists entirely of doubly dehydrogenated bitartrate species [20–22], bonded strongly to the surface via the four oxygen atoms of the carboxylate groups. These discrete and rigidly bonded bitartrate units give rise to a highly organized chiral phase at the surface. The molecular model constructed from RAIRS, LEED and STM data shows that chiral "trimers" of (R,R)-bitartrate molecules assemble to form long chiral chains aligned along the [1 $\bar{1}$4] crystallographic direction, thus breaking the mirror symmetry elements of the Cu(1 1 0) surface. The chains are propagated across large length scales and the macroscopic surface organization is, therefore, also chiral and belongs to the C_2 chiral space group and is nonsuperimposable on its mirror image. When the opposite enantiomer, (S,S)-tartaric acid, is adsorbed, the chirality of the adlayer is switched to give the mirror organization, with chains growing along the mirror [$\bar{1}$14] direction. Periodic density functional theory (DFT) calculations [33] reveal that adsorption via all four carboxylate oxygen atoms creates a concomitant chiral distortion in the molecule. This distortion is twofold, first along the carbon–carbon backbone and, second, from the O–C–O plane in the <110> direction of the surface, Figure 6.8. Importantly, the distortions are enantiomer-specific, with the (R,R) enantiomer distorting one way and the (S,S) enantiomer distorting in the mirror configuration. This adsorbate distortion has subsequently been verified experimentally [34] via X-ray photoelectron diffraction. From the description above, it can be seen that chirality at the single-adsorbate level is bestowed both from the inherent molecular structure of (R,R)-tartaric acid and from the actual adsorption process. This adsorption motif then assembles in a chiral organization that is adopted across the entire surface, demonstrating clearly that chirality is transferred with high fidelity from the individual adsorption motif, to the nanoscale trimer organization to the macroscale trimer-chain organization. Periodic-DFT calculations show that this chiral transfer from adsorption motif to surface organization is governed by repulsive molecule–molecule interactions within the array that combine to produce an energy difference of 10 kJ mol^{-1} between (R,R)-bitartrate packed in the preferred (1 2, −9 0) structure compared to accommodation within the mirror (9 0, −1 2) enantiomorph. This energy difference is sufficient to ensure that, at 300 K, over 95% of adsorption would result in the preferred chiral organization. For the (S,S)-bitartrate adsorbate, the mirror structure is similarly favored. A point worth noting is that the ability of a chiral molecule, such as tartaric acid, to form a chiral arrays is not only critically dependent on the particular adsorption phase created at a surface but is also very dependent on the actual metal surface. For example, tartaric acid does not form an ordered 2D array structure on the Ni(1 1 0) surface [17]. However, RAIRS, STM and TPD work by Jones et al. [19] on the adsorption of (R,R)-tartaric acid on the smoother Ni(1 1 1) surface does show the creation of chiral arrays.

Globally organized homochiral arrays have also been obtained from other functional molecules. For example, the simplest chiral amino acid, alanine, is strongly chemisorbed on the Cu(1 1 0) surface where it dehydrogenates to create the alaninate species in which the carboxylic acid functionality has been converted into the carboxylate upon adsorption. As is often the case for such multifunctional molecules at surfaces, the system is highly polymorphic and shows a range of adsorption phases as coverage and temperature are altered [35–37]. One of the most interesting

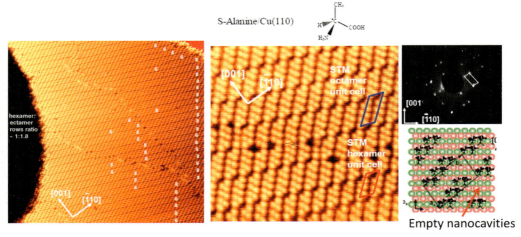

Figure 6.9 Large area 50 nm × 49 nm STM image (top) of the chiral phase of (S)-alaninate on the Cu(1 1 0) surface obtained after annealing to 403 K; (bottom left) Close-up 10 × 10 nm STM image shows assembly of size-selected chiral clusters; (right) The schematic of the overlayer consisting of hexamer clusters and stress breaks that lead to empty nanocavities at the surface. The LEED yields a (2 −2, 5 3) overlayer unit cell, with a repeat unit comprising an individual cluster and associated nanocavities. (R)-alaninate on Cu(1 1 0) creates the mirror phase. Reproduced with permission of Elsevier, from [37].

and unusual phases comprises of chiral clusters that become part of a global macroscopic chiral array, as confirmed by the LEED pattern obtained, Figure 6.9. Here, individual alaninate molecules self-assemble into size-selected, chirally organized clusters of six or eight molecules that are not aligned along any of the main symmetry axes of the Cu(1 1 0) surface [36, 37]. The STM image in Figure 6.9 shows a typical arrangement of clusters for (S)-alaninate growing broadly parallel to the [11̄2] direction. RAIRS analysis of this adsorption phase suggests the presence of two differently oriented alaninate species: one species is bound to the copper surface through both oxygen atoms of the carboxylate group and the nitrogen atom of the amino group while the other is bonded to the copper through the amino group and only one of the carboxylate oxygen atoms. For both alaninate species, the methyl group is held away from the surface. This orientation results in direct chirality transfer into the footprint of the adsorbed alanine molecules with the (S)-alaninate adopting a right-handed kink in the CCN backbone and the (R)-alanine molecules a left-handed kink, when viewed from above. A model for the clusters has been suggested, Figure 6.9, possessing both alaninate species, woven together with a two-tier network of hydrogen bonds between the N−H and O groups held close to the surface and the C−H and O groups located at a higher level [36, 37]. These intermolecular hydrogen-bonding interactions are similar to those exhibited by solid crystals of alanine. Finally, when this phase is created from the mirror R-enantiomer, the mirror enantiomorph results, that is, there is direct chirality transfer from the individual nanoscale molecules to the footprint of the adsorbed species to the nanoscale size-selected clusters to the macroscale organization.

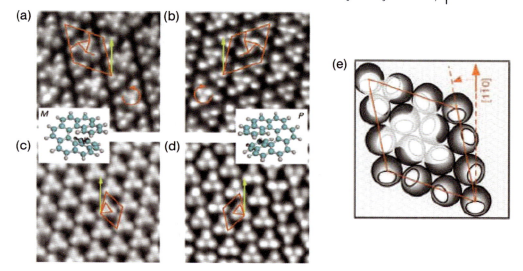

Figure 6.10 Adsorption of heptahelicene on the Cu(1 1 1) surface. STM images (10 × 10 nm) showing clusters of (a) (M)-[7]-helicene and (b) (P)-[7]-helicene at $\vartheta = 0.95$, (c) (M)-[7]-helicene and (d) (P)-[7]-Helicene at $\vartheta = 1$. The model in (e) shows the structure of the M-[7]-helicene "6 and 3" clusters seen at $\vartheta = 0.95$, based on a hexagonal close packing of the molecules and systematically varying azimuthal orientations. Reproduced with permission of Wiley-VCH GmbH from Figures 2 and 4 [38].

Turning to another functionality type, chiral self-assembled monolayers have also been observed for (M)- or (P)-[7]-helicene on Cu(1 1 1)[38, 39], where the driving force dictating assembly is steric repulsion. In contrast to the tartaric acid and amino acid systems, this chiral helical molecule is weakly bound to the substrate and intermolecular repulsive forces dominate the lateral interactions so ordered structures are only formed when the molecules are closely packed. Two ordered structures are seen, with clusters of 6 and 3 molecules imaged with STM at coverages of 95% of a monolayer, and clusters with 3 molecules observed for a saturated monolayer as in Figure 6.10. For both structures, the unit cells and the arrangement of molecules within the cells for the P-enantiomer are chiral and are the mirror image of those seen for the M-enantiomer. X-ray photoelectron diffraction (XPD) studies have shown that the molecule is adsorbed with its terminal phenanthrene ring parallel to the surface and the rest of the molecule spiralling away from the surface in one of 6 possible azimuthal orientations. Assuming that the brightest features in the STM images are due to tunnelling into the uppermost parts of the [7]-helicene, the molecules can be represented by a circular disk with an off-center protrusion in one of these 6 azimuthal orientations. Thus, it has been possible to recreate the STM images by using a model where the molecules are hexagonally close packed but give rise to a "cluster" appearance in the STM images due to the different orientations adopted by the molecules. In particular, the imaged handedness or chirality of the organizations is a direct reflection of the orientational differences adopted within the close-packed layer, as depicted in Figure 6.10. Molecular modelling calculations have shown that molecules adopt the different azimuthal orientations in order to minimize the repulsive interactions, that is, in this case the supramolecular driving force is steric repulsion.

6.3.4
Chiral Nanocavity Arrays

An important manifestation of chirality in the 2D arrays is not simply where the molecules are organized, but rather where empty cavities are present since these can also be chiral by virtue of the arrangements of the surrounding molecules that form their boundaries. These chiral nanochannels are potentially very interesting in terms of technological applications, since they provide a confined environment within which enantioselective recognition, catalysis, separation or sensing could take place. The bitartrate on Cu(110) structure discussed earlier creates an empty nanochannel after every third bitartrate molecule, that is, each long timer chain is separated from the adjacent trimer chain by a vacant channel, Figure 6.8. These nanochannels are also directed along the chiral <114> direction and, therefore, are inherently chiral spaces on the copper substrate. DFT calculations [40] reveal that the origin of these channels is the surface stress created by adsorption of the bitartrate species to the surface; the Cu–Cu distance of the bonding metal atoms is increased from their bulk truncation value of 2.58–2.62 Å, leading to a strong compressive strain along the close-packed [1–10] direction. This stress builds up with each additional carboxylate functionality bonding to a particular row and can only be tolerated for three molecules in a row before a break in the organization is necessary to relieve it.

Similarly, the alaninate/Cu(1 1 0) system creates size-selected clusters, Figure 6.9, a phenomenon attributed directly to system stresses that arise from balancing molecule–molecule interactions and molecule–metal bonding [36, 37]. Thus, the stresses and strains of maintaining both optimum adsorption sites for the alaninate molecules and maximizing the intermolecular interactions become too great once a critical cluster size is reached, leading to a fracture in the assembly. In this particular example, the clusters plus their stress fractures, become "synthons" for a globally organized macroscopic array with a unique chiral arrangement of molecular clusters interspersed with chiral channels and spaces, as shown in the model in Figure 6.9. The size of the LEED unit cell suggests that each defined cluster and its surrounding space acts as the repeat unit for scattering, Figure 6.9. Overall, the size-defined chiral clusters further self-assemble into a defined chiral array with nanochannels of bare metal left between the chiral clusters that are themselves chiral. This suggests that the stresses that induce the channels must also be chiral, in a manner reminiscent of that reported for individual molecules at surfaces [17].

The creation of nanoporous 2D supramolecular structures has also been demonstrated [30, 41, 43] for robust metalorganic frameworks created by the supramolecular assembly of Fe(trimesic acid)$_4$ clusters on Cu(1 0 0) discussed in Section 6.3.1. Here, Fe(TMA)$_4$ clusters first combine with other Fe(TMA)$_3$ units to give a nanogrid that then self-assembles to create an extended array. The juxtapositions of the nanogrids within this array leaves chiral cavities, which expose underlying Cu surface atoms, decorated by surrounding carboxylate groups of the TMA.

6.4
Chiral Recognition Mapped at the Single-Molecule Level

Chiral molecular recognition plays a pivotal role in controlling key events in biological systems and in technological applications such as enantioselective chemistry, catalysis, chiral separations and sensors. However, chiral interactions are difficult to probe generally and it is here that STM studies at surface have made a pivotal contribution, because of its unique ability to observe single molecules, very often with submolecular resolution.

6.4.1
Homochiral Self-Recognition

The loss of symmetry elements at a surface means that racemic mixtures of molecules may be more prone to segregate into homochiral conglomerates, compared to their 3D counterparts. Any surface system showing this behavior necessarily must have undergone large-scale chiral recognition. Single-molecule events that underpin such processes have now been mapped at the nanoscale by STM. For example, chiral self-recognition by the cysteine./Au(1 1 0) system [42] leads to STM images, Figure 6.11, for the enantiopure system showing dimer pairs of molecules with the L-cysteine dimers rotated 20° clockwise with respect to the [1 $\bar{1}$ 0] direction and D-cysteine dimers rotated 20° anticlockwise that is, the adsorption footprints of the dimers are chiral, breaking the mirror symmetry of the gold surface. When a

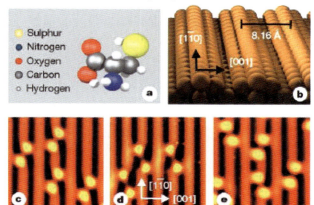

Figure 6.11 Homochiral recognition of cysteine molecules on Au(1 1 0) showing: (a) Schematic drawing of cysteine molecule; (b) Model of the (1 × 2) Au(1 1 1) surface; (c) STM image of L-cysteine pairs showing the main axis of the pair rotated by 20° clockwise (49 × 53 Å); (d) D-cysteine pairs rotated anticlockwise (same size); (e) Homochiral molecular pairs from DL-cysteine (same size). (Reprinted by permission from Macmillan Publishers Ltd from reference [42], copyright 2002).

racemic mixture of cysteine is deposited on the surface, only homochiral dimers, identical in appearance to those formed from pure enantiomers, are observed showing that chiral self-recognition has prevailed in the racemic system. DFT calculations indicate that the preferred formation of homochiral dimers is driven by the optimization of three bonds with the cysteine molecules. According to the simulation, the most stable conformation involves hydrogen bonding between adjacent carboxylic acid groups and bonding to the gold surface via the sulfur and amino groups. It has been widely postulated that such "three-point" bonding is a necessary condition for chiral recognition but, as the next example shows, the phenomenon is much more complex.

Recently, dynamic chiral recognition events have been reported for the chiral dipeptides di-D- and di-L-phenylalanine (D-Phe-D-Phe and L-Phe-L-Phe) adsorbed on Cu(1 1 0) [43]. The larger size of the dipeptide molecule compared with the single amino acid enables STM to map submolecular features, revealing two bright protrusions for the phenyl rings and a central dimmer part associated with the peptide backbone, as shown in Figure 6.12. STM images from the enantiopure systems reveals that each enantiomer transcribes a preferred chiral adsorption footprint, so that the main axis connecting the two phenyl rings is rotated 34° clockwise from the [1 $\bar{1}$ 0] substrate direction for L-Phe-L-Phe but 34° counterclockwise for D-Phe-D-Phe. The dipeptide molecules further self-assemble into chains that grow along nonsymmetry directions of the surface, with the self-assembly

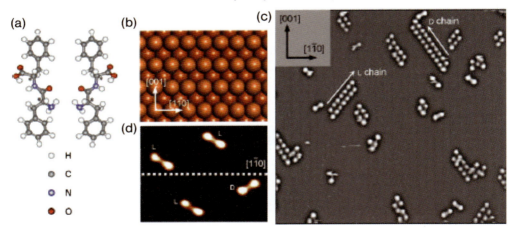

Figure 6.12 D-Phe-D-Phe and L-Phe-L-Phe adsorption on Cu(1 1 0) showing: (a) schematic diagrams of L-Phe-L-Phe (left) and D-Phe-D-Phe (right); (b) Ball model of Cu(1 1 0) surface; (c) STM image (36 × 34 nm) of self-assembled chains of L-Phe-L-Phe (left) and D-Phe-D-Phe from a racemic mixture on Cu(1 1 0). The arrows indicate the growth directions of the homochiral chains; (d) STM images (8.3 nm × 6.4 nm) of individual di-phenylalanine molecules showing that the two enantiomers are mutual mirror reflections with respect to the plane perpendicular to the surface through the [1–10] axis. (Reprinted from reference [43] with permission from Wiley-VCH).

process incurring a further rotation of the dipeptide molecular axis that now lies 74° clockwise from the [1 1̄ 0] substrate direction for L-Phe-L-Phe and 74° counter clockwise for D-Phe-D-Phe. When a racemic mixture is adsorbed, dynamic chiral self-recognition processes are recorded in time-lapse STM images, with only homochiral chains being formed, while heterochiral assemblies do not lead to stable structures. DFT calculations show that individual dipeptide molecules bind to the surface through the nitrogen of the amino group, one oxygen atom of the carboxyl group and the oxygen atom of the carbonyl group, with the carboxylic and amino groups located on the same side of the main molecular axis. However, when the dipeptide is assembled into chains, the conformation of the molecules changes and the carboxylic and amino groups now point in opposite directions, which allows head-to-tail bonding. In addition, the bonding oxygen atoms change their relative positions with respect to the [1 1̄ 0] direction. The main intermolecular interaction is now a strong hydrogen bond between the carboxyl group of one molecule and the amino group of an adjacent molecule, with proton transfer occurring to give a chain of zwitterionic molecules. Thus, the chiral recognition process leading to the formation of these supramolecular homochiral chains is shown to be highly dynamic and does not just result from a static "lock-and-key" fit of the molecules.

For prochiral molecules that have become chiral upon adsorption at surfaces, it is sometimes possible to flip chirality in the adsorbed phase. Figure 6.13 shows such a molecule that can, depending on the rotation around the C–C bonds, create LR/RL, RR and LL stereoisomers at a Au(1 1 1) surface [44], that is, the chirality of the adsorbed species is now a dynamic property. This flexibility makes the molecule highly receptive to the chirality of its surrounding and plays an important part in accommodation of molecules within an enantiopure chiral array. This effect is illustrated when LL and SS enantiomers self-organize into homochiral windmill structures containing four molecules, which then undergo further assembly to create a highly organized 2D chiral array. Time-dependent STM data show that the expansion of this domain often involves chiral switching by arriving molecules in order to conform to domain chirality. As an example, Figure 6.13, shows superimposed time-lapse STM images showing a RL isomer at the edge of a RR extended domain, undergoing conformational switching to attain RR chirality in order to be accommodated within the 2D array. This chiral recognition and dynamical switching events offer interesting opportunities to steer overall system chirality into desired directions.

6.4.2
Diastereomeric Chiral Recognition

6.4.2.1 Diastereomeric Chiral Recognition by Homochiral Structures
Homochiral recognition events of the type discussed in the previous section are now becoming well documented. However, chiral recognition between dissimilar molecules is far less investigated. Here, the creation of diastereoisomeric interactions is central to the chiral discrimination. The first example of such recognition at surfaces was provided by the adenine/Cu(1 1 0) system [23] discussed in Section 6.2, where

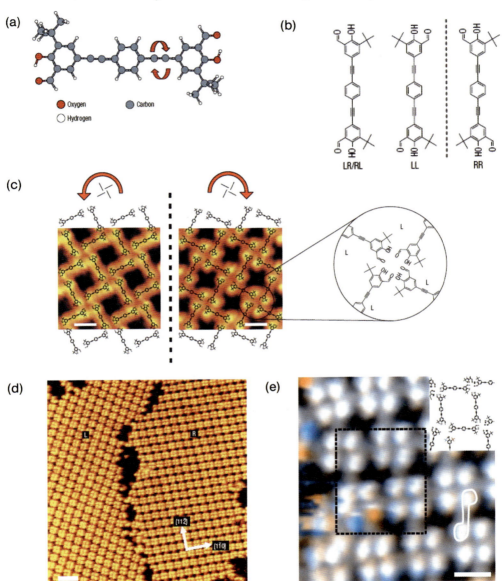

Figure 6.13 Dynamical switching of molecular chirality at surfaces: (a) Schematic of molecule; (b) Rotation around C–C bonds enabling LR/RL, RR and LL isomers to be created at the Au(1 1 1) surface; (c) Detail of STM image to show homochiral windmill structures that further assemble to give (d) large coexisting 2D homochiral arrays; (c) Overlay of two STM images separated by 19 s, showing transformation of an RL conformer into a RR conformer at a domain boundary in order to be accommodated into the domain (Adapted from reference [44] with permission from Macmillan Publishers Ltd.

S-Phenylglycine/Adenine/Cu(110)   R-Phenylglycine/Adenine/Cu(110)

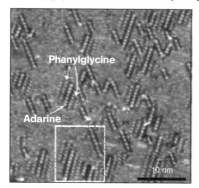

Figure 6.14 Chiral recognition by dissimilar species. (Left) Coadsorbed (S)-phenylglycine with adenine on Cu(1 1 0) showing adsorption only next to 1D adenine chains growing in the +19.5° (1,2) direction, with adenine chains of the opposite chirality remaining bare. (Right) Coadsorbed (R)-Phenylglycine with adenine on Cu(1 1 0) is found to only decorate adenine chains of the mirror chirality growing in the −19.5° (−1, 2) direction. From reference [23] with permission from Macmillan Publishers Ltd.

homochiral 1D chains of both surface enantiomers are observed. When this system is subsequently exposed to phenylglycine, the covalently bound amino acid was found to show high chiral selectivity in adsorption, with enantiopure (S)-phenylglycine coadsorbing only next to 1D adenine chains growing in the +19.5° (1,2) direction, with adenine chains of the opposite chirality remaining bare, Figure 6.14. Conversely, when (R)-phenylglycine is introduced, it is found to only decorate adenine chains of the mirror chirality growing in the −19.5° (−1, 2) direction. Interestingly, the chiral selection occurs over a 20 Å distance. Recent DFT calculations [45] suggest that this chiral recognition process is governed by surface-mediated Coulomb repulsion between the phenylglycine amino group and the DNA base rather than any direct intermolecular interactions.

6.4.2.2 Diastereomeric Chiral Recognition by Heterochiral Structures

Chiral recognition abilities are generally associated with homochiral superstructures. However, recently, the first STM observation of chiral recognition within a two-dimensional *heterochiral* racemic assembly of succinic acid (SU) molecules on Cu(1 1 0) was reported [46], which leads to highly enantiospecific substitution of individual tartaric acid (TA) guest molecules. This process represents an important stepping-stone towards "mirror-symmetry breaking" in intrinsically racemic architectures at solid surfaces, and detailed insights into its nature have been obtained by STM, LEED, RAIRS and periodic DFT [46].

When SU is adsorbed on Cu(1 1 0) at a coverage of 0.25 monolayer, it creates a heterochiral p(4 × 2) structure, in which the doubly dehydrogenated bisuccinate species adopts a chiral motif that can exist in two distinct mirror orientations, L-SU and D-SU, aligned along asymmetric directions of the substrate, Figure 6.15.

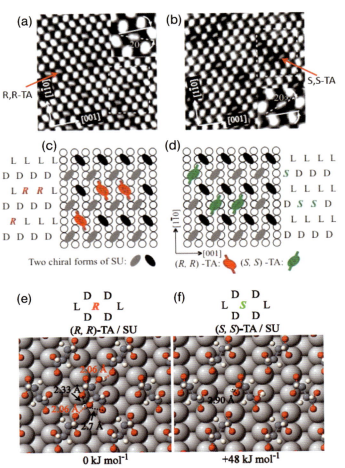

Figure 6.15 A, B 80 × 80 Å STM topographic images of (R,R) and (S,S)-TA substitution in the p (4 × 2) SU/Cu(1 1 0) structure, respectively, The angle of the long axis of TA with respect to [001] direction is denoted in the close-up images, 16 × 16 Å, for (R,R)-TA. C Schematic of the adsorption sites occupied by (R,R)-TA (R in the figure) in the area defined by the dashed rectangle in A; D Schematic of the adsorption sites occupied by (S,S)-TA (S in the figure) in the area defined by the dashed rectangle in B; (E) DFT structural model of (R,R)-TA substituted at an L site compared with (F) (S,S)-TA substituted at an L site, which is energetically unstable by 48 kJ mol^{-1}. (Reprinted from reference [46] with permission from Wiley-VCH.)

When enantiopure chiral guest molecules of (R,R)-tartaric acid are coadsorbed with SU molecules, STM images show that the doubly dehydrogenated bitartarte, (R,R)-TA, is incorporated within the structure at highly specific sites. In contrast to the oval shape imaged SU, the TA molecules are resolved as much thinner "slots," which pinpoints their exact locations from the STM image. What clearly emerges from the STM image in Figure 6.15a is that (R,R)-TA is only substituted in sites originally occupied by L-SU. When the opposite enantiomer (S,S)-TA is coadsorbed, it only substitutes at the D-SU sites, as shown in Figure 6.15b. The structures of the mixed overlayers, which are effectively quasiracemates, can be mapped directly from the STM data, and are shown in Figures 6.15c and 15D. Theoretical DFT modelling, using detailed information from the RAIRS, LEED and STM data, shows that the enantiospecificity of site substitution is essentially dictated by the architecture of the vacancy created within the heterochiral structure, which leads to significant enantiospecific differences in inter- and intramolecular hydrogen-bonding interactions and molecular backbone distortion costs, Figures 6.15e and f. This combination leads to a large enantiospecific preference of some 48 kJ mol^{-1}, with the major contributor to the enantiospecific preference being the backbone distortion penalty that arises when the guest molecule is forced into the wrong site. Significantly, this work signals that ordered heterochiral assemblies at solid surfaces can be efficiently desymmetrized via enantiospecific insertion of enantiopure guests within homochiral chains in the structure, leading to stochastic creation of a range of diastereoisomers. Such processes, leading to departures from the racemic state, have been suggested [11] as important stepping stones in the creation of biological homochirality.

6.5 Summary

A wide range of rigorous surface science and theoretical techniques have now been deployed to capture the nucleation, expression and transfer of chirality at the nanoscale level. These first mapping have enabled the first molecular and atomic level models of chiral induction at surfaces to be constructed, and this review has highlighted some of the ways in which chirality is created at surfaces by the adsorption of organic molecules. Examples of structures ranging in size from 1D chains, to small 2D clusters and, finally, to large 2D chiral arrays are presented. The phenomenon of chiral recognition, inherent in any kind of chiral assembly or separation at surfaces, is also discussed. These pioneering studies will, undoubtedly, form the foundation from which more sophisticated chiral phenomena can be studied and, more importantly, exploited to create sophisticated functions at surfaces and interfaces. There is also little doubt that the insights garnered by such surface-science studies will illuminate chiral phenomena in liquid, gel and solid media.

Glossary of Terms Used

(STM)   Scanning Tunnelling Microscopy
(LEED)   Low-Energy Electron Diffraction
(RAIRS) Reflection Absorption Infrared Spectroscopy
(DFT)   Density Functional Theory

Coverage at the surface is given in terms of fractional monolayers (ML), quoted with respect to the number density of surface metal atoms.

The adlayer unit meshes are given in standard matrix notation as follows and quoted in the text as $(G_{11}G_{12}, G_{21}G_{22})$:

$$\begin{pmatrix} a' \\ b' \end{pmatrix} = \begin{pmatrix} G_{11} & G_{12} \\ G_{21} & G_{22} \end{pmatrix} \begin{pmatrix} a \\ b \end{pmatrix}$$

where **a′**, **b′** are the overlayer net vectors and the underlying metal surface mesh is defined by **a** and **b**, using standard conventions.

References

1 Lehn, J.-M. (1995) *Supramolecular Chemistry*, VCH, Weinheim.
2 Philp, D. and Stoddart, J.F. (1996) *Angewandte Chemie-International Edition*, **35**, 1155.
3 Whitesides, G.M., Mathias, J.P. and Seto, C.T. (1991) *Science*, **254**, 1312.
4 Cornelissen, J.J.L.M., Rowan, A.E., Nolte, R.J.M. and Sommerdijk, N.A.J.M. (2001) *Chemical Reviews*, **101**, 4039.
5 Barlow, S.M. and Raval, R. (2003) *Surface Science Reports*, **50**, 201.
6 Raval, R. (2001) *Cattech*, **5**, 12.
7 Raval, R. (2002) *Journal of Physics-Condensed Matter*, **14**, 4119.
8 Raval, R. (2003) *Current Opinion in Solid State & Materials Science*, **7**, 67.
9 De Feyter, S. and Schryver De, F.C. (2003) *Chemical Society Reviews*, **32**, 139.
10 Hazen, R.M. and Sholl, D. (2003) *Nature Materials*, **2**, 367.
11 Weissbuch, I., Leiserowitz, L. and Lahav, M. (2005) *Biocatalysis – from Discovery to Application*, **259**, 123.
12 Ernst, K.-H. (2006) *Biocatalysis – from Discovery to Application*, **265**, 209.
13 Perez-Garcia, L. and Amabilino, D.B. (2007) *Chemical Society Reviews*, **36**, 941.
14 Barlow, S.M. and Raval, R. (2008) *Current Opinion in Colloid & Interface Science*, **13**, 65.
15 Barth, J.V., Weckesser, J., Cai, C.Z., Gunter, P., Burgi, L., Jeandupeux, O. and Kern, K. (2000) *Angewandte Chemie-International Edition*, **39**, 1230.
16 Weckesser, J., De Vita, A., Barth, J.V., Cai, C. and Kern, K. (2001) *Physical Review Letters*, **8709**, art. no. 096101.
17 Humblot, V., Haq, S., Muryn, C., Hofer, W.A. and Raval, R. (2002) *Journal of the American Chemical Society*, **124**, 503.
18 Hofer, W.A., Humblot, V. and Raval, R. (2004) *Surface Science*, **554**, (2–3), 141–149.
19 Jones, T.E. and Baddeley, C.J. (2002) *Surface Science*, **513**, 453.
20 Lorenzo, M.O., Baddeley, C.J., Muryn, C. and Raval, R. (2000) *Nature*, **404**, 376.
21 Lorenzo, M.O., Haq, S., Bertrams, T., Murray, P., Raval, R. and Baddeley, C.J. (1999) *Journal of Physical Chemistry B*, **103**, 10661.
22 Lorenzo, M.O., Humblot, V., Murray, P., Baddeley, C.J., Haq, S. and Raval, R. (2002) *Journal of Catalysis*, **205**, 123.
23 Chen, Q. and Richardson, N.V. (2003) *Nature Materials*, **2**, 324.
24 Blankenberg, S. and Schmidt, W.G. (2005) *Physical Review Letters*, **94**, 236102.
25 McNutt, A., Haq, S. and Raval, R. (2003) *Surface Science*, **531**, 131.
26 Böhringer, M., Morgenstern, K., Schneider, W.D. and Berndt, R. (1999) *Angewandte Chemie-International Edition*, **38**, 821.
27 Böhringer, M., Morgenstern, K., Schneider, W.D., Berndt, R., Mauri, F., De Vita, A. and Car, R. (1999) *Physical Review Letters*, **83**, 324.

28 Pasteur, L. (1848) *Annals of Physics*, **24**, 442.
29 Messina, P., Dmitriev, A., Lin, N., Spillmann, H., Abel, M., Barth, J.V. and Kern, K. (2002) *Journal of the American Chemical Society*, **124**, 14000.
30 Spillmann, H., Dmitriev, A., Lin, N., Messina, P., Barth, J.V. and Kern, K. (2003) *Journal of the American Chemical Society*, **125**, 10725.
31 Blum, M.C., Cavar, E., Pivetta, M., Patthey, F. and Schneider, W.D. (2005) *Angewandte Chemie-International Edition*, **44**, 5334.
32 In't Veld, M., Iavicoli, P., Haq, S., Amabilino, D.B. and Raval, R. (2008) *Chemical Communications*, 1536–1538.
33 Barbosa, L. and Sautet, P. (2001) *Journal of the American Chemical Society*, **123**, 6639.
34 Fasel, R., Wider, J., Quitmann, C., Ernst, K.-H. and Greber, T. (2004) *Angewandte Chemie-International Edition*, **116**, 2913.
35 Williams, J., Haq, S. and Raval, R. (1996) *Surface Science*, **368**, 303.
36 Barlow, S.M., Louafi, S., Le Roux, D., Williams, J., Muryn, C., Haq, S. and Raval, R. (2004) *Langmuir*, **20**, 7171–7176.
37 Barlow, S.M., Louafi, S., Le Roux, D., Williams, J., Muryn, C., Haq, S. and Raval, R. (2005) *Surface Science*, **590**, 243–263.
38 Fasel, R., Parschau, M. and Ernst, K.-H. (2003) *Angewandte Chemie-International Edition*, **42**, 5178.
39 Ernst, K.H., Kuster, Y., Fasel, R., Muller, M. and Ellerbeck, U. (2001) *Chirality*, **13**, 675.
40 Hernse, C.G.M., Van Bavel, A.P., Jansen, A.P.J., Barbosa, L.A.M.M., Sautet, P. and Van Santen, R.A.J. (2004) *Journal of Physical Chemistry B*, **108**, 11035–11043.
41 Dmitriev, A., Spillmann, H., Lingenfelder, M., Lin, N., Barth, J.V. and Kern, K. (2004) *Langmuir*, **20**, 4799.
42 Kühnle, A., Linderoth, T.R., Hammer, B. and Besenbacher, F. (2002) *Nature*, **415**, 891.
43 Lingenfelder, M., Tomba, G., Constantini, G., Ciachchi, L.C., De Vita, A. and Kern, K. (2007) *Angewandte Chemie International Edition*, **46**, 4492–4495.
44 Weigelt, S., Busse, C., Petersen, L., Rauls, E., Hammer, B., Gothelf, K.V., Besenbacher, F. and Linderoth, T.R. (2006) *Nature Materials*, **5**, 112.
45 Blankenberg, S. and Schmidt, W.G. (2007) *Physical Review Letters*, **99**, 196107.
46 Liu, N., Haq, S., Darling, G. and Raval, R. (2007) *Angewandte Chemie-International Edition*, **46**, 7613.

7
Expression of Chirality in Physisorbed Monolayers Observed by Scanning Tunneling Microscopy
Steven De Feyter, Patrizia Iavicoli, and Hong Xu

7.1
Introduction

Chirality abounds in natural and synthetic systems [1–3], and also it is therefore not surprising that it is manifested in self-assembled monolayers at the liquid/solid interface, although observing it directly has been virtually impossible until relatively recently. As we will show, it is actually hard not to induce chirality on surfaces by adsorp- tion of molecules, even for achiral molecules. Chirality on surfaces is both expressed at the level of the molecular organization and at the relation of this molecular organization with respect to the symmetry of the substrate underneath (when the latter is crystalline).

Why bother with a dedicated chapter on chirality of physisorbed self-assembled monolayers at the liquid/solid interface? What is different with respect to those studies carried out for chemisorbed systems? Is there a difference at all? For sure, there are many similarities: the arguments to discuss chirality in monolayers formed under UHV conditions or at the liquid/solid interface are the same [4–7]. There are some important differences though, which on the one hand relate to technical aspects, and on the other hand to some more fundamental issues.

Under both conditions, scanning tunneling microscopy (STM) [8, 9] is one of the preferred techniques to investigate the ordering and properties of these self-assembled layers. In STM, a metallic tip is brought very close to a conductive substrate and by applying a voltage between the conductive media, a tunneling current through a classically impenetrable barrier results between the two electrodes. The direction of the tunneling depends on the bias polarity. The exponential distance dependence of the tunneling current leads to excellent control of the distance between the probe and the surface and very high resolution (atomic) on atomically flat conductive substrates can be achieved. For imaging purposes, the tip and substrate are scanned precisely relative to one another and the current is accurately monitored as a function of the lateral position. The contrast in STM images reflects both topography and electronic effects. In the constant height mode, the absolute vertical position of the probe remains constant during raster scanning and the tunneling current is plotted as a function of the lateral position. In the more popular constant current mode, the

Chirality at the Nanoscale: Nanoparticles, Surfaces, Materials and more. Edited by David B. Amabilino
Copyright © 2009 WILEY-VCH Verlag GmbH & Co. KGaA, Weinheim
ISBN: 978-3-527-32013-4

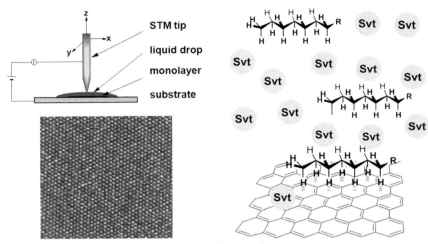

Figure 7.1 Top left: Schematic representation of STM at the liquid/solid interface. Bottom left: STM image of highly oriented pyrolytic graphite. Right: Molecules at the liquid/solid interface.

absolute vertical position of the probe (or sample) changes, and these changes are plotted as a function of the lateral position.

The way molecules are brought onto the surface is different. Under UHV conditions, molecules are "evaporated" onto the substrate with a superior control on the surface coverage. To induce adsorption of molecules and order at the liquid/solid interface (Figure 7.1), a drop of the solution with the compound of interest is deposited on top of the substrate [10–12]. A major difference between the UHV conditions and the liquid/solid interface is … indeed the solvent. The choice of solvent can be tuned as a function of the particular solute and/or substrate. Typically, the solvent has a low vapor pressure, is nonconductive (electrochemically inert), and shows a lower affinity for the substrate than the solute. On the one hand, the dynamic exchange of molecules adsorbed on the surface and in the liquid phase promotes repair of defects in the self-assembled layers. On the other hand, these solvent-mediated dynamics, in combination with the fact that the temperature window at the liquid/solid interface is much smaller than under UHV conditions, typically leads to the visualization of monolayers, rather than submonolayers or clusters that can be easily probed at low temperature under UHV conditions or in chemisorbed systems where equilibria are slow [13]. Despite the fact that the effect of solvent has not been probed in a systematically way yet, its role is anticipated to become crucial in directing the expression of chirality at the liquid/solid interface. So far, a number of studies have focused on the effect of solvent (polarity, viscosity, hydrogen bonding, etc.) on monolayer formation [14, 15] but comprehensive studies on its role in directing/tuning chirality at the liquid/solid interface are still awaited [16].

In this chapter, rather than giving an exhaustive overview of the many chirality studies at the liquid/solid interface, selected examples will highlight the different aspects of chiral self-assembly at the liquid/solid interface.

7.2
How to Recognize Chirality at the Liquid/Solid Interface

7.2.1
Chirality at the Level of the Monolayer Symmetry

Basically, chirality is expressed at two different levels in self-assembled monolayers. The first level is the molecular ordering or packing itself. The second level is the orientation of the monolayer with respect to the substrate underneath. Let us first focus on the symmetry aspects of the molecular ordering. In two dimensions, molecules can self-assemble in 17 plane groups, 5 of which are chiral: *p*1, *p*2, *p*3, *p*4, and *p*6. Matzger *et al.* made a tremendous effort to evaluate the data of all known systems that self-assemble in 2D patterns at the liquid/solid interface so far [17]. More specifically, they have compiled those data in what they call a two-dimensional structural database (2DSD). This compilation is a 2D analogue of the 3D crystal structure databases. They described the 2DSD as *"providing the unified view of interfacial self-assembly essential for investigation of 2D crystallization and comparison with bulk crystals to uncover the basic similarities underlying all forms of self-assembly and the differences due to the presence of an interface."* 876 monolayers are included in this database. For each entry in the 2DSD, a structural description that matched the STM image was developed including the plane group, the number of molecules in the asymmetric unit, and the symmetry element on which the molecules reside.

Figure 7.2 is a schematic representation of the 17 plane groups used to describe monolayer symmetry [17]. What is striking is the preference of a few plane group symmetries out of the numerous possibilities and it resembles the space-group preference apparent in the crystal structures of organic and metallo-organic compounds. 5 of the 230 space groups describe more than 75% of all crystal structures [17]. These space groups have in common that they allow the densest packing and maximize intermolecular interactions. In 2D, the plane groups *p*1, *p*2, *pg*, and *p*2*gg* are predicted to enable objects of any shape to contact the largest number of neighboring objects [17]. And indeed, these space groups are favored. Together *p*2 (58%), *p*1 (17%) and *p*2*gg* (10%) represent 85% of these structures. Therefore, it is safe to conclude that the structures are in general influenced by the tendency toward minimization of empty space. This has important implications for chirality. In three dimensions, inversion centers are the most favored symmetry elements because they generate the least amount of surrounding empty space and, furthermore, they limit the like–like interactions that interfere with close packing [18]. In 3D, this leads to a preference for centrosymmeric space groups. As a major consequence, achiral molecules crystallize generally in achiral space groups and unit cells of racemic mixtures contain both enantiomers. Crystals that contain both enantiomers in the unit cell tend to be denser [19] and more stable than those composed of the pure enantiomers [20]. In contrast, inversion centers are normally incompatible with the surface-confined monolayers because of the noncentrosymmetricity at interfaces. In two dimensions, twofold rotations provide the closest packing, while mirror planes hinder close packings. Therefore, the propensity of twofold rotations in packing

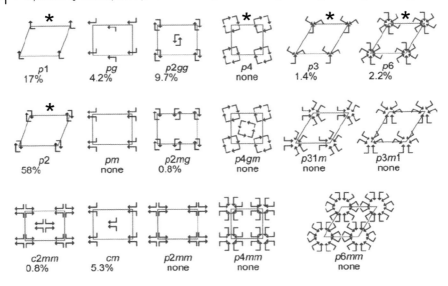

Figure 7.2 Schematic representation of the 17 plane groups used to describe monolayer symmetry for entries in the 2DSD, and their abundances. The chiral plane groups are indicated by a star. Reprinted with permission from [17].

motifs on a surface, lacking mirror planes, leads to a preference for 2D chiral crystal formation. Indeed, *p*2 and *p*1 are the most abundant space groups and they are chiral: mirror-related, nonsuperimposable domains can be formed on the surface. 79% of the structures in the 2DSD are chiral, despite the fact that most of the molecules investigated are achiral. Indeed, most molecules, chiral and achiral, and even racemic mixtures form enantiomerically pure domains (see also Section 7.6). This propensity for forming chiral 2D patterns cannot be explained by a change in the percentage of chiral symmetry groups in two and three dimensions. It is due to the fact that close packing in two dimensions is most commonly satisfied in chiral plane groups *p*1 and *p*2. An important practical consequence is that such chiral 2D layers can be used to induce formation of enantioenriched oligomers [21] and produce crystals that vary in orientation with respect to the interface by inducing face-selective nucleation [22]. So, in those cases where STM imaging is of sufficient quality, the formation and observation of chiral patterns is often obvious, just by evaluating the symmetry of the unit cell.

As we will stress later, though both achiral and chiral molecules form chiral patterns, there is a distinct difference between both. While achiral molecules form mirror-related domains at achiral surfaces, pure enantiomeric samples typically form enantiomorphous domains: the patterns of "left-handed" and "right-handed" molecules are mirror-related, but an enantiomerically pure compound, being "left-handed" or "right-handed" does not form both mirror-related domains. This is the general observation (rule) that will be discussed further in Section 7.3.

7.2.2
Chirality at the Level of the Monolayer – Substrate Orientation

Monolayer chirality though is also (often) expressed at a different level, namely at the orientation of the monolayer with respect to the substrate underneath. Let us discuss this issue for graphite, which is the most popular surface studied so far at the liquid/solid interface. Figure 7.3 is the structure of the upper carbon layer of the graphite substrate, forming a honeycomb motif. The symmetry directions, according to the Weber notation, are also indicated. Those along the equivalent <–12–10> directions (indicated by thick solid arrows) are often called the "major symmetry directions." The set of equivalent <–1100> directions (indicated by dashed arrows) are running perpendicular to the corresponding "major symmetry directions." Let us take a simple example, for instance molecules that stack in rows. Often it is observed that these rows do not match the direction of the main symmetry axes or those running normal to them (see for instance the thin solid arrow). A straightforward description of the orientation of a monolayer, or a domain there of, with respect to the graphite substrate underneath is by evaluating the orientation of a unit-cell vector with respect to a symmetry axis of the substrate, that is, the angle between both. For practical reasons, often that particular symmetry axis is selected (a major symmetry axis or a "normal") that makes the smallest angle with the unit-cell vector. Therefore, for those systems where the unit-cell vector does not run parallel with respect to one of the substrate symmetry axes, the selected unit-cell vector can be rotated clockwise (positive angle) or anticlockwise (negative angle) with respect to the reference axis. As we will illustrate later, achiral systems form *both* positive (unit-cell vector rotated

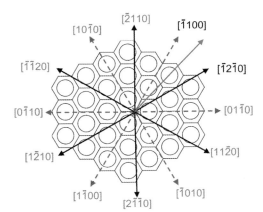

Figure 7.3 Upper honeycomb lattice of highly oriented pyrolytic graphite. The "main symmetry directions" are indicated by thick solid arrows; those running perpendicular to them are indicated by dashed arrows. The direction of the vectors is indicated using the Weber notation (in grey). The two in black highlight a "main symmetry direction" (belonging to the set of equivalent <–12–10> directions) and a "normal" (belonging to the set of equivalent <–1100> directions. The thin solid arrow illustrates the orientation of a unit-cell vector of the monolayer which typically deviates in orientation from the symmetry directions of graphite.

clockwise) and negative (unit-cell vector rotated anticlockwise) domains, while an enantiomeric pure molecule typically self-assembles in *one* of both types of domains.

Obviously, the kind of analysis stated above only makes sense if the orientation of the monolayers with respect to the substrate underneath is not random. Many of the molecules, both alkylated and nonalkylated, investigated at the liquid/solid interface often show a nonrandom orientation with respect to the graphite substrate. Molecule–substrate interactions that for organic molecules on graphite are of the van der Waals type are weak but non-negligible. Take, for example, alkylated molecules. Most of the molecules investigated at the liquid/graphite interface carry alkyl chains. This is because of the very good match between the structural parameters of the alkyl chains (the distance between next-neighbor methylene groups measures 2.51 Å) and the distance between adjacent hexagons in the graphite lattice (2.46 Å) [9, 23]. The correspondence between the zigzag alternation of methylene groups and the distance between the honeycomb centers of graphite leads to a stabilization energy of about 6.2 kJ mol^{-1} per methylene group [24]. As a result, alkyl chains run typically parallel to one of the major symmetry axes of graphite, and this effect is the more pronounced the longer the alkyl chains. Note that alkanes themselves do not form a chiral pattern on graphite. They are stacked in a row: the long axes of the alkanes run parallel to a main symmetry axis, while the row axis runs parallel to the corresponding normal. In many cases though, rows formed of substituted alkanes will not run parallel to a graphite's symmetry axis. Any substitution that induces a lateral offset of the alkyl chains will lead to a chiral pattern (see, for instance, Section 7.3).

7.2.3
Determination Absolute Configuration

With STM, it is occasionally possible to determine the absolute chirality of molecules, for instance of individual adsorbed molecules (but not always by any means). The geometric configuration (*cis* or *trans*) of several simple alkenes chemisorbed on the silicon (100) surface has been determined under UHV conditions, through the ability of the STM to identify individual methyl groups [25].

The direct determination of the absolute configuration has also been demonstrated for a larger organic molecule with a single stereogenic center, that is, (*R*)/(*S*)-2-bromohexadecanoic acid ($CH_3(CH_2)_{13}CHBrCOOH$), which forms 2D crystals at the liquid/solid interface [26]. In the high-resolution images of the monolayers formed by this compound, the relative position of the bromine atom (bright) and the carboxyl group (dark) can be discerned and as such, the orientation of the rest of the alkyl chain is also determined (Figure 7.4). The long alkyl chain of the molecule must lie down on the opposite site of the bromine atom from the carboxyl group. So, it is assumed that the "bulky" Br group is facing away from the substrate. Given that three of the four groups attached to the chiral carbon atom have been directly determined, the fourth, which is a hydrogen atom, is right beneath the bromine atom. Therefore, the atomic resolution enables to determine the absolute configuration of a single organic molecule directly. Note, however, that molecular dynamics calculations (see Section 7.6.1) indicate that the interpretation of the STM data might not be that straightforward.

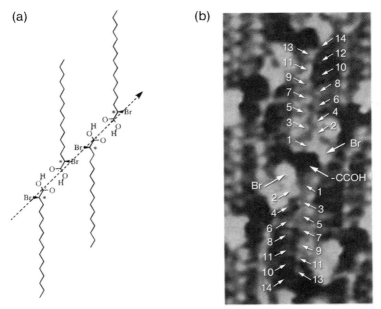

Figure 7.4 Schematic of the molecular structure of (R)-2-bromohexadecanoic acid physisorbed on graphite. Detailed assignment of individual atoms on a zoom of an STM image of a chiral pair of (R)-2-bromohexadecanoic acid molecules. The numbers 1–14 point to the positions of 14 hydrogen atoms on the 14 methylene groups. Reprinted with permission from [26].

Though the absolute chirality was determined in both cases discussed here, the systems differ significantly. In the first example, individual molecules are chemisorbed, while in the last case, physisorbed molecules are forming a 2D crystal. Chemisorption can strongly affect the nature of the adsorbed molecule, for instance by creating chiral centers as a result of a change in the hybridization state of the carbon atoms. In contrast, physisorption does not affect the molecular properties (such as valence) substantially upon adsorption, only the conformation is expected to by influenced to some extent upon physisorption of flexible molecules to a surface.

7.3
Chirality in Monolayers Composed of Enantiopure Molecules

Walba *et al.* were the first to report on the ordering of chiral molecules and the expression of 2D chirality observed by STM [27, 28]. They highlighted the direct observation of enantiomorphous monolayer crystals at the interphase between HOPG and liquid-crystalline (smectic A phase) enantiomers at room temperature. The images of monolayer crystals grown from both enantiomers exhibited well-defined rows of tilted, rod-shaped bright regions, but with the bright rods tilted in opposite directions for the enantiomers. Since then, many reports have appeared on the self-assembly of enantiopure molecules, especially at the liquid/solid interface. Examples include terephthalic acid derivatives [29] carboxylic acid derivatives [30], dihydroxyalkanes [31], 4-phenylene-vinylene oligomers [32–35], and many others. In all these systems, the enantiomers form enantiomorphous monolayers.

As representative examples, Figures 7.5A and B show STM images of physisorbed monolayer structures of, respectively, the (S)-enantiomer and (R)-enantiomer of a chiral terephthalic acid derivative, 2,5-bis[10-(2-methylbutoxy)decyloxy]-terephthalic acid (TTA) with two identical stereogenic centers [29]. The aromatic terephthalic acid groups appear as the larger bright spots. The monolayers are characterized by two different spacings between adjacent rows of (R,R)-TTA or (S,S)-TTA terephthalic acid groups. For both enantiomers, the width of the broader lamellae ($\Delta L_1 = 2.54 \pm 0.05$ nm) corresponds to the dimension of fully extended alkoxy chains, which are lying flat on the graphite surface and almost parallel to a main graphite axis. The width of the narrow lamellae ($\Delta L_2 = 1.9 \pm 0.1$ nm) indicates that the terminal 2-methylbutoxy groups are bent away from the surface, while the decyloxy groups are lying flat on the graphite surface adopting an all *anti* conformation. For this system, monolayer chirality is expressed in several ways. In regions of the monolayer where the alkoxy chains are fully extended, the unit cells for the (S,S)- and (R,R)-enantiomer are clearly chiral (plane group p2). Moreover, the STM images exhibit a clear modulation of the contrast along the lamellae. This superstructure (Moiré pattern) is attributed to the incommensurability of the monolayer with the underlying graphite lattice. The unit cells of this contrast modulation, indicated in red, are mirror images for both enantiomers, which means that each enantiomer forms its characteristic enantiomorphous monolayer structure. This enantiomorphism is also expressed by the orientation of the lamella axes with respect to the graphite lattice: the angle θ between a lamella axis (any line parallel to a row formed by terephthalic acid groups) and a graphite reference axis (i.e. one of the symmetry-equivalent <–1100> directions), which is (nearly) perpendicular to the alkoxy chains, takes a value of $-3.7° \pm 0.3°$ and $+3.7° \pm 0.3°$ for the (S,S)-enantiomer and (R,R)-enantiomer, respectively. In addition to the effect of chirality on the 2D ordering of these monolayers outlined above, monolayer images reveal elongated discontinuous features (arrows), both in narrow and wide lamellae. In the narrow lamellae, the position of those features can be assigned to the location of 2-methylbutoxy groups, which are pointing away from the graphite surface. The discontinuous fuzzy character of the observed features is due to the mobility of the nonadsorbed chain ends and the interaction with the STM tip during the scanning process. However, these streaky features are also observed in the wide lamellae, and are attributed to the interaction between the scanning tip and the protruding methyl unit on the chiral carbon atom, which allows the visualization of the location of stereogenic centers in a direct way. Further support for this hypothesis was provided by the observation that an increase of the bias voltage that results in a slight retraction of the tip only leads to the disappearance of the spots correlated with the stereogenic centers, while the spots related to the 2-methylbutoxy groups are still visible.

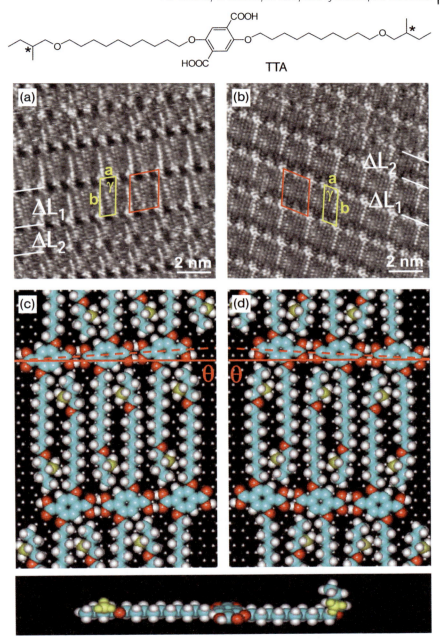

Figure 7.5 STM images at the solid(HOPG)/liquid interface and model of (A,C) S, S-TTA and (B,D) R, R-TTA. Both enantiomers form mirror-image-type patterns. Bottom: alkyl chains are not always extended: the 2-methylbutoxy group is often raised up from the HOPG surface (ΔL_2). The monolayer unit cell (for the fully extended alkyl chains (ΔL_1)) is indicated in yellow. The cell parameters a, b, and γ measure 0.96 ± 0.01 nm, 2.6 ± 0.1 nm and $73 \pm 3°$. The red unit cell refers to the epitaxy with the HOPG surface. Reprinted with permission from [29].

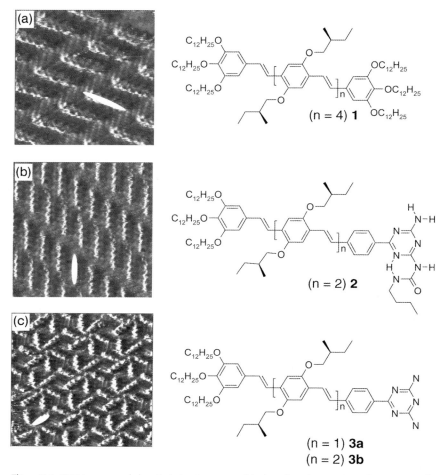

Figure 7.6 STM images and chemical structures of (A) **1** ($n=4$) at the 1,2,4-trichlorobenzene/graphite interface. Image size: $10.7 \times 10.7 \, \text{nm}^2$. (B) **2** ($n=2$) at the 1,2,4-trichlorobenzene/graphite interface. Image size: $12.1 \times 12.1 \, \text{nm}^2$ and (C) **3b** ($n=2$) at the 1-phenyloctane/graphite interface. Image size: $18.4 \times 18.4 \, \text{nm}^2$. Hydrogen bonding has a strong effect on the supramolecular architecture. The white ovals indicate a conjugated backbone. Reprinted with permission from [32–34].

Among the many enantiomerically pure systems that have been investigated, some contain multiple chiral centers, such as the oligo-p-phenylene vinylenes (Figure 7.6) [32–35]. They all carry (S)-2-methyl butoxy groups along the backbone. In the first type, oligo-*para*-phenylene vinylenes that are functionalized at both termini with three dodecyl chains (**1**, Figure 7.6A), self-assemble in highly organized 2D crystals on graphite by spontaneous self-assembly. The molecules form stacks as shown by the STM images of the monolayers in which the bright features correspond to the conjugated backbones, of which one is indicated by a white oval in the figure.

Alkyl chains are interdigitated [32]. In the second and the third type, molecules are functionalized by hydrogen-bonding groups such as ureido-s-triazine (type 2) (**2**, Figure 7.6B) [33] or 2,5-diamino-triazine groups (type 3) (**3**, Figure 7.6C) [34, 35]. The ureido-s-triazine derivatized oligo-*para*-phenylene vinylenes show *linear dimerization* via self-complementary hydrogen bonding, as expected for this supramolecular synthon. The molecules are stacked in parallel, though not equidistant rows. The fact that the conjugated backbones forming a dimer are slightly shifted is in line with the hydrogen-bonding pattern formed. In contrast, the 2,5-diamino-triazine derivatized oligo-*para*-phenylene vinylenes show *cyclic hexamer* formation. Both **1** and **2** self-assemble according to the *p*2 symmetry group, while **3** according to the *p*6 plane group. The domains formed by these compounds are again enantiomorphous too: mirror-image related patterns were never observed, except of course for their mirror-image enantiomers (not shown).

Within these series of oligo-*para*-phenylene vinylene derivatives, compound **3a** and **3b** take a special place. For both molecules, the arrangement of the molecules in individual rosettes is unidirectional but different for the two oligomers discussed. The **3a** rosettes appear exclusively to rotate clockwise (CW) while in **3b** rosettes, the molecules are exclusively arranged in a counterclockwise (CCW) fashion. No rosettes of opposite chirality were found. Molecular chirality is transferred to the rosette structures that in turn form chiral 2D crystalline patterns. Intuitively, one would expect that the rosettes formed would show the same "rotation" direction, independent of the number of stereogenic centers or conjugated oligomer length. The difference in the virtual rotation direction of the **3a** and **3b** rosettes can be explained by balanced molecule–molecule and molecule–substrate interactions. Molecules tend on the one hand to minimize the free surface area (favored by a CCW rotation) but in order to minimize steric interactions a less dense packing might be more favorable (CW rotation). The final 2D pattern is the result of a delicate balance between hydrogen bonding, van der Waals interactions between the alkyl chains and between the stereogenic centers and the substrate, and the free surface area, which is to a large extent affected by the ratio between the length of the conjugated backbone and the alkyl chains.

The champion as far as the number of stereogenic centers is concerned is compound **4** [36]. Typical STM images (Figure 7.7) show six-armed stars. The unit-cell vectors *a* and *b* are identical in length (5.56 ± 0.07 nm) and the angle between both unit-cell vectors measures $61 \pm 2°$. Though at a first glance it may be not that obvious, but the STM images are two-dimensionally chiral. Consider, for instance, the orientation of the longer dashed and shorter solid white lines that connect the terminal phenyl groups of similarly oriented OPV units along unit-cell vector *b* in Figure 7.7C. Note that these dashed and solid marker lines are not collinear. Their relative orientation can be described as dashed-up and solid-down. **4** self-assembles into a chiral pattern in accordance with the plane group *p*6. The surface pattern formed by this covalent star shows many similarities with **3b**. The similarities as far as the expression of molecular chirality on the level of the monolayer structure and symmetry are concerned indicate that the chiral substituents play the same role in both systems.

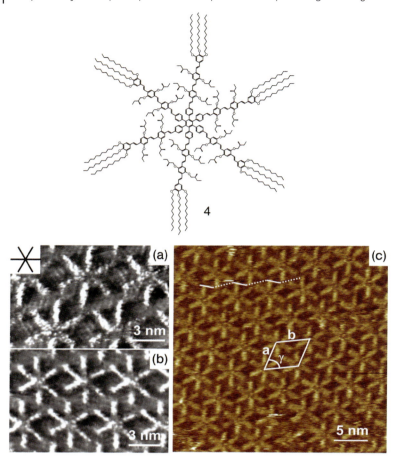

Figure 7.7 (A) and (B) Small-scale STM images of **4** at the TCB/graphite interface. The bright rods are the OPV-units. Alkyl chains can clearly be identified as the gray stripes between the OPV-rods of adjacent molecules. Inset: the orientation of the main symmetry axes of graphite. (C) Large-scale image. The unit cell is indicated. The longer dashed and shorter solid white lines connect the terminal phenyl groups of similarly oriented OPV units along unit-cell vector b. Reprinted with permission from [36].

So far, the question why do enantiopure molecules not form both mirror-image related patterns was not discussed. The chiral organization of a chiral enantiopure tetra meso-amidophenyl-substituted porphyrin containing long hydrophobic tails (**5**) at the periphery of the conjugated π-electron system has been studied with the aim of exploring the influence of stereogenic centers on the self-assembly in two dimensions [37]. STM images of the compound at the graphite/heptanol interface reveal a chiral arrangement of the molecules with the porphyrin rows tilted by 12.0° with respect to the normal to the graphite axes (Figure 7.8). The porphyrin molecules self-assemble into an interdigitated structure where the alkyl chains align along one of the main symmetry axes of graphite and the porphyrin cores are slightly shifted with

7.3 Chirality in Monolayers Composed of Enantiopure Molecules

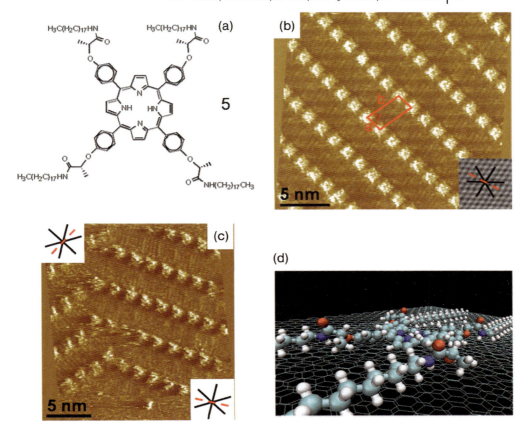

Figure 7.8 (A) Chemical structure of **5**. (B, C) STM images of **5** physisorbed at the interface between graphite and 1-heptanol. The black lines in the insets indicate the direction of the major symmetry axes of graphite. The red dashed line, running perpendicular to one of the main symmetry axes, is a reference axis selected to evaluate the orientation of the monolayer with respect to the substrate. (B) The unit cell is indicated in red. The inset shows an STM image of graphite (not to scale). (C) Image of a domain boundary. Each inset relates to the domain underneath. (D) Side view of a model of porphyrin **5** adsorbed on graphite. Note that the methyl group on the chiral center is directed towards the substrate. Reprinted with permission from [37].

respect to one another. The direction of this shift, which defines the chirality of the monolayer, is set by the chirality of the stereogenic centers. Importantly, molecular modeling shows that the methyl groups of the stereogenic center point toward the graphite surface. Note that this type of arrangement arises from the peculiar architecture of the system studied here, namely the presence of the chiral center close to the porphyrin core. Another conformation where the methyl group is pointing out of the surface has been reported in a molecular dynamics study for a chiral mesogen containing a 1-methylheptyloxy chain [38]. Experimental evidence also led to the conclusion that the methyl group is pointing away from the surface for self-assemblies of chiral terephtalic acid derivatives where the chiral center on the terminal 2-

methylbutoxy group is far away from the aromatic core [29]. In fact, it is counterintuitive to expect that a "bulky" methyl group will face the graphite substrate, unless it is merely there to fill space. The difference with the other systems where the methyl group is pointing out of the surface, is the specific location of the chiral center in **5** (next to an amide functionality) and the fact that the phenyl group close to the chiral center is tilted, favoring therefore the interaction of the methyl group with the graphite surface.

Because of the presence of a "bulky" group on the stereogenic center, at the level of the monolayer, it will matter if these "bulky" groups point up or down. Whatever the preferred orientation, the other one, obtained by a C2 orientation along an axis parallel to the substrate, would lead to a decrease in stability of the interface layer because of nonoptimized interactions with the substrate. Therefore, enantiomeric pure compounds show only one pattern and not the mirror-image one (ignoring the absolute chirality of the molecules).

7.4
Polymorphism

Polymorphism is a phenomenon that is quite normal in 3D crystals but is less explored in self-assembled monolayers. Physisorbed monolayer films of a chiral terephthalic acid (**6**) derivative have been imaged on graphite (HOPG) at the solution/substrate interface using STM [39]. The molecule comprises a nonchiral aromatic moiety and a chiral handle. It is found to form several 2D polymorphs, all corresponding to the plane group *p2* (Figure 7.9). The STM images are characterized by rather large bright spots corresponding to the location of aromatic terephthalic acid groups, which are aligned in rows and define the lamella axis. The rows of smaller spots perpendicular to the lamella axis reflect the orientation of the extended eicosyloxy groups. The 2-octyloxy groups at the other side of the terephthalic acid row appear with a different contrast, indicating a nonoptimal packing and increased dynamics on the STM timescale. They are slightly clockwise rotated with respect to the long axis of the eicosyloxy groups. A simplified model is indicated in yellow. This molecule self-assembles in polymorphous patterns (see STM image). As the eicosyloxy groups are oriented parallel to a main graphite axis the lamella structure can be characterized unequivocally in terms of its relation to the underlying graphite substrate. These data are summarized for both enantiomers in the histograms in Figure 7.9. The angle between the terephthalic acid rows and the appropriate substrate symmetry axis has been determined for all domains composed of (*S*)-**1** and the characteristic angles are approximately $-3°$ and $-12°$ and $+3°$. Similarly, for (*R*)-**1** the images were analyzed in the same way and depending on the domain, this angle takes the values $-3°$, $+3°$ or $+12°$. In agreement with symmetry considerations, the similar absolute value of the row to graphite symmetry axis angle for both positive and negative domains must be fortuitous and for the same enantiomer, domains with positive and negative values cannot be truly enantiomorphous. Such domains that form apparently mirror images are diastereomeric and are different 2D polymorphs. Polymorphism and quasi-enantiomorphism have also been reported for some liquid-crystalline compounds at their interface with graphite [28].

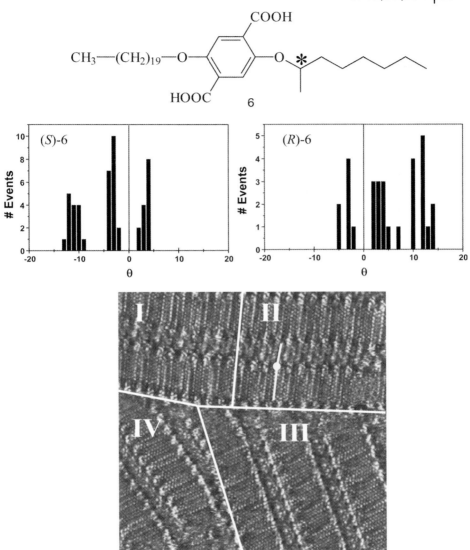

Figure 7.9 Top: Chemical structure of **6**. Center: Histograms reflecting the number of observations as a function of the value of the angle between a row of terephthalic acid groups and the substrate's appropriate symmetry axis (reference axis). Left: (S)-enantiomer. Right: (R)-enantiomer. Bottom: STM image containing several domain boundaries of (S)-**6** physisorbed at the 1-phenyloctane/graphite interface. $20 \times 20\,\text{nm}^2$. The white lines define the several domain areas. These domains are polymorphous and are characterized by different values for the angle between a row of terephthalic acid groups and the substrate's appropriate symmetry axis. The value of this angle in the domains I, II, III and IV is $+3.5°$, $-3.5°$, $+4.5°$ and $-11°$, respectively. The orientation of the 2-octyloxy groups can only be distinguished in the upper two domains. They appear to be rotated a few degrees clockwise with respect to the normal on the lamella axis. Reprinted with permission from [39].

7.5
Is Chirality Always Expressed?

In all these cases where chiral molecules are adsorbed on the surface the chiral groups feel or interact with the surface. However, what is the fate of those systems where the chiral groups do not touch the surface as a result of conformational or packing constraints? We have already indicated that in case of the terephthalic acid derivatives with one or two chiral handles, chirality is expressed even if the chiral handle itself is desorbed from the surface, which was attributed to the fact that there are no domains where all chiral handles are adsorbed. In other words, the chirality of the monolayer is directed by those molecules that have their chiral groups in contact with the substrate.

Modeling studies indicated that for compound **7** (Figure 7.10) the combination of a diacetylene and 2-methylbutoxy group can cause strain in the monolayer packing upon interdigitation of the alkyl chains. Only in 1-octanol spontaneous monolayer formation was observed. In other solvents – such as 1-phenyloctane, tetradecane, and 1-octanoic acid – no monolayer formation could be observed at all [40]. Figure 7.10 reveals a 2D arrangement of molecules in an image with submolecular resolution. The aromatic isophthalic acid groups appear as bright spots and the distance between these aromatic groups along a lamella axis is 0.96 ±0.03 nm. Two different spacings

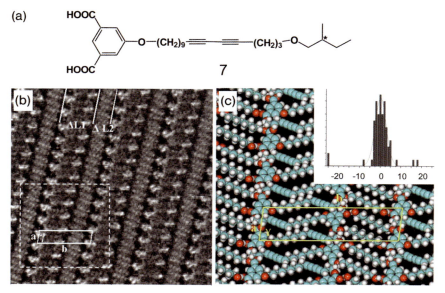

Figure 7.10 (A) Chemical structure of **7**. (B) STM image of an ordered monolayer of (S)-**7** formed by physisorption at the 1-octanol/graphite interface. The image size is 14 × 14 nm². $I_{set} = 0.56$ nA, and $V_{bias} = -0.744$ V. ΔL_1 indicates a lamella formed by (S)-**7**. ΔL_2 corresponds to the width of a lamella built up by 1-octanol molecules. (C) Tentative molecular model for the 2D packing of (S)-**7** molecules in the area shown in B. Inset: Histogram of the angle between a row of isophthalic acid groups and the substrate's relevant symmetry axis.

are found between adjacent rows of (S)-**7** head groups. The smaller spacing (ΔL2 1.09 ±0.07 nm) corresponds to the dimension of coadsorbed 1-octanol solvent molecules. For this particular molecule, coadsorption of the 1-octanol molecules is essential for monolayer formation that indicates that the typical lamella-type structure for alkylated isophthalic acid derivatives involving interdigitation of alkyl chains, is not a stable configuration for this system. No diacetylene groups could be visualized. The width of the larger (S)-**7** lamellae: (ΔL1 = 2.39 ±0.07 nm) is much smaller than the dimension of interdigitating (S)-**7** molecules with fully extended alkyl chains (3.05 nm). However, the experimentally obtained value is in good agreement with the distance between the isophthalic acid group and the oxygen atom in the alkoxy chain (2.43 nm). This is a strong indication that part of the chain, namely, the 2-methylbutoxy group is bent away from the surface, while the other part of the molecule is lying flat on the graphite surface, adopting an extended conformation.

The chiral center in the 2-methylbutoxy groups may still influence the 2D packing of the molecules, even though this part of the chain is bent away from the substrate's surface. However, the angle between a row of isophthalic acid groups and the substrate's relevant symmetry axis is symmetrically distributed around $\theta = 0$. The chiral nature of the molecules is not expressed by the ordering of the molecules on the surface, which is most probably due to the fact that all the 2-methylbutoxy groups are bent away from the surface, pointing up into the liquid phase.

7.6
Racemic Mixtures: Spontaneous Resolution?

An interesting case of surface stereochemistry is that of the fate of a solution of a racemate on a substrate. Based upon symmetry considerations, it should be easier to induce racemic conglomerate – each domain contains only one of the two enantiomers – formation rationally at surfaces than in three-dimensional (3D) systems [41]. Most studies reported so far for self-assembled monolayers on graphite – the case that is pertinent here – demonstrate racemic conglomerate formation, while in 3D crystals, for instance, racemic compound – in which the two enantiomers are present in equal proportions in the unit cell – formation is dominant. The reasons for the preferred formation of racemic conglomerates in 2D were already discussed in Section 7.2.1. We will limit ourselves here to present some systems, always at the liquid/solid interface.

7.6.1
Chiral Molecules

As stated before, most of the reports of molecules self-assembled by physisorption at the liquid/solid interface indicate racemic conglomerate formation [2], including a study by Flynn et al. on the self-assembly of (R)/(S)-2-bromohexadecanoic acid ($CH_3(CH_2)_{13}CHBrCOOH$), which forms 2D crystals at the liquid/graphite interface and allowed the determination of the absolute configuration of the molecules [26]. Recent molecular-dynamics simulation studies carried out in "UHV" conditions, neglecting solvent effects and image-charge interactions suggest, however, that racemic (R)/(S)-2-bromohexadecanoic acid on graphite is likely to form enantiomixed domains though, and highlight the subtleties that can lead to the formation of enantiomixed rather than enantiopure domains [42]. In particular, the calculations indicate that the difference in height between bromine atoms in a hydrogen-bonded racemic or heterochiral dimer between (R)-2-bromohexadecanoic acid (bromine atom facing down) and (S)-2-bromohexadecanoic acid (bromine atom facing up) is quite small (only 0.3 Å), and is less than intuitively anticipated.

Only in a few cases has pseudoracemate formation been reported. In the STM images of **PB**, physisorbed at the 1-phenoloctane or 1-heptanol/graphite interface (Figure 7.11), the phenyl benzoate moieties show up as the bright structures corresponding to high tunneling current through the aromatic groups [43, 44]. The alkyl chains are located in between the rows of the phenyl benzoate groups, and sometimes their orientation can be discerned. The location and orientation of a number of phenyl benzoate groups are indicated by sticks. The molecules form chains that in turn form dimers. The X-ray data of their 3D crystals indicate hydrogen-bonding interactions along the chains and between them through the formamide groups. The molecular chirality of the pure enantiomers of **PB** is expressed at the 1-heptanol/graphite interface at two different levels: (1) at the level of the monolayer structure as expressed by the orientation of the phenyl benzoate moieties (α) with respect to the lamella normal and (2) at the level of the orientation of the adlayer with respect to the underlying graphite lattice as expressed by the direction of the lamella axis of the monolayer with respect to the symmetry axes of graphite (θ) (Figures 7.11B and C). The histograms in Figure 7.11D represent the number of domains for which the lamella axis is rotated a given angle θ with respect to the graphite's reference axis and they reflect this second level of expression of chirality for (R)-**PB**, (S)-**PB** and (rac)-**PB**, respectively. Despite the considerable spread, in none of the three cases is the orientation of the lamella axis with respect to the substrate's symmetry completely random. (R)-**PB** has a strong tendency to form domains with $\theta > 0$ (Figure 7.11Da) while (S)-**PB** forms preferentially domains with $\theta < 0$ (Figure 7.11Db). In addition, both enantiomers form a substantial fraction of domains for which the angle θ is close to zero (Figures 7.11Da and b). In contrast to the enantiopure forms, the racemate forms exclusively domains with the angle θ_1 close to zero (Figure 7.11Dc). So, the patterns formed by the racemate are not a mere reflection of the adsorbate layers formed by the pure enantiomers. Therefore no racemic conglomerate formation takes place. The exclusive formation of $\theta = 0$ domains for the racemate suggests that there is a preferred interaction between both enantiomers, as opposed to the formation of homochiral domains.

Another example of nonconglomerate formation is the self-assembly of a racemic mixture of (9R,10R)-9,10-diiodooctadecan-1-ol and (9S,10S)-9,10-diiodooctadecan-1-ol [45]. These are relatively flat molecules and when they approach the surface, they could expose two different faces. In these molecules, the two iodine atoms are on the same side of the backbone chain. As a consequence, upon adsorption, the two iodine atoms

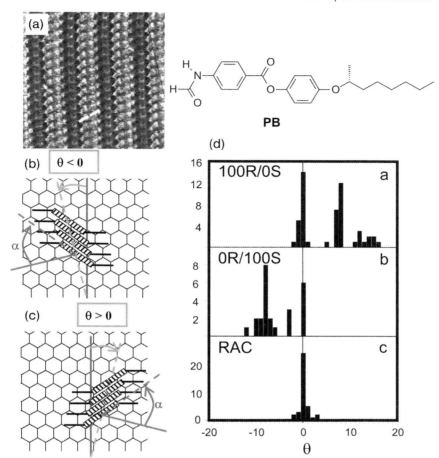

Figure 7.11 (A) STM image of (R)-**PB** at the 1-heptanol/graphite interface. Image size: 11.0 × 11.0 nm². Two phenyl benzoate groups are indicated by a white bar. (B, C) Schematic representations of the orientation of "dimers" of **PB** on graphite. The vertical line indicates the reference axis of graphite. The dashed line represents a lamella axis. θ: angle between the reference axis and the lamella axis. α: angle between the normal on the lamella axis and the (R)-**PB** dimer. (D) Histograms of the angle θ observed for physisorbed monolayers of (a) (R)-**PB** (b) (S)-**PB** and (c) (rac)-**PB**. Reprinted with permission from [44].

are facing up, above the alkyl chain, or they are facing down, buried underneath the alkane chain (Figure 7.12). Due to the presence of the chiral centers, for a given enantiomer, the interaction of one of the faces with the substrate will be favored. The molecules assemble in rows and each row is composed of only one type of optical isomer. Molecules in adjacent rows are the opposite optical isomers. Despite the fact that (R)- and (S)-rows appear in an alternating fashion, adjacent rows don't form mirror images because for one enantiomer, both iodine atoms are facing up while in the adjacent row for the other optical enantiomer, both iodine atoms are facing down. The entire domain, although composed of both the S and R enantiomer, is chiral.

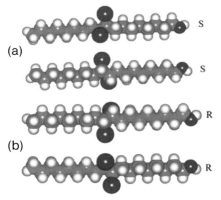

Figure 7.12 Models of (a) (9S,10S)-9,10-diiodooctadecan-1-ol and (b) (9R,10R)-9,19-diiodooctadecan-1-ol, showing different faces of the two molecules. Reprinted with permission from [45].

In contrast to studies in UHV [46], only few examples of nonracemic mixtures have been investigated and no firm conclusions concerning the possibility of "majority rules" effects could be drawn [44].

7.6.2
Achiral Molecules

Certain molecules that are achiral in the gas phase or solution phase can be described as prochiral – one desymmetrizing step away from chirality – if they become asymmetric when they are constrained to a surface because of the nonequivalence of the two faces. Others remain achiral upon deposition on the surfaces but, nevertheless, form chiral structures. Why do achiral molecules form chiral 2D structures, at least locally, even if they are nonprochiral? The event of forming a chiral 2D pattern is not a consequence of (just) molecular asymmetry. It is the result of intermolecular and molecule–substrate interactions [47]. Asymmetry is easily introduced in the 2D packing motif.

Consider for instance 4′-alkyl-4-cyanobiphenyl derivatives [48]. Though these molecules are achiral in solution, upon adsorption on an atomically flat substrate, the alkyl chains make an angle with the biphenyl axis, breaking the symmetry of the molecule. As a result, 2D chiral patterns are formed (Figure 7.13).

Also, molecules with threefold symmetry often undergo symmetry breaking upon adsorption [49]. For instance, alkoxylated triphenylene derivatives (**Tn**) with short alkoxy chains ($n = 5$) (**T5**) assemble in nonchiral hexagonal lattices (symmetry $p6m$) (Figure 7.14) [50, 51]. Upon increasing the alkyl chain length ($n = 11$) (**T12**), the substrate–alkyl chain and alkyl chain–alkyl chain interactions start to dominate the packing and the arrangement of the alkyl chains is locally chiral (symmetry $p6$). Further increasing the alkyl chain length (**T16**) often leads to molecules organizing in (chiral) dimer rows, again the result of increasing alkyl chain – alkyl chain and alkyl

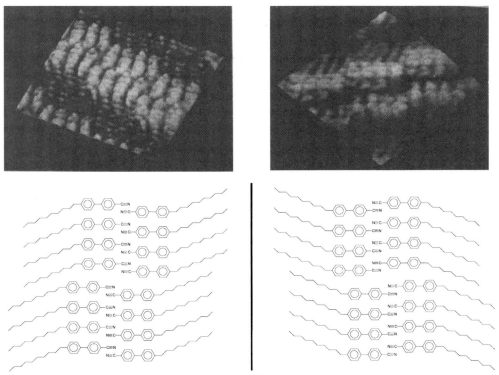

Figure 7.13 Top: STM images of a 4'-alkyl-4-cyanobiphenyl derivative (10CB) adsorbed on HOPG. Tentative model of the ordering of the two-possible, mirror-image related, orientations of 4'-alkyl-4-cyanobiphenyl compound on a surface. Reprinted with permission from [48].

chain–substrate interactions (symmetry $p2$) [52]. In addition, the substrate plays an important role in mediating the molecule–substrate interactions. While for **T11** a monolayer of $p6$ symmetry is obtained, upon adsorption on graphite, a $p2$ lattice is formed on Au (111) [53].

Similar to chiral molecules, most achiral systems that become 2D chiral upon adsorption on a surface show conglomerate formation: packing of one (2D) stereoisomer with copies of itself is thermodynamically more favorable (or faster kinetically) than packing with other molecules [54, 55].

An elegant system has been presented where the stereochemical morphology of monolayers formed from alkylated prochiral molecules on HOPG switches from a 2D racemate to a 2D conglomerate by the addition of a single methylene unit to each side chain [56]. The two anthracene derivatives (1,5-bis-(3'-thiaalkyl) anthracenes) contain linear alkyl chains at the anthracene 1 and 5 positions, the alkyl chains of the longer analogue contain one additional CH_2 unit. (Figure 7.15) A striking difference is observed between both compounds. For the smaller compound ($n = 11$), the orientation of the anthracene moieties alternates from row to row, leading to a

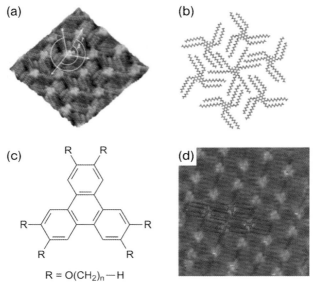

Figure 7.14 (A) STM image of **T12** on HOPG and (B) corresponding molecular model. (C) Chemical structure of alkoxylated triphenylene **Tn**. (D) STM image of **T16** on graphite. Reprinted with permission from [52].

racemic monolayer with *pg* plane group symmetry, while for the longer analogue ($n = 12$), the orientation of the anthracene units remains constant within a given domain, reflecting the formation of a 2D conglomerate monolayer with *p2* plane group symmetry. Isolated mirror image 2D enantiomers have identical energy, while

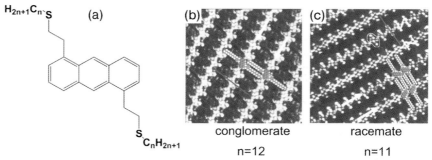

Figure 7.15 (A) Structures of the anthracene derivatives. (B) STM-image (11×11 nm^2) of the longer alkyl chain compound ($n = 12$) adsorbed on HOPG. From row to row, the anthracene units are parallel. A 2D conglomerate is formed. (C) STM-image (12×12 nm^2) of the shorter alkyl chain compound ($n = 11$) adsorbed onto HOPG. From row to row, the anthracene units (indicated by arrows) are twisted. A 2D racemate is formed. Reprinted with permission from [56].

pairs of interacting 2D isomers are diastereomeric and have distinct energies. For both compounds, the lowest-energy row contains molecules of identical 2D chirality. However, the difference between both compounds is a result of the relative orientation of adjacent rows. What determines this difference? The monolayer structure is determined both by molecule–molecule and molecule–substrate interactions. Alkyl chains tend to align along one of the main symmetry axes of graphite and methylene groups of adjacent chains align in registry so as to maximize intermolecular van der Waals interactions. The different stereomorphologies arise from the different relative orientation of the CH_2-CH_3 and the C-aryl-C1' bonds within the same side chain (see Figure 7.15). For the shorter compound, both bonds are parallel, while for the longer compound, these bonds make an angle of about 110° within the same chain. Optimizing the interchain van der Waals interactions, in combination with the all-*trans* conformation of the alkyl chains, leads to the difference between interrow orientations of the anthracenes.

Interestingly, by using crystal-engineering principles, the controlled formation of 2D conglomerates and racemates was extended, by using alkylated anthracene derivatives with one or two oxygen atoms in each alkyl chain. The placement and orientation of the ether dipoles in the alkyl chains determine the monolayer morphology and overrule the odd-even effect [57].

7.7
Multicomponent Structures

A number of studies deal with the self-assembly of multicomponent structures, where achiral and chiral molecules are mixed. In a first example the components are very similar. A 1:1 mixture of hexadecanoic acid with racemic 2-bromohexadecanoic acid physisorbed onto a graphite surface. These compounds coadsorb onto the surface, and the mixed monolayer film resolves into domains of hexadecanoic acid with either (*R*)- or (*S*)-2-bromohexadecanoic acid [58]. As observed for other fatty-acid molecules with an even number of carbon atoms, achiral hexadecanoic acid forms domains exhibiting nonsuperimposable mirror-image domains – enantiomorphous domains. The molecular axis is oriented perpendicular to the lamella axis. As demonstrated by the random appearance of bright spots in the images, which are assigned to bromine atoms, the 2-bromohexadecanoic acid molecules sporadically appear in the monolayer. The relative orientation of the bromine atoms to the carboxylic groups remains consistent throughout the STM images of a single domain that is mainly composed of achiral hexadecanoic acid molecules. In a given domain, all of the bromine atoms lie either above and to the left of the carboxyl group or below and to the right of the carboxyl group. This means that racemic 2-bromohexadecanoic acid coadsorbed with hexadecanoic acid segregates into separate chiral domains (Figure 7.16). The formation of enantiomorphous domains by hexadecanoic acid and the segregation of 2-bromohexadecanoic acid in hexadecanoic acid domains was explained in the following way. As in hexadecanoic acid the two-dimensional self-assembly is controlled by hydrogen bonding, the hexadecanoic acid molecules appear as homochiral

(a) (b)

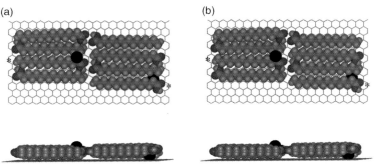

Figure 7.16 (A) Top and side view of a possible model for a domain consisting of both (R)- and (S)-2-bromomhexadecanoic acid mixed with hexadecanoic acid molecules physisorbed on graphite. In A (B), the conformer on the left-hand side indicated with a star is an (R)-enantiomer ((S)-enantiomer)), the one on the right-hand side is an (S)-enantiomer ((R)-enantiomer). In A (B), the (R)-enantiomer ((S)-enantiomer) has a bromine protruding up and away from the graphite substrate, while the (S)-enantiomer ((R)-enantiomer) has the bromine pushing into the surface. This last configuration is energetically unfavorable and therefore is rejected as a model for the mixed monolayer. Reprinted with permission from [58].

pairs on the substrate. As a consequence, the domains composed of these chiral pairs are also chiral. Assuming that the most stable interaction of a 2-bromohexadecanoic has the bromine groups directed to the liquid phase instead of the substrate, (R)-2-bromohexadecanoic acid fits only in an "A" domain (Figure 7.16A) and (S)-2-bromohexadecanoic acid coadsorbs in an "B" domain (Figure 7.16B).

To investigate the presence of an odd/even effect, similar experiments have been carried out but now in the presence of heptadecanoic acid [59]. Determination of the absolute configuration of the racemic 2-bromohexadecanoic acid molecules coadsorbed in the heptadecanoic acid matrix suggests that each lamella is composed of alternating coadsorbed (R)- and (S)-2-bromo-hexadecanoic acid molecules. These experiments indicate a clear odd-even effect: even-numbered acids induce domains of high enantiomeric purity, while odd-numbered acids cause coadsorption of both enantiomers within the domains.

Few other examples of multicomponent structures have been reported. Figures 7.17A and B represent STM images of a monolayer of the (S)-enantiomer (S)-ISA and the (R)-enantiomer (R)-ISA, respectively, of a chiral isophthalic acid derivative (5-[10-(2-methylbutoxy)decyloxy]isophthalic acid) (ISA) physisorbed from 1-heptanol [60]. As far as the packing of the isophthalic acid molecules is concerned, each enantiomer forms its characteristic enantiomorphous lamella structure. The orientation of both enantiomers with respect to the graphite surface is shown in Figures 7.17C and D. The orientation of the coadsorbed 1-heptanol molecules depends on the enantiomeric character of the domain in which those molecules are coadsorbed. For (S)-ISA, the alcohol molecules are rotated clockwise with respect to the normal on the isophthalic acid lamellae while for (R)-ISA monolayers, this rotation is counterclockwise. As such, the orientation of achiral coadsorbed molecules is influenced by the chiral character of the domains. Coadsorption of aliphatic alcohols (1-heptanol, 1-octanol, 1-undecanol) is a typical phenomenon in the

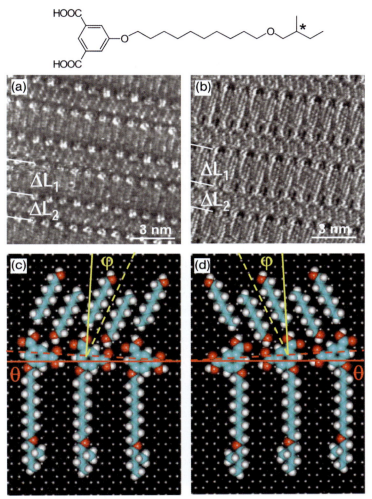

Figure 7.17 Monolayer image and model of (S)-ISA (A and C) and (R)-ISA (B and D). ΔL_1 and ΔL_2 indicate lamellae composed of ISA and 1-octanol molecules, respectively. θ is the angle between the lamella axis and the graphite reference axis. φ is the angle between the 1-octanol axis and the normal on the lamella axis of ISA. Reprinted with permission from [60].

presence of 5-alkyl(oxy)-isophthalic acid derivatives [61]. The investigation of the racemic solution indicates conglomerate formation.

Self-assembled monolayers of stearic acid, (R,S)- and (S)-16-methyloctadecanoic acid complexed with 4,4'-bipyridine in a 2 : 1 ratio have been observed at the liquid/graphite interface too [62]. Of special interest is the fact that the pure enantiomers self-assemble into a pattern that can hardly be called chiral. The alkyl chains run perpendicular to the row director. It is even suggested that the chiral unit itself is desorbed from the substrate. However, upon adding bipyridine to the enantiomer, a supramolecular complex is formed with a clear expression of enanotiomorphous 2D chirality. Adding bipyridine to (R,S)-16-methyloctadecanoic acid creates an identical pattern too, exhibiting the mirror-image-related organizations, indicating spontaneous resolution (Figure 7.18).

7.8
Physical Fields

Patrick *et al.* introduced a new method that allows the preparation of macroscopically homochiral films with twisted but achiral inputs [63]. The chiral symmetry of the films is systematically broken by the use of a liquid crystal (LC) solvent and a magnetic field to induce a preferred inplane orientation of adsorbed molecules on an achiral single-crystal substrate. The molecular packing can be selected and controlled so that left-handed, right-handed, or a mixture of left- and right-handed molecular domains can be prepared (Figure 7.19). Crystallization of an 4-cyano-4′-octylbiphenyl (8CB) monolayer on graphite leads to racemic surfaces on a macroscopic scale. Figures 7.19b and c show two possible packing arrangements. By applying a magnetic field (1.2 T) parallel to the substrate as a droplet of 8CB was placed on the surface, the orientational order in the bulk supernatant LC becomes imprinted on the monolayer, producing a molecular film with macroscopic uniform in plane alignment. By rotating the magnetic field direction one packing can be more favored while another one is less favored. Thus, the symmetry between two molecular packings was broken, producing a film with a net excess of one enantiomorph over the other.

7.9
Outlook

Chirality at the liquid/solid interface is without any doubt a fashionable topic. It is fascinating to discover how molecules self-assemble on surfaces and how the outcome of the self-assembly is directed by the presence or absence of chiral centers. At the liquid/solid interface, a lot of progress has been made in evaluating the relation between molecular structure and the 2D chirality of the formed patterns. However, is that all there is to be done? Definitely not! What is needed are more and targeted systematic studies to address specific aspects such as the role of the location of a stereogenic center along an alkyl chain, the effect of the size of the stereogenic center, the effect of the number of stereogenic centers... to name just a few.

The most popular substrate has been graphite – because of the ease of preparation of the surface – and as a consequence the interaction between the molecules and the substrate can be regarded to be physisorption in nature. The popularity of graphite has, however, also limited the extent to which the effect of substrate–molecule interactions has been probed. There are only a few systematic studies of the effect of substrate and clearly much more effort is needed.

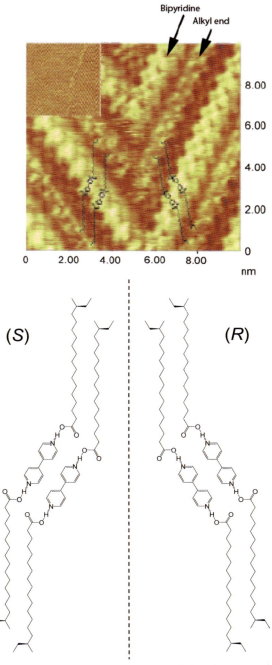

Figure 7.18 (A) STM image of (R,S)-16-methyloctadecanoic acid in the presence of 4,4′-bipyridine. (B) Structural drawing of the mirror-image-related domains. Reprinted with permission from [62].

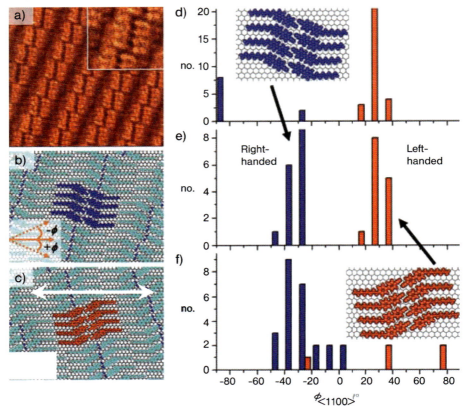

Figure 7.19 Monolayers of 8CB on graphite: (a) STM image of a right-handed domain (20 × 20 nm²). (b) Model illustration of right-handed and (c) left-handed forms, which represent two of the six possible arrangements of molecules in the monolayers; note that the two are mirror images. (d)–(f) Histograms of molecular headgroup orientations determined by STM for samples prepared with three representative field alignments with respect to the arrow indicated in image (c): (d) field at −19°; (e) field at 0°; and (f) field at +19°. Reprinted with permission from [63].

The great unknown so far is the role played by the solvent. It is well known that the solvent can have a major effect on the outcome of the self-assembly process. In some cases, the solvent is coadsorbed, that is, it is part of a complex 2D multicomponent crystal, while for other systems, the 2D patterns formed depend on the nature of the solvent though the solvent itself apparently is not incorporated in the 2D self-assembled pattern. For chiral systems, the role of solvent has not been investigated so far. Also, what about the role chiral solvents might play in directing the self-assembly of achiral or chiral solutes at the liquid/solid interface?

One of the key topics will remain the investigation of how spontaneous resolution can be achieved, controlled or templated at the liquid/solid interface, a major issue for the pharmaceutical industry. Amplification of chirality effects, such as the "majority-rules" effect and the "sergeants-and-soldiers" principle [64], at the liquid/solid interface also need to be explored. What is striking so far is the limited number of theoretical studies addressing topics such as the interaction of chiral molecules on surfaces and chiral separation in two-dimensional systems [38, 42, 47]. This should not come as a surprise as such calculations require a lot of effort and time. However, the ever increasing computer power will allow computational probing of the self-assembly of at least relative simple systems at the liquid/solid interface. For sure, there is still a lot to discover and to learn ahead of us.

Acknowledgment

We thank all our collaborators who have contributed to our own research in this area. This work was supported by the European Union Marie Curie Research Training Network CHEXTAN (MRTN-CT-2004-512161), by the Interuniversity Attraction Pole program of the Belgian Federal Science Policy Office (PAI 6/27), by the Fund of Scientific Research – Flanders (FWO), and the Dirección General de Investigación, Ciencia y Tecnología (MEC, Spain), under the project CTQ2006-06333/BQU, and the DGR, Catalonia (Project 2005 SGR-00591).

References

1 Wagnière, G.H. (2007) *On Chirality and the Universal Asymmetry*, Wiley-VCH, Weinheim.
2 Perez-Garcia, L. and Amabilino, D.B. (2002) *Chemical Society Reviews*, **31**, 342–356.
3 Perez-Garcia, L. and Amabilino, D.B. (2007) *Chemical Society Reviews*, **36**, 941–967.
4 Barlow, S.M. and Raval, R. (2003) *Surface Science Reports*, **50**, 201–341.
5 Ernst, K.H. (2006) *Topics in Current Chemistry*, **269**, 209–252.
6 De Feyter, S. and De Schryver, F.C. (2003) *Chemical Society Reviews*, **32**, 139–150.
7 De Feyter, S. and De Schryver, F.C. (2006) *Scanning Probe Microscopies beyond Imaging: Manipulation of Molecules and Nanostructures*, Wiley-VCH, Weinheim.
8 Binnig, G., Rohrer, H., Gerber, C. and Weibel, E. (1982) *Physical Review Letters*, **49**, 57–61.
9 Cyr, D.M., Venkataraman, B. and Flynn, G.W. (1996) *Chemistry of Materials*, **8**, 1600–1615.
10 McGonigal, G.C., Bernhardt, R.H. and Thomson, D.J. (1990) *Applied Physics Letters*, **57**, 28–30.
11 Rabe, J. and Buchholz, S. (1991) *Science*, **253**, 424–427.
12 De Feyter, S. and De Schryver, F.C. (2005) *Journal of Physical Chemistry B*, **109**, 4290–4302.
13 Böhringer, M., Morgenstern, K., Schneider, W.D. and Berndt, R. (1999) *Angewandte Chemie-International Edition*, **38**, 821–823.
14 Kampschulte, L., Lackinger, M., Maier, A.K., Kishore, R.S.K., Griessl, S., Schmittel, M. and Heckl, W.M. (2006) *Journal of Physical Chemistry B*, **110**, 10829–10836.
15 Mamdouh, W., Uji-i, H., Ladislaw, J.S., Dulcey, A.E., Percec, V., De Schryver, F.C. and De Feyter, S. (2006) *Journal of the American Chemical Society*, **128**, 317–325.
16 Li, C.J., Zeng, Q.D., Wang, C., Wan, L.J., Xu, S.L., Wang, C.R. and Bai, C.L. (2003) *Journal of Physical Chemistry B*, **107**, 747–750.

17 Plass, K.E., Grzesiak, A.L. and Matzger, A.J. (2007) *Accounts of Chemical Research*, **40**, 287–293.
18 Brock, C.P. and Dunitz, J.D. (1994) *Chemistry of Materials*, **6**, 1118–1127.
19 Brock, C.P., Schweizer, W.B. and Dunitz, J.D. (1991) *Journal of the American Chemical Society*, **113**, 9811–9820.
20 Jacques, J., Collet, A. and Wilen, S.H. (1981) *Enantiomers, Racemates, and Resolutions*, John Wiley & Sons, New York, pp. 23–31.
21 Zepik, H., Shavit, E., Tang, M., Jense, T.R., Kjaer, K., Bobach, G., Leiserowitz, L., Weissbuch, I. and Laval, M. (2002) *Science*, **295**, 1266–1269.
22 Weissbuch, I., Berfeld, M., Bouwman, W., Kjaer, K., Als-Nielsen, J., Lahav, M. and Leiserowith, L. (1997) *Journal of the American Chemical Society*, **119**, 933–942.
23 Groszek, A.J. (1970) *Proceedings of the Royal Society of London. Series A, Mathematical and Physical Sciences*, **314**, 473.
24 Gellman, A.J. and Paserba, K.R. (2002) *Journal of Physical Chemistry B*, **106**, 13231–13241.
25 Lopinski, G.P., Moffatt, D.J., Wayner, D.D.M. and Wolkow, R.A. (1998) *Nature*, **392**, 909–911.
26 Fang, H., Giancarlo, L. and Flynn, G.W. (1998) *Journal of Physical Chemistry B*, **102**, 7311–7315.
27 Stevens, F., Dyer, D.J. and Walba, D.M. (1996) *Angewandte Chemie-International Edition*, **35**, 900–901.
28 Walba, D.M., Stevens, F., Clark, N.A. and Parks, D.C. (1996) *Accounts of Chemical Research*, **29**, 591–597.
29 De Feyter, S., Gesquière, A., Meiners, C., Sieffert, M., Müllen, K. and De Schryver, F.C. (1999) *Langmuir*, **15**, 2817–2822.
30 Yablon, D.G., Guo, J.S., Knapp, D., Fang, H.B. and Flynn, G.W. (2001) *Journal of Physical Chemistry B*, **105**, 4313–4316, and references therein.
31 Qian, P., Nanjo, H., Yokoyama, T. and Suzuki, T.M. (1998) *Chemistry Letters*, **11**, 1133–1134.
32 Gesquière, A., Jonkheijm, P., Schenning, A.P.H.J., Mena-Osteritz, E., Bäuerle, P., De Feyter, S., De Schryver, F.C. and Meijer, E.W. (2003) *Journal of Materials Chemistry*, **13**, 2164–2167.
33 Gesquière, A., Jonkheijm, P., Hoeben, F.J.M., Schenning, A.P.H.J., De Feyter, S., De Schryver, F.C. and Meijer, E.W. (2004) *Nano Letters*, **4**, 1175–1179.
34 Jonkheijm, P., Miura, A., Zdanowska, M., Hoeben, F.J.M., De, Feyter, S., Schenning, A.P.H.J., De, Schryver, F.C. and Meijer, E.W. (2004) *Angewandte Chemie-International Edition*, **43**, 74–78.
35 Miura, A., Jonkheijm, P., De Feyter, S., Schenning, A.P.H.J., Meijer, E.W. and De Schryver, F.C. (2005) *Small*, **1**, 131–137.
36 Tomović, Ž., van Dongen, J., George, S.J., Xu, H., Pisula, W., Leclère, P., De Feyter, S., Meijer, E.W. and Schenning, A.P.H.J. (2007) *Journal of the American Chemical Society*, **129**, 16190–16196.
37 Linares, M., Iavicoli, P., Psychogyiopoulou, K., Beljonne, D., De Feyter, S., Lazzaroni, R. and Amabilino, D.B. (2008) *Langmuir*, **24**, 9566–9574.
38 Yoneya, M. and Yokoyama, H. (2001) *Journal of Chemical Physics*, **14**, 9532–9538.
39 De Feyter, S., Gesquière, A., De Schryver, F.C., Meiners, C., Sieffert, M. and Müllen, K. (2000) *Langmuir*, **16**, 9887–9894.
40 Zhang, J., Gesquière, A., Sieffert, M., Klapper, M., Müllen, K., De Schryver, F.C. and De Feyter, S. (2005) *Nano Letters*, **5**, 1395–1398.
41 Kuzmenko, I., Weissbuch, I., Gurovich, E., Leiserowitz, L. and Lahav, M. (1998) *Chirality*, **10**, 415–424.
42 Ilan, B., Berne, B.J. and Flynn, G.W. (2007) *Journal of Physical Chemistry C*, **111**, 18243–18250.
43 De Feyter, S., Gesquière, A., Wurst, K., Amabilino, D.B., Veciana, J. and De Schryver, F.C. (2001) *Angewandte Chemie-International Edition*, **40**, 3217–3220.
44 Mamdouh, W., Uji-i, H., Gesquière, A., De Feyter, S., Amabilino, D.B., Abdel-Mottaleb, M.M.S., Veciana, J. and De

Schryver, F.C. (2004) *Langmuir*, **20**, 9628–9635.

45 Cai, Y. and Bernasek, S.L. (2003) *Journal of the American Chemical Society*, **125**, 1655–1659.

46 Ernst, K.-H. (2008) *Current Opinion in Colloid and Interface Science*, **13**, 54–59.

47 Paci, I., Szleifer, I. and Ratner, M.A. (2007) *Journal of the American Chemical Society*, **129**, 3545–3555.

48 Smith, D.P.E. (1991) *Journal of Vacuum Science & Technology B*, **9**, 1119–1125.

49 Li, C., Zeng, Q., Wu, P., Xu, S., Wang, C., Qiao, Y., Wan, L. and Bai, C. (2002) *Journal of Physical Chemistry B*, **106**, 13262–13267.

50 Charra, F. and Cousty, J. (1998) *Physical Review Letters*, **80**, 1682–1685.

51 Perronet, K. and Charra, F. (2004) *Surface Science*, **551**, 213–218.

52 Wu, P., Zeng, Q., Xu, S., Wang, C., Yin, S. and Bai, C. (2001) *ChemPhysChem*, **2**, 750–754.

53 Katsonis, N., Marchenko, A. and Fichou, D. (2003) *Journal of the American Chemical Society*, **125**, 13682–13683.

54 Tao, F. and Bernasek, S.L. (2005) *Journal of Physical Chemistry B*, **109**, 6233–6238.

55 Merz, L., Guntherodt, H., Scherer, L.J., Constable, E.C., Housecroft, C.E., Neuburger, M. and Herrmann, B.A. (2005) *Chemistry – A European Journal*, **11**, 2307–2318.

56 Wei, Y., Kannappan, K., Flynn, G.W. and Zimmt, M.B. (2004) *Journal of the American Chemical Society*, **126**, 5318–5322.

57 Wei, Y., Tong, W., Wise, C., Wei, X., Armbrust, K. and Zimmt, M. (2006) *Journal of the American Chemical Society*, **128**, 13362–13363.

58 Yablon, D.G., Giancarlo, L.C. and Flynn, G.W. (2000) *Journal of Physical Chemistry B*, **104**, 7627–7635.

59 Yablon, D.G., Wintgens, D. and Flynn, G.W. (2002) *Journal of Physical Chemistry B*, **106**, 5470–5475.

60 De Feyter, S., Grim, P.C.M., Rücker, M., Vanoppen, P., Meiners, C., Sieffert, M., Valiyaveettil, S., Müllen, K. and De Schryver, F.C. (1998) *Angewandte Chemie-International Edition*, **37**, 1223–1226.

61 Tao, F. and Bernasek, S.L. (2005) *Journal of the American Chemical Society*, **127**, 12750–12751.

62 Qian, P., Nanjo, H., Yokoyama, T., Suzuki, T.M., Akasaka, K. and Orhui, H. (2000) *Chemical Communications*, **20**, 2021–2022.

63 Berg, A.M. and Patrick, D.L. (2005) *Angewandte Chemie-International Edition*, **44**, 1821–1823.

64 Palmans, A.R.A. and Meijer, E.W. (2007) *Angewandte Chemie-International Edition*, **47**, 8948–8968.

8
Structure and Function of Chiral Architectures of Amphiphilic Molecules at the Air/Water Interface

Isabelle Weissbuch, Leslie Leiserowitz, and Meir Lahav

8.1
An introduction to Chiral Monolayers on Water Surface

Monolayers of self-assembled amphiphilic chiral molecules at the air/water interface attract great interest since they can serve as useful models for biological membranes and the design of surfaces in the material sciences. In particular, such systems, in which polar head-groups are those that are directed towards the water phase, while the hydrophobic groups (normally alkyl chains) are directed away from it, can provide an important insight on the enantioselective interactions taking place between molecules. The interactions include those between the hydrophobic tails of homochiral and heterochiral molecules as well as those of their polar head-groups with molecules present in the aqueous solution.

The first study on chiral monolayers was reported in the 1950s by Zeelen in his Ph. D. thesis supervised by Professor Havinga [1]. Years later, the Swedish scientist Monica Lundquist [2] reported comparative surface-pressure–area (Π-A) isotherm studies as function of temperature on monomolecular films composed from either enantiopure or racemic amphiphilic molecules deposited on water. She complemented these studies by X-ray powder diffraction measurements on the highly compressed films after their transfer on solid support. On the basis of these studies she inferred that the amphiphilic molecules in the compressed state are packed in crystalline arrays so that in some of these films the hydrocarbon tails are aligned vertical on the water surface, whereas in others the chains are tilted. Additional Π–A isotherm studies performed during the 1980s by Arnett and coworkers demonstrated that certain racemic mixtures of amphiphilic molecules spread at the air/aqueous interface undergo spontaneous resolution in two-dimensions. These early investigations were summarized in two comprehensive reviews by Arnett's group [3, 4].

The advent of modern surface-sensitive analytical tools coupled with theoretical computations on 2D films has recently revolutionized the surface sciences. In particular, with the availability of intense, monochromatic and low-divergence synchrotron X-rays, and the application of epifluorescence, Brewster angle and scanning force microscopy methods, it became possible to determine the morphology, at near nanometer level, and the structure at near atomic level, of ordered 2D molecular self-assemblies, directly on liquid surfaces. Of particular interest are the reports that describe the dynamics of the structural changes that take place during phase separation in 2D nonracemic films or during their reactivity at the air/water interface.

Here, we present results on comparative structural and functional properties of monolayers composed from self-assembled enantiopure and racemic amphiphilic molecules. We focus on model systems that can shed light on the early stages of crystal nucleation at the molecular level and on the role played by such architectures as templates for epitaxial crystallization, as well as on the use of such architectures for "mirror-symmetry breaking" and amplification of chirality in mother Nature. A special issue in the journal of Current Opinions in Colloids and Interface Science (2008, Volume 13) was dedicated to recent progress to this field.

8.2
Two-Dimensional Crystalline Self-Assembly of Enantiopure and Racemates of Amphiphiles at the Air/Water Interface; Spontaneous Segregation of Racemates into Enantiomorphous 2D Domains

Upon crystallization, racemates can form either (i) a racemic compound, in which the two enantiomers – that are related by an inversion center and/or by glide symmetry – pack in the same crystal, or (ii) undergo spontaneous segregation into a racemic mixture of enantiomorphous crystals, in which the symmetry elements mentioned for case (i) are absent. According to Kitaigorodskii's hypothesis [5], an inversion center leads to a denser phase and therefore most racemates tend to crystallize as three-dimensional (3D) racemic compounds. Moreover, the inversion center and the glide plane provide additional elements of symmetry, which are not available to enantiopure systems.

In contradistinction, racemic mixtures of amphiphilic molecules composed of a hydrophilic head-group and a hydrophobic chain, which form crystalline self-assembled monolayers at a liquid surface, cannot pack via a glide plane parallel to the surface or across a center of symmetry. Therefore, the probability of racemates to self-assemble into enantiomorphous two-dimensional domains is increased provided molecular packing via a glide plane perpendicular to the surface can be avoided. This situation generally does not occur for the racemates, since the hydrocarbon tails that emerge from the water surface, invariably pack in a herring-bone motif via glide symmetry in the nonchiral plane group *pg*, as represented in Figure 8.1.

In order to induce spontaneous separation of racemates into a mixture of enantiomorphous 2D crystallites, the glide symmetry plane perpendicular to the surface or possible formation of enantiomerically disordered solid solutions must be prevented. Accordingly, the insertion of functional groups within the hydrocarbon tails has been reported to have a tendency to prevent the formation of the herring-bone motif and, at the same time, to promote chain packing by translation symmetry only. The presence of such a group may induce the racemate to self-segregate into enantiomorphous domains of plane group *p1*, (Figure 8.1). Reasoning along the lines

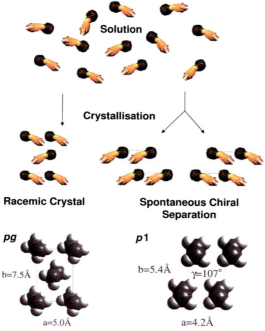

Figure 8.1 Schematic representation of the crystallization of a racemic mixture of amphiphilic molecules into either racemic crystals packing in the plane group *pg*, or into racemic mixtures of enantiomorphous crystals packing in the plane group *p*1.

described above, the insertion of polar groups – such a secondary amide that can replace the herring-bone interactions by $N-H\cdots O=C$ hydrogen bonds – was applied as illustrated below for the packing arrangement of amphiphilic α-amino acids.

8.3
Langmuir Monolayers of Amphiphilic α-Amino Acids

The self-assembly of several enantiopure and racemic long-chain α-amino acids (C_nH_{2n+1}-$CH(NH_3^+)CO_2^-$ with $n = 12-16$, and labeled C_n-Gly) spread at the air/water interface to generate a monolayer was investigated by grazing-incidence X-ray diffraction (GIXD) method. (The method is described in Reference [6], and the diffraction data is consistent with a monolayer) The measured GIXD diffraction patterns – presented as 2D intensity contour plots – of racemic and enantiopure (S)- or (R)-) 2D crystallites of C_{16}-Gly at the air/water interface are shown in Figure 8.2.

The GIXD patterns indicate that the essentially two-dimensional crystallites of racemic C*n*-Gly amphiphiles $n = 16$ are of plane space group *pg* and contain both enantiomeric molecules related to one another by a glide plane perpendicular to the surface of the monolayer, in a herring-bone motif. The crystalline packing

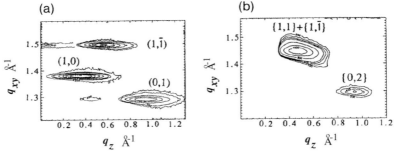

Figure 8.2 GIXD patterns measured from self-assembled 2D crystallites of (a) enantiopure and (b) racemic C_{16}-Gly amphiphiles on water at 4 °C, where q_{xy} and q_z are the horizontal and vertical components of the X-ray scattering vector, respectively.

arrangement is shown in Figure 8.3. By contrast, the enantiopure molecules self-assemble in an oblique unit cell of plane group $p1$ in which the molecules are related by translation symmetry (Figure 8.4) [7].

In order to design racemates that undergo spontaneous segregation into enantiomorphous two-dimensional domains, secondary amide groups were incorporated within the molecules, between the long hydrocarbon chains that provide hydrophobic character and the head-group that orients them toward the water surface. The structure of the monolayer films of several enantiopure and racemic α-amino acids bearing various alkyl chains $C_nH_{2n+1}CONH(CH_2)_4$-$CH(NH_3^+)CO_2^-$, $n = 15, 17, 21$ (C_n-Lys) was determined by GIXD [7, 8]. The enantiopure and the racemic mixtures of

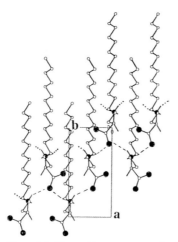

Figure 8.3 The packing arrangements of racemic 2D crystallites of the (R,S) C_{16}-Gly monolayer on water, viewed perpendicular to the layer. For clarity, the hydrogen atoms of the n-alkyl chains are omitted.

(a)

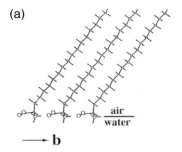

(b)

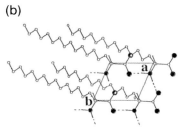

Figure 8.4 (a, b) The packing arrangement of enantiopure C_{16}-Gly crystallites viewed parallel and perpendicular to the water surface, respectively. For clarity the nitrogen and oxygen atoms of the headgroups in (b) are filled and the hydrogen atoms of n-alkyl chains are omitted.

molecules yielded very similar GIXD patterns (shown in Figure 8.5 for $n = 17$) indicative of the packing into an oblique unit cell with translational symmetry only. The 2D packing arrangement determined from the GIXD measurements is shown in Figure 8.6 [7].

Increasing the length of the alkyl chain attached to the carbonyl group of the amide moiety to $n = 29$ (C_{29}-Lys) enhanced the tendency for their packing in the

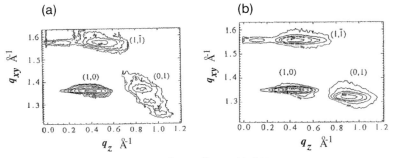

Figure 8.5 GIXD patterns measured from self-assembled 2D crystallites of enantiomerically pure (a) and racemic (b) $C_{17}H_{35}CONH(CH_2)_4$-$CH(NH_3^+)CO_2^-$ amphiphiles on water at 4 °C, where q_{xy} and q_z are the horizontal and vertical components of the X-ray scattering vector, respectively.

(a)

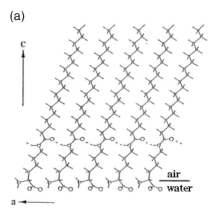

(b)

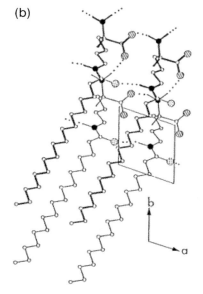

Figure 8.6 The packing arrangement of the enantiomorphous 2D crystallites self-assembled from either enantiopure or racemic $C_{17}H_{35}CONH(CH_2)_4\text{-}CH(NH_3^+)CO_2^-$ amphiphiles viewed: (a) parallel and, (b) perpendicular to the water surface.

herring-bone motif thus overruling the effect of the hydrogen bonding between the amide groups and leading to the formation of racemic two-dimensional crystallites.

Spontaneous segregation of racemic $C_{17}H_{35}CONH(CH_2)_4\text{-}CH(NH_3^+)CO_2^-$ was also obtained when these molecules were spread within a monolayer of a phospholipids, di-palmitoylphosphatidyl-ethanolamine (DPPE). By contrast, racemic $C_{18}H_{37}OCO(CH_2)_2\text{-}CH(NH_3^+)CO_2^-$ α-amino acid crystallizes in the form of a racemic compound. Furthermore, a nonracemic mixture of composition 7:3 R:S of these molecules undergoes a phase separation into a racemic phase and an enantiomorphous phase of the excess of R-enantiomer, both on water as well as within a phospholipid monolayer [9].

8.3.1
Domain Morphology and Energy Calculations in Monolayers of N-acyl-α-Amino Acids

Two optical microscopy methods that enable *in-situ* imaging of the morphology of monolayer domains at the air/water interface are epifluorescence microscopy, first reported by McConnell [10, 11] and Möhwald [12, 13], and Brewster angle microscopy, first reported for monolayer imaging by Hönig and Möbius [14].

Applying the latter method, Vollhardt and coworkers recently reported detailed studies on the chiral discrimination between enantiomers in monolayers of N-palmitoyl-aspartic acid, N-stearoylserine methyl ester, and N-tetradecyl-γ-δ-dihydroxypentanoic acid amide as well as of 1-O-hexadecyl glycerol [15, 16]. For a recent review, see references by Nandi and Vollhardt [17, 18]. The Brewster microscopy images of the condensed-phase domains of enantiopure N-stearoyl serine methyl ester demonstrated the formation of enantiomorphous morphologies, as shown in Figure 8.7. Furthermore, the images obtained by spreading the racemate on the water

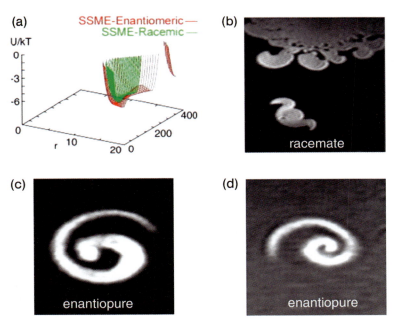

Figure 8.7 Chiral discrimination in condensed-phase domains in monolayer of N-stearoyl serine methyl ester as calculated theoretically (a) from pair potential theory and from the different handedness of the pure enantiomeric domains and racemic domain and the mirror-symmetry breaking as measured by Brewster angle microscopy images (b of the racemate and c and d the enantiomers) that show the homochiral preference that is a signature of self-recognition. (with the permission of *Current Opinion in Colloid & Interface Science*)

demonstrated the formation of pairs of enantiomorphous domains where two domains of opposite handedness are stitched to form a meso-morphology (Figure 8.7). The computed enantiomeric discrimination energy between pairs of molecules show that in those cases where racemates undergo spontaneous segregation the close-packed enantiomeric pairs have a lower minimum pair potential as compared to the racemic pair (Figure 8.7). The pair potential was computed using a coarse-grained molecular model and it has been shown that at intermediate separation between molecules, the hetero- or racemic pair recognize each other better and form the meso-type architectures, while the same type of enantiomers prefer each other at shorter separation. On the basis of these studies, the authors were able to provide a direct correlation between the absolute configuration of the amphiphilic molecules and the 2D enantiomorphous morphology of the crystallites.

Enantiomerically pure amphiphiles of the long-chain α-amino acids bearing an amide group, such as N^ε-stearoyl-lysine, C_{17}-Lys, were reported to form stable Langmuir–Blodgett (LB) films arranged in a polar (Z-type) configuration [19]. Polarizable groups such as p-nitro-aniline and merocyanine, when they were attached covalently to molecules such as C_{17}-Lys, preserved their property to yield polar LB films and to display second-harmonic generation. The Z-type films of enantiopure C_{17}-Lys, deposited on a glass coated with gold, could be used to generate highly polarized electron beams when scattered through these films [20].

8.4
Stochastic Asymmetric Transformations in Two Dimensions at the Water Surface

The ability to convert achiral molecules into homochiral polymers or supramolecular architectures of single handedness had played a ubiquitous role in the origin of homochirality on Earth. Such a process may be simpler in two-dimensional than in three-dimensional crystalline assemblies, leading to an "absolute" asymmetric transformation performed on surfaces. Thus, for example, whereas the pure alkanoic acids self-assemble on water in achiral space groups when they are exposed on aqueous solutions containing Cd(II) salts, these films self-assemble at the interface as mixtures of enantiomorphous crystalline domains as demonstrated by GIXD [21] and atomic force microscopy (AFM) [22] studies.

Recently, Liu *et al.* [23] reported that a variety of nonchiral amphiphilic diacetylenes, nonchiral barbituric acids [24], aryl-benzimidazoles [25, 26], achiral amphiphile of 4-octadecyloxylcoumarin [27], self-assemble as mixtures of enantiomorphous clusters at the air/water interface or on aqueous solutions containing Ag(I) ions, as demonstrated by circularly polarized light measurements. More recently, this group also reported the adsorption and the aggregation of a selenacarbocyanine dye (3,3-disulfopropyl-9-methyl-selenacarbocyanine, onto Langmuir monolayers of a series of gemini amphiphiles [28], where the chains are separated via a spacer containing a different numbers of methylene groups. The transfer of some of the films onto a solid support displayed circular dichroism indicating the formation of 2D nonracemic mixture composed of enantimorphous domains of opposite handedness [29].

8.5
Self-Assembly of Diastereoisomeric Films at the Air/Water Interface

Structural studies were also performed on monolayers of diastereoisomeric salts of different composition on water. Thus, for example diastereoisomeric salts of mandelic acid-phenethylamine, where both molecules were functionalized with alkyl chains in the 4-positions of their benzene rings, were analyzed by GIXD [30]. The Langmuir film composed from 1:1 mixtures of R- (or S-) 4-tetradecylphenethylamine, $C_{14}H_{29}$-C_6H_4-$CH(CH_3)NH_3$, C_{14}-PEA, and R'- (or S'-) 4-pentadecylmandelic acid, $C_{15}H_{31}$-C_6H_4-CH(OH)COOH, C_{15}-MA, form stable crystalline monolayer films. The GIXD pattern of the diastereoisomeric (R,S') crystallites is different from that of (R,R') crystallites, whereas equimolar mixtures of the four components give rise to diffraction patterns almost identical to that of the (R,R') mixture (Figure 8.8). Studies on different chiral compositions of the four components in the system proved

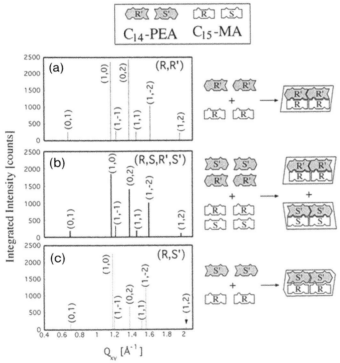

Figure 8.8 (a–c, left) GIXD patterns for the three diastereomeric films: (a) R,R' (thin line), (b) R,S, R', S' (thick line) and (c) R, S' (dotted line), measured at 5 C and $\Pi = 1$ mN/m. The observed Bragg peaks represented by lines of corresponding integrated intensities with assigned (h,k) indices. (a–c, right) Schematic views of the molecular assembly into crystals corresponding to the three diffraction patterns.

the formation of chiral domains containing enantiomeric disorder. The unit cells of the (R,R') and (R,S') films are oblique and contain one diastereomeric pair of molecules. Studies on different chiral compositions of the four-component system demonstrated the formation of chiral domains containing enantiomeric disorder. A similar kind of disorder of the α-OH group of mandelic acid has been observed in 3D crystals of analogous structures of phenethylamine-mandelate salts. This example shows that a *p*1 lattice symmetry (one molecule per unit cell) cannot provide unambiguous proof of a Pasteur-type separation in two-dimensions unless enantiomeric disorder is prevented. Further support for the presence of disorder in the present system was provided by the observation of a similar disorder between the enantiomers in the 3D crystal structure of the salt 45 : 55 S-C_{15}-MA:R-PEA, where the structure refinement revealed about 50% chiral disorder of the -OH group of MA [30].

8.6
Interactions of the Polar Head Groups with the Molecules of the Aqueous Environment

Many biological processes, among them enzyme recognition and reactivity, self-replication of nucleic acids, antibody–antigen interactions, comprise molecular recognition by complementary hydrogen bonding between host and guest systems taking place at soft surfaces. Molecular-recognition between monolayer components and an aqueous guest have also been extensively investigated as artificial models for shedding light on these processes [31]. Kitano and Ringsdorf reported a recognition process, at the air/water interface based on the effect of a nucleobase binding to a host monolayer [32]. This study was followed by several research groups, among them Ariga and Kunitake who made a systematic investigation of a variety of combinations of the host monolayers and guests in the aqueous solution [33]. Most interestingly, they found that molecular recognition at the air/water interface generally occurs with enhanced binding constants relative to those in bulk media [34]. Other studies comprise numerous examples of enantioselective interactions, where the polar group was either a cyclodextrin group or a crown-ether-macrocycle interacting with α-amino acids [35–37], or enantiopure phospholipids interacting enantioselectively with the odorants (+) and (−) carvone [38]. Early studies were performed also on the recognition of peptides by amphiphilic peptides [39, 40]. For a recent review on the cyclodextrin systems we suggest that of Shahgaldian [41]. Some of these systems were interpreted theoretically using quantum-chemical calculations [42, 43].

The elucidation of the organization of such combined systems at the molecular level became feasible with the application of the GIXD methodology. By such means we may study intercalation of amphiphilic guest molecules into a monolayer film, or attachment of guest amphiphiles to the periphery of monolayer crystalline islands of the host molecule. The common Π–A isotherms cannot unambiguously differentiate between these two cases, and therefore one has to apply other methods such as GIXD.

The following system provides an example that illustrates the attachment of the guest molecules between enantiomorphous crystallites. When monolayers of enantiopure S- (or R-) or racemic R,S-C_{15}-Lys or C_{16}-Gly were spread on copper acetate aqueous solutions in the absence or in the presence of enantiopure S – or R-alanine, serine or valine (hereafter labelled R or S), they exhibited a more expanded Π–A isotherm when alanine, serine or valine of the same absolute configuration as the monolayer was injected into the aqueous subphase (Figure 8.9) [44]. The diastereoisomers assume different packing arrangements, as demonstrated by GIXD measurements on the films directly formed at the air/solution interface.

Furthermore, on the basis of comparative X-ray photoelectron spectroscopy (XPS) studies of the films of the two types of heterocomplexes and 3D crystals of (S-serine)

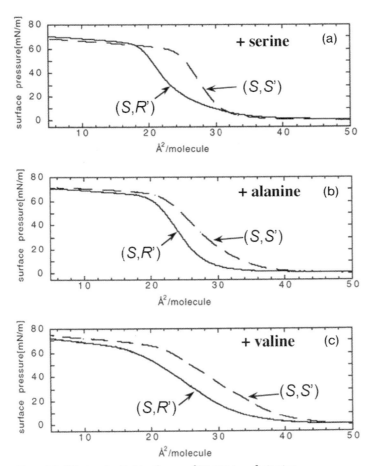

Figure 8.9 Effect on the Π–A isotherms of (S)-C15-Lys of injecting of (a) serine, (b) alanine, and (c) valine into the copper acetate solution. Labels (S, S'), (S,R') point to isotherm of the (S) amphiphile on the subphase containing the injected (S') or (R') water-soluble α-amino acid, respectively.

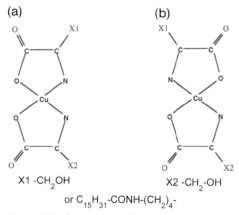

X1 -CH$_2$OH   X2 -CH$_2$-OH
                or C$_{15}$H$_{31}$-CONH-(CH$_2$)$_4$-

Figure 8.10 The syn- (a) and anti- (b) configurations of copper (II) serine complexes

Cu and (R, S -serine)Cu complexes, the configuration of the α-amino acid ligand around the Cu(II) ions in such complexes indicated a structural similarity. In the thin films, the S–Cu–S complex displays a *syn* configuration, whereas the S–Cu–R complex adopts an anti configuration (Figure 8.10). The more expanded Π–A isotherms of the amphiphile and solute of the same absolute configuration were rationalized by proposing that water soluble S–Cu–S complex molecules of *syn* configuration can bind more strongly to the periphery of the polar head-groups of the monolayer islands of S–Cu–S of the same absolute configuration by virtue of a better molecular fit. Because of steric hindrance, the molecules of the S–Cu–S complex molecules may bind only loosely to the periphery of the S–Cu–R domains that assume an *anti* configuration.

We now consider a system in which intercalation of the guest molecule takes place within the host monolayer lattice. It is composed of the amphiphilic molecules that incorporate two moieties of different size, a chiral hydrophobic tail that has a larger cross-section than the hydrophilic head-group. When this type of molecules is spread on water, they create a corrugated surface at the monolayer/solution interface that has chiral grooves that may interact enantioselectively with chiral molecules present in the aqueous solution. Thus, for example, cholesteryl-S-glutamate (CLG), which is composed of a hydrophobic steroid moiety with a cross-sectional area of 38 Å^2 and a hydrophilic glutamate head-group with an area of 24 Å^2 forms such corrugated Langmuir films. When the film is formed on an aqueous solution containing hydrophobic S α-amino acids, such as leucine or phenylalanine, which are of the same sense of chirality as the glutamate head-groups, they are easily intercalated within the grooves of the monolayer film (Figure 8.11) [45]. On the other hand, when the CLG amphiphile was spread on a solution of R-leucine, the crystallinity of the film was destroyed, as determined by the GIXD (Figure 8.12). The hydrophilic α-amino acids were not incorporated within the film.

Another interesting example reported recently is the monolayer of a cholesterol-armed cyclen Na$^+$ complex host molecule where the homochiral film interacts

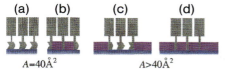

$A=40\text{Å}^2$ $A>40\text{Å}^2$

Figure 8.11 Schematic representation of the packing of CLG molecules on: (a) water, (b) aqueous solutions of hydrophobic S-α-amino acids that can be intercalated between CLG molecules, (c) at the periphery of the domains, or (d) at both sites. CLG are in gray and amino acids are in purple.

preferentially at low surface pressure with S-Leu and S-Val, whereas at higher pressures it interacts with the R-amino acids [46]. Dynamic formation of a cavity structure has been reported for a monolayer of steroid cyclophane host molecules (Figure 8.13) that have a large binding constant to water-soluble naphthalene guests

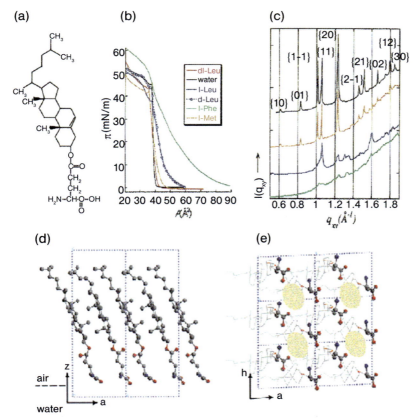

Figure 8.12 (a) Molecular structure of CLG; (b) Π–A isotherms of CLF spread on aqueous solutions of α-amino acids; (c) GIXD patterns; (d, e) Packing arrangements of the 2D crystallites of CLG, without and with intercalates water-soluble α-amino acid molecules, viewed perpendicular to the liquid surface.

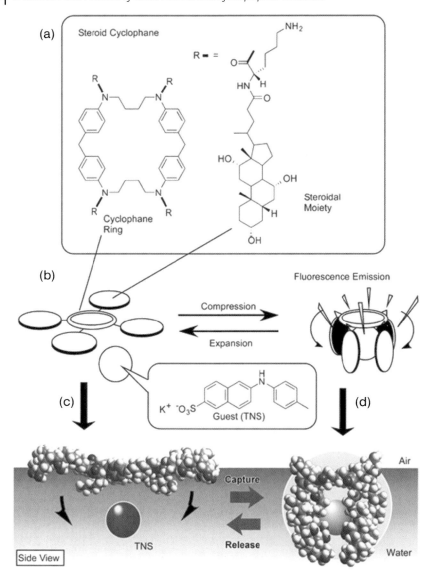

Figure 8.13 (a) Chemical formula of the steroid cyclophane; (b) Plausible models for the conformational change of the steroid cyclophane upon monolayer compression and reversible guest (TNS) capture: (c) open conformation at low pressures; (d) cavity conformation at high pressures (with permission of The Royal Chemical Society).

such as 6-(p-toluidino)naphthalene-2-sulfonate (TNS) in a 1:1 stoichiometry. In a monolayer at low pressures, the host molecules adopt an open conformation. Compression of the monolayer induces shrinkage of the steroid cyclophane molecule resulting in a cavity conformation that is more closed. Using this conformational change, the reversible binding of the fluorescent TNS guest by a repeated compression and expansion of the monolayer was demonstrated [33, 47, 48].

8.7
Interdigitated Bi- or Multilayer Films on the Water Surface

Interdigitation between hydrocarbon chain moieties of certain classes of molecules is an effective way to ensure cohesive contacts between the molecules, forming ordered layer-like systems provided certain structural conditions are met. Partial interdigitation is found in some membranes in Nature composed from bilayers of phospholipids by virtue of the difference in the length of the two chains constituting the phospholipid molecule. Another class of molecules or two-component systems that have a tendency to self-assemble spontaneously or by compression into bi- or multilayers films on water, via a process of interdigitation, are those composed of two different moieties with different surface areas.

Thus, for example, derivatives of cholesterol bearing a long hydrocarbon ester chain with a cross-sectional area of 18–20 Å^2 and a cholesterol group with a cross-sectional area twice that of the ester 38–40 Å^2. This molecule self-assembles spontaneously into an interdigitated multilayer at the air/water interface, as demonstrated by GIXD. The unit-cell parameters of such bilayers at the water surface almost match those of this molecule in three-dimensional crystals [49].

Chain interdigitation can also be achieved in crystalline thin films at the air/solution interface, making use of binary systems composed from molecules A and B exhibiting an acid–base complementarity [50]. When A is a water-insoluble amphiphile bearing a long hydrocarbon chain substituent and B is water soluble without a chain, they spontaneously form two-dimensional arrays composed of -A-B-A-B-A-B- chains at the water surface. Upon compression of the monolayer beyond its collapse point, the film is transformed into a multilayer incorporating interdigitated chains. This process has been demonstrated in several systems. For example, use was made of the diastereoisomeric salts of mandelic acid, $C_6H_5CH(OH)CO_2H$ (MA), and phenylethylamine, $C_6H_5CH(CH_3)NH_2$ (PEA), as chiral functional groups for interdigitation, where either the acid or the amine are modified by attaching an alkyl chain $C_{15}H_{31}$- in the 4-position of the phenyl ring. A film of R-$C_{15}H_{31}$-$C_6H_4CH(OH)CO_2H$, labeled R-$C_{15}H_{31}$-MA, spread on a solution of S-PEA did not diffract at any point along the compression isotherm. In contrast, replacing the solution by R-PEA an intense GIXD pattern was obtained after high compression. The film structure (Figure 8.14) is an interdigitated crystalline trilayer whose top layer of alkyl chains exposed to air is disordered. Indeed, the system will only pack well in a crystalline bilayer containing interdigitated chains.

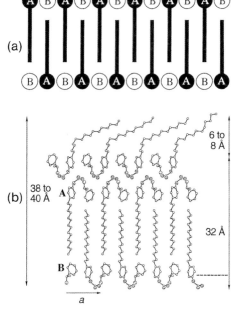

Figure 8.14 (a) Schematic representation of the interdigitated ABAB system; (b) Packing arrangement of the interdigitated bilayer, as determined by GIXD.

The effect of chirality was explained in terms of the 3D crystal structures of MA and PEA; only the chiral groups MA with the same handedness, e.g., R-MA and R-PEA, are compatible with the formation of the interdigitated bilayer but not the groups PEA and MA of opposite handedness.

Ward and coworkers reported the interdigitation in a molecular system obtained upon the spreading of sodium octadecanesulfonate, $C_{18}H_{37}SO_3^-Na^+$, amphiphile over an aqueous solution containing guanidinium sulfonate [51]. This acid–base system yielded a surface pressure–area isotherm compatible with intercalation of the guanidinium cations between the $C_{18}H_{37}SO_3^-$ anions into a two-dimensional hydrogen-bonded network. The shape of the isotherm after monolayer collapse is in keeping with interdigitation, as observed in the 3D crystal structures of guanidinium arenesulfonate. The presence of interdigitation has also been demonstrated in a system composed of amphiphilic acids deposited on aqueous solutions bearing short-chain benzamidium bases [52].

From the above examples it appears that the basic requirement for chain interdigitation is the spontaneous formation of an initial mixed monolayer comprising the complementary acid–base components in proper registry, despite the loss in effective chain packing. In the next step, the system will pack into a crystalline architecture only if the chains are properly structured for interdigitation. A model for the general process of the molecular reorganization as a function of film compression

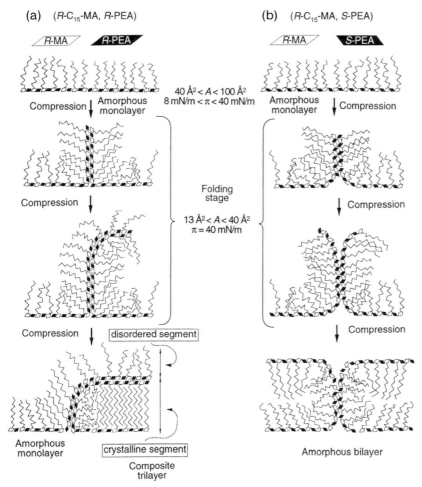

Scheme 8.1 Schematic model for the general process of the molecular reorganization as a function of film compression with a tendency towards corrugation, followed by interdigitation, for different diastereomers of $C_{15}MA$ and PEA.

with a tendency towards corrugation, followed by interdigitation, is depicted in Scheme 8.1.

8.8
Structural Transfer from 2D Monolayers to 3D Crystals

Chiral amphiphilic self-assemblies were successfully used as templates for oriented 3D crystallization and for the resolution of enantiomers. The concept was

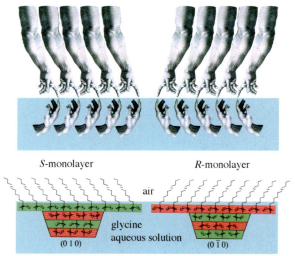

Figure 8.15 Schematic representation of the stereochemical correlation between the α-amino acid amphiphile molecules in the monolayer and the structure of the nucleated face of α-glycine crystals. S-monolayer-induced oriented nucleation of α-glycine from the (0–10) face whereas the R-monolayer is from the (010) face.

demonstrated in an early study of the oriented crystallization of 3D crystals of α-glycine induced by enantiopure amphiphilic α-amino acids monolayers at the air/aqueous glycine solution interface or induced by their Langmuir–Blodgett films in contact with the solution [53]. These studies have shown a clear cut stereochemical correlation between the absolute configuration of the α-amino acid head-groups of the monolayer, packed in enantiomorphous 2D crystals as determined by GIXD (Figures 8.5 and 8.6) and the structure of the chiral face of the α-polymorph of the α-glycine crystals nucleated at the interface (Figures 8.15 and 8.16) [6, 54].

This oriented crystallization method has been extended in recent years to other systems, in particular, on gold surfaces coated with self-assembled monolayers of rigid thiols that were used to induce 3D crystallization of variety of 3D crystals of amino acids [55]. Very recently, gold surfaces coated with enantiopure cysteine, $HS\text{-}CH_2\text{-}CH(NH_3^+)CO_2^-$, molecules attached to the surface via their SH group and exposing to the aqueous solution the α-amino acid polar groups were used to reportedly induce resolution of the racemic glutamic acid into enantiomers [56].

Another interesting application of the interaction between the homochiral amino acid groups organized on gold was achieved in the enantioselective reduction of the drug DOPA on gold electrodes covered by homocysteine [57]. Enantioselective crystal growth of leucine on self-assembled monolayers of leucine molecules covalently bound to the gold surface has been reported [58].

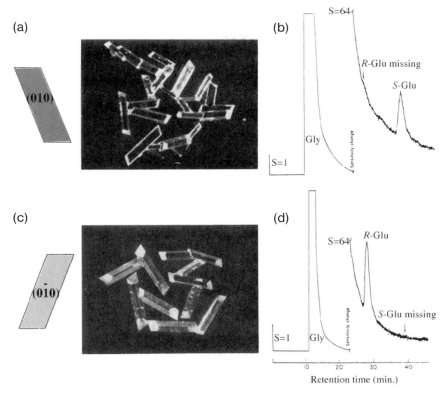

Figure 8.16 (a, c) Crusts of α-glycine crystals grown in the presence of racemic glutamic acid (Glu) in solution, under R- or S-monolayers, respectively. All the crystals are attached with their (010) or (0–10) faces, respectively, to the monolayer and exhibit enantiomorphous morphology (schematically shown in dark grey and light grey). Note that the crystals attached with (010) or (0-10) to the monolayer are not superimposable by rotation in the plane of the air/water interface; (b, d) Enantiomeric HPLC analysis of crystals grown as in (a, c), respectively.

8.9
Homochiral Peptides from Racemic Amphiphilic Monomers at the Air/Water Interface

One of the mists in the field of origin of life is the emergence of the biopolymers that are composed of α-amino acids and sugars of the same handedness (isotactic) in an achiral prebiotic world. Since the polymerization of racemates in ideal solutions should result in the formation of atactic polymers composed of both enantiomers, it is imperative to design alternative synthetic routes for the preparation of those of homochiral sequence. A way to override the random polymerization process is to self-assemble the racemic monomers into crystalline domains either prior to reaction or to form supramolecular architectures, in the course of the polymerization reaction that can engender enantioselectivity in the ensuing processes of chain elongation. As we have seen above, the air/water interface provides an arranging medium where amphiphilic molecules may self-assemble spontaneously into crystalline domains. Once the molecules are organized at the air/water interface, the addition of a catalyst in the aqueous subphase can bring on the polymerization of the polar head-groups. Furthermore, the ability to determine the structures of these clusters at the molecular level allows us to select and to design particular molecules that can be appropriated for the generation of homochiral peptides of single handedness.

Earlier reports, by Fukuda [59], on the polymerization of amphiphilic esters at the air/water interface have claimed, on the basis of IR studies, that such esters undergo polymerization to yield polypeptides. Recent reinvestigation of these reactions by mass spectrometry has demonstrated, however, that the peptides formed are no longer than dipeptides [60]. For this reason, such esters cannot be regarded as realistic prebiotic model systems for the formation of long oligopeptides. On the other hand, polymerization of amphiphilic N^{α}-carboxyanhydrides or thioesters of amphiphilic α-amino acids yield longer oligopeptides [61]. GIXD studies have demonstrated that racemic N^{ε}-alkanoyl-lysine ($n = 12$–21) and their corresponding N^{α}-carboxyanhydrides undergo spontaneous segregation of the enantiomers into enantiomorphous 2D crystalline domains at the water surface. Polymerization reactions within such enantiomorphous crystallites, using Ni(II) ions as catalyst, yielded mixtures of oligopeptides up to six repeating units with some excess of the homochiral oligopeptides [62]. Longer homochiral oligopeptides were obtained in the polymerization N^{ε}-stearoyl-lysine-thio-ethyl ester (C_{18}-TELys) that crystallizes into a racemic plane group [61]. Polymerization with Ag(I) ions as catalysts yielded a library of diastereoisomeric peptides of various lengths where those of homochiral sequence represent the dominant fraction, in agreement with a preferred reaction between molecules of the same handedness along a translation axis. Furthermore, an enhanced formation of homochiral peptides is anticipated from the polymerization of nonracemic mixtures of the above amphiphilic molecules, as a result of phase separation into a mixture of racemic crystallites and the enantiomorphous crystallites of the enantiomer in excess. In such systems, we anticipated that the enantiomer in excess should be converted into peptides of homochiral sequence [63].

GIXD studies, demonstrated the occurrence of phase separation of nonracemic O^{γ}-stearylglutamic acid (C_{18}-Glu) [9]. On the basis of a phase separation of nonracemic N^{α}-carboxyanhydride of O^{γ}-stearyl-glutamic acid (C_{18}-Glu-NCA) amphiphiles into racemic crystallites accompanied by enantiomophous ones comprising the enantiomer in excess, polymerization of racemic crystallites yielded heterochiral oligopeptides containing up to six residues of both absolute configuration, whereas the enantiomorphous crystallites yielded, upon polymerization, oligopeptides of homochiral sequence of up to ten residues, as demonstrated by matrix-assisted laser desorption-ionization time-of-flight mass spectrometry (MALDI-TOF MS) analysis (Figure 8.17) [64].

Similar phase separations were also observed when nonracemic C_{18}-Glu-NCA within a phospholipids monolayer since the diastereoisomeric distribution of the oligopeptides within the latter monolayers was similar to that observed on water.

8.9 Homochiral Peptides from Racemic Amphiphilic Monomers at the Air/Water Interface

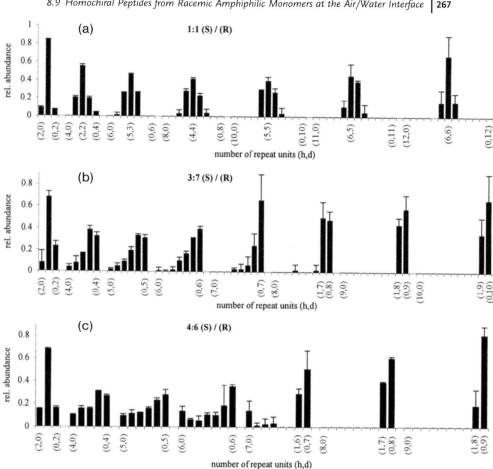

Figure 8.17 MALDI-TOF MS analysis of the oligopeptides obtained from racemic and nonracemic mixtures of C_{18}-Glu-NCA monomer: (a) Racemic (b) 3:7 (S:R) and (c) 4:6 (S:R). The R monomer contained deuterated hydrocarbon chains.

When DPPE, which contains an amine group, was used as "the membrane-like monolayer" it operates as a surface initiator and could be detected at the C-terminus of the peptide chains [63]. A very efficient desymmetrization of the racemic mixtures was accomplished when the polymerization reactions of the thio-esters were performed in the presence of various amounts of the corresponding esters molecules. Since the latter form *quasi*racemates with the thio-esters, the enantiopure ester can be inserted enantioselectively within specific chiral sites of the thio-ester crystallites. Consequently, polymerization within the mixed crystallites of the esters and the thio-esters yield nonracemic mixtures of the formed peptides [65].

Finally, the lessons learned on the organization and the reactivity of amphiphilic molecules at the air/water interface were applied for the generation of homochiral peptides from racemic mixtures of water-soluble N^α-carboxyanhydrides of hydrophobic α-amino acids and successfully performed in the polymerization of amino acids in aqueous solutions [66].

8.10
Conclusions

The ability to probe the structures of Langmuir films with modern analytical tools complemented by theoretical calculations opens new directions in the utilization of such films as model systems in the investigation of the structure and dynamics of the process of interfacial molecular recognition of artificial systems in general and in the understanding of biological recognition processes in biological membranes in particular. Studies on chiral films expanded in recent years to complex organic and biological systems, such as in the reactivity of phospholipases that induce hydrolysis of phospholipids in order to provide messenger molecules for signal transmission. Furthermore, monitoring the reactivity of these enzymes within a phospholipid membrane provides a useful route for the elucidation of the local structures of biomembranes [67].

Chiral Langmuir films are also finding potential applications in material sciences such as in the preparation of new sensor systems [29] as well as in the control of the early stages in crystal nucleation in general and as models for the understanding the process of biomineralization in particular. Research along such directions is anticipated to continue in the years to come.

Acknowledgement

This work was supported by the Israel Science Foundation.

References

1 Zeelen, F.J. (1956) *Steroyl Aninozuren: Synthese, Spreiding en Photochemie*, State University of Leiden, Doctoral Thesis.
2 Lundquist, M. (1978) *Progress in the Chemistry of Fats and Other Lipids*, **16**, 101.
3 Stewart, M.V. and Arnett, E.M. (1982) *Topics in Stereochemistry*, Vol. 13 (eds. Eliel, E.L., Wilen, S.H. and Allinger, N.L.), John Wiley & Sons Inc, New York.
4 Arnett, E.M., Harvey, N.G. and Rose, P.L. (1989) *Accounts of Chemical Research*, **22**, 131–138.
5 Kitaigorodskii, A.I. (1961) *Organic Chemical Crystallography*, Consultants Bureau Press, New York.
6 Kuzmenko, I., Rapaport, H., Kjaer, K., Als-Nielsen, J., Weissbuch, I., Lahav, M. and Leiserowitz, L. (2001) *Chemical Reviews*, **101**, 1659–1696.
7 Weissbuch, I., Berfeld, M., Bouwman, W.G., Kjaer, K., Als-Nielsen, J., Lahav, M. and Leiserowitz, L. (1997) *Journal of the American Chemical Society*, **119**, 933–942.

8 Nassoy, P., Goldmann, M., Bouloussa, O. and Rondelez, F. (1995) *Physical Review Letters*, **75**, 457.

9 Weissbuch, I., Rubinstein, I., Weygand, M.J., Kjaer, K., Leiserowitz, L. and Lahav, M. (2003) *Helvetica Chimica Acta*, **86**, 3867–3874.

10 von Tscharner, V. and McConnell, H.M. (1981) *Biophysical Journal*, **36**, 409–419.

11 Weiss, R.M. and McConnell, H.M. (1984) *Nature*, **310**, 47.

12 Lösche, M. and Möhwald, H. (1984) *Colloids and Surfaces*, **10**, 217–224.

13 Möhwald, H. (1993) *Reports on Progress in Physics*, **56**, 653–685.

14 Hönig, D. and Möbius, D. (1991) *The Journal of Physical Chemistry*, **95**, 4590.

15 Nandi, N. and Vollhardt, D. (2003) *Journal of Physical Chemistry B*, **107**, 3464–3475.

16 Nandi, N., Vollhardt, D. and Brezesinski, G. (2004) *Journal of Physical Chemistry B*, **108**, 327–335.

17 Nandi, N. and Vollhardt, D. (2003) *Chemical Reviews*, **103**, 4033–4075.

18 Nandi, N. and Vollhardt, D. (2008) *Current Opinion in Colloid & Interface Science*, **13**, 40–46.

19 Popovitz-Biro, R., Hill, K., Shavit, E., Hung, D.J., Lahav, M., Leiserowitz, L., Sagiv, J., Hsiung, H., Meredith, G.R. and Vanherzeele, H. (1990) *Journal of the American Chemical Society*, **112**, 2498–2506.

20 Ray, K., Ananthavel, S.P., Waldeck, D.H. and Naaman, R. (1999) *Science*, **283**, 814–816.

21 Weissbuch, I., Leiserowitz, L. and Lahav, M. (2008) *Current Opinion in Colloid & Interface Science*, **13**, 12–22.

22 Viswanathan, R., Zasadzinski, J.A. and Schwartz, D.K. (1994) *Nature*, **368**, 440–443.

23 Huang, X., Jiang, S. and Liu, M. (2004) *Journal of Physical Chemistry B*, **109**, 114–119.

24 Huang, X., Li, C., Jiang, S., Wang, X., Zhang, B. and Liu, M. (2004) *Journal of the American Chemical Society*, **126**, 1322–1323.

25 Yuan, J. and Liu, M. (2003) *Journal of the American Chemical Society*, **125**, 5051–5056.

26 Yuan, J. and Liu, M. (2006) *International Journal of Nanoscience*, **5**, 689–695.

27 Guo, Z., Jiao, T. and Liu, M. (2007) *Langmuir*, **23**, 1824–1829.

28 Zhai, X., Zhang, L. and Liu, M. (2004) *Journal of Physical Chemistry B*, **108**, 7180–7185.

29 Guo, P., Zhang, L. and Liu, M. (2006) *Advanced Materials*, **18**, 177–180.

30 Kuzmenko, I., Kjaer, K., Als-Nielsen, J., Lahav, M. and Leiserowitz, L. (1999) *Journal of the American Chemical Society*, **121**, 2657–2661.

31 Leblanc, R.M. (2006) *Current Opinion in Chemical Biology*, **10**, 529–536.

32 Kitano, H. and Ringsdorf, H. (1985) *Bulletin of the Chemical Society of Japan*, **58**, 2826–2828.

33 Ariga, K., Michinobu, T., Nakanishi, T. and Hill, J.P. (2008) *Current Opinion in Colloid & Interface Science*, **13**, 23–30.

34 Onda, M., Yoshihara, K., Koyano, H., Ariga, K. and Kunitake, T. (1996) *Journal of the American Chemical Society*, **118**, 8524–8530.

35 Kawabata, H. and Shinkai, S. (1994) *Chemistry Letters*, **2**, 375–378.

36 Badis, M., Tomaszkiewicz, I., Joly, J.-P. and Rogalska, E. (2004) *Langmuir*, **20**, 6359–6267.

37 Shahgaldian, P., Pieles, U. and Hegner, M. (2005) *Langmuir*, **21**, 6503–6507.

38 Pathirana, S., Neely, W.C., Myers, L.J. and Vodyanoy, V. (1992) *Journal of the American Chemical Society*, **114**, 1404–1405.

39 Cha, X., Ariga, K., Onda, M. and Kunitake, T. (1995) *Journal of the American Chemical Society*, **117**, 11833–11838.

40 Cha, X., Ariga, K. and Kunitake, T. (1996) *Journal of the American Chemical Society*, **118**, 9545–9551.

41 Shahgaldian, P. and Pieles, U. (2006) *Sensors*, **6**, 593–615.

42 Sakurai, M., Tamagawa, H., Inoue, Y., Ariga, K. and Kunitake, T. (1997) *The Journal of Physical Chemistry. B*, **101**, 4810–1816.

43 Tamagawa, H., Sakurai, M., Inoue, Y., Ariga, K. and Kunitake, T. (1997) *The Journal of Physical Chemistry. B*, **101**, 4817–1825.

44 Berfeld, M., Kuzmenko, I., Weissbuch, I., Cohen, H., Howes, P.B., Kjaer, K., Als-Nielsen, J., Leiserowitz, L. and Lahav, M. (1999) *Journal of Physical Chemistry B*, **103**, 6891.

45 Alonso, C., Eliash, R., Jensen, T.R., Kjaer, K., Lahav, M. and Leiserowitz, L. (2001) *Journal of the American Chemical Society*, **123**, 10105–10106.

46 Michinobu, T., Shinoda, S., Nakanishi, T., Hill, J.P., Fujii, K., Player, T.N., Tsukube, H. and Ariga, K. (2006) *Journal of the American Chemical Society*, **128**, 14478–14479.

47 Ariga, K., Nakanishi, T., Terasaka, Y., Tsuji, H., Sakai, D. and Kukuchi, J. (2005) *Langmuir*, **21**, 976–981.

48 Ariga, K., Nakanishi, T. and Hill, J.P. (2006) *Soft Matter*, **2**, 465–477.

49 Alonso, C., Kuzmenko, I., Jensen, T.R., Kjaer, K., Lahav, M. and Leiserowitz, L. (2001) *Journal of Physical Chemistry B*, **105**, 8563–8568.

50 Kuzmenko, I., Buller, R., Bouwman, W., Kjaer, K., Als-Nielsen, J., Lahav, M. and Leiserowitz, L. (1996) *Science*, **274**, 2046–2049.

51 Martin, S.M., Kjaer, K., Weygand, M.J., Weissbuch, I. and Ward, M.D. (2006) *Journal of Physical Chemistry B*, **110**, 14292–14299.

52 Kuzmenko, I., Kindermann, M., Kjaer, K., Howes, P.B., Als-Nielsen, J., Granek, R., Kiedrowski, G.v., Leiserowitz, L. and Lahav, M. (2001) *Journal of the American Chemical Society*, **123**, 3771.

53 Landau, E.M., Levanon, M., Leiserowitz, L., Lahav, M. and Sagiv, J. (1985) *Nature*, **318**, 353.

54 Landau, E.M., Woilf, S.G., Levanon, M., Leiserowitz, L., Lahav, M. and Sagiv, J. (1989) *Journal of the American Chemical Society*, **111**, 1436.

55 Lee, A.Y., Ulman, A. and Myerson, A.S. (2002) *Langmuir*, **18**, 5886–5898.

56 Dressler, D.H. and Mastai, Y. (2007) *Chirality*, **19**, 258–365.

57 Nakanishi, T., Matsunaga, N., Nagasaka, M., Asahi, T. and Osaka, T. (2006) *Journal of the American Chemical Society*, **128**, 13322–13323.

58 Banno, N., Nakanishi, T., Matsunaga, M., Asahi, T. and Osaka, T. (2004) *Journal of the American Chemical Society*, **126**, 428–429.

59 Fukuda, K., Shibasaki, Y., Nakahara, H. and Liu, M. (2000) *Advances in Colloid and Interface Science*, **87**, 113–145.

60 Eliash, R., Weissbuch, I., Weygand, M.J., Kjaer, K., Leiserowitz, L. and Lahav, M. (2004) *Journal of Physical Chemistry B*, **108**, 7228–7240.

61 Zepik, H., Shavit, E., Tang, M., Jensen, T.R., Kjaer, K., Bolbach, G., Leiserowitz, L., Weissbuch, I. and Lahav, M. (2002) *Science*, **295**, 1266–1269.

62 Weissbuch, I., Bolbach, G., Zepik, H., Shavit, E., Tang, M., Frey, J., Jensen, T.R., Kjaer, K., Leiserowitz, L. and Lahav, M. (2002) *Journal of the American Chemical Society*, **124**, 9093–9104.

63 Rubinstein, I., Bolbach, G., Weygand, M.J., Kjaer, K., Weissbuch, I. and Lahav, M. (2003) *Helvetica Chimica Acta*, **86**, 3851–3866.

64 Weissbuch, I., Zepik, H., Bolbach, G., Shavit, E., Tang, M., Jensen, T.R., Kjaer, K., Leiserowitz, L. and Lahav, M. (2003) *Chemistry – A European Journal*, **9**, 1782–1794.

65 Rubinstein, I., Kjaer, K., Weissbuch, I. and Lahav, M. (2005) *Chemical Communications*, 5432–5434.

66 Rubinstein, I., Eliash, R., Bolbach, G., Weissbuch, I. and Lahav, M. (2007) *Angewandte Chemie-International Edition*, **46**, 3710–3713.

67 Wagner, K. and Brezesinski, G. (2008) *Current Opinion in Colloid & Interface Science*, **13**, 47–53.

9
Nanoscale Stereochemistry in Liquid Crystals
Carsten Tschierske

9.1
The Liquid-Crystalline State

The liquid-crystalline (LC) state is a state of matter between solids and liquids that possesses properties reminiscent of both. Like a crystalline solid it has long-range order and like a liquid it is fluid. Long-range order in liquid crystals leads to direction-dependent (anisotropic) properties, whereas fluidity allows LC systems to respond to external stimuli (e.g., electric, magnetic fields and mechanical forces) by changing their configuration. Hence, liquid crystals (LCs) represent soft matter that combines order and mobility and this unique combination is the basis of numerous technical applications [1] as well as being a prerequisite for the formation of life on Earth. There are two distinct types of long-range order in LC systems, orientational order and positional order. Orientational order develops by the parallel alignment of mesogenic units with an overall anisometric shape, driven by the desire to minimize the excluded volume and maximizing the van der Waals interactions between them. The shape of the mesogenic units is either rod-like (calamitic) or disc-like (discotic) in most cases (Figure 9.1), but also other shapes, such as board-like (sanidic), "banana"-like (bent-core mesogens) and cone-like are common [2a]. Positional order has different reasons. In particular, nanoscale segregation (microsegregation) of incompatible segments of the molecules (e.g., polar vs. nonpolar or rigid vs. flexible) contributes to positional long-range order [2, 3].

Mobility is due to conformational, rotational and translational motions. It is often provided by soft and flexible segments (e.g., alkyl or oligooxyethylene chains), which are connected with the rigid and anisometric cores, or by solvent molecules that are located either between the anisometric units, or that specifically interact with distinct parts of the mesogens. Accordingly, LC systems can be subdivided into thermotropic [2] and lyotropic [4]. Thermotropic LCs exhibit phase transitions from the crystalline via the LC state to the isotropic liquid state as temperature is increased, whereas lyotropic LCs exhibit phase transitions as a function of temperature and as a function of concentration in a solvent.

Chirality at the Nanoscale: Nanoparticles, Surfaces, Materials and more. Edited by David B. Amabilino
Copyright © 2009 WILEY-VCH Verlag GmbH & Co. KGaA, Weinheim
ISBN: 978-3-527-32013-4

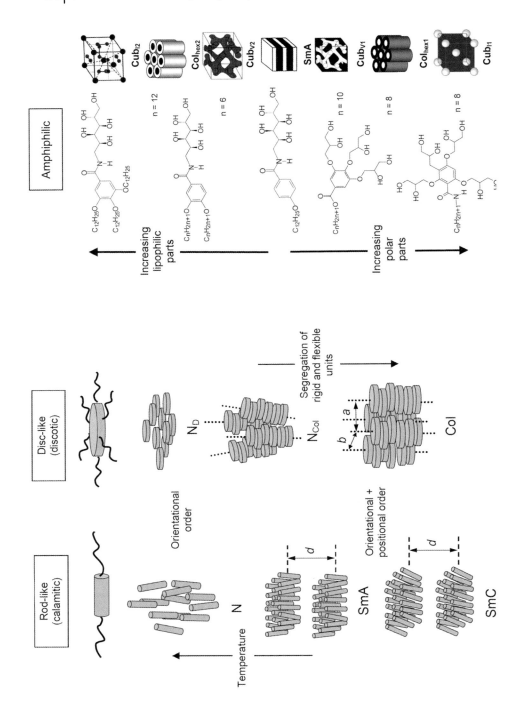

There are four fundamental modes of organization in LC systems (Figure 9.1).

1. Nematic LC phases (N) have exclusively orientational long-range order.
2. Smectic LC phases (Sm) represent layer structures (lamellar phases) with a long-range positional order between the layers (1D periodicity). In the layers the molecules can be disordered or organized either on average perpendicular to the layer planes (SmA phases) or tilted with respect to the layer normal (SmC phases).
3. Columnar LC phases (Col) have a long-range positional order in two directions, leading to a 2D periodicity. There are different subtypes with distinct plane groups. Within the columns disc-like units can be organized on average perpendicular or tilted to the column long axis [5].
4. Cubic (Cub) and noncubic mesophases with a long-range periodic order on a 3D lattice are often built up by bicontinuous networks (Cub_V) or spheroidic entities (Cub_I, see Figure 9.1, right) [2a,3a–c,4,6]. Mesophases with 3D periodicity can also be formed if in lamellar or columnar phases the molecules have a positional order within the layers or columns, respectively, and if there is a coupling of these periodicities in adjacent layers/columns [2a,5]. Due to the 3D lattice, these materials are not fluid and therefore it is recommended to designate them more generally as mesophases. All mesophases are characterized by disorder in at least one of the compartments and hence a diffuse scattering is observed in the wide-angle region of the X-ray diffraction pattern (usually around 0.45 nm as typical for the mean distance between fluid alkyl chains). LCs represent a special type of mesophase characterized by the absence of positional long-range order in at least one direction.

The mesogens considered herein represent mostly molecules or supramolecular aggregates with a specific (rod-like, disc-like) shape. However, flexible amphiphilic molecules composed of at least two incompatible molecular parts, but without specific shape, can also form LC phases composed of lamellar, columnar or spheroidic aggregates (Figure 9.1, right), depending on the size of the distinct molecular parts [6].

9.2
Chirality in Liquid Crystals Based on Fixed Molecular Chirality

In LC systems there are at least two levels of chirality. The first level is the chirality of the molecule that can be configurational and conformational and both can be either fixed or fluxional in nature. Under ambient conditions configurational chirality is fixed in most cases, whereas conformational chirality is fluxional. However, there are exceptions, such as binaphthyls and biphenyls with bulky substituents close to the

Figure 9.1 The fundamental modes of organization of rod-like, disc-like and amphiphilic mesogens in LC phases (N = nematic, SmA nontilted smectic, SmC = synclinic tilted smectic, N_D discotic nematic, N_{Col} = columnar nematic, Col = columnar, Cub_V = bicontinuous cubic, Cub_I = micellar cubic) [2–4, 6].

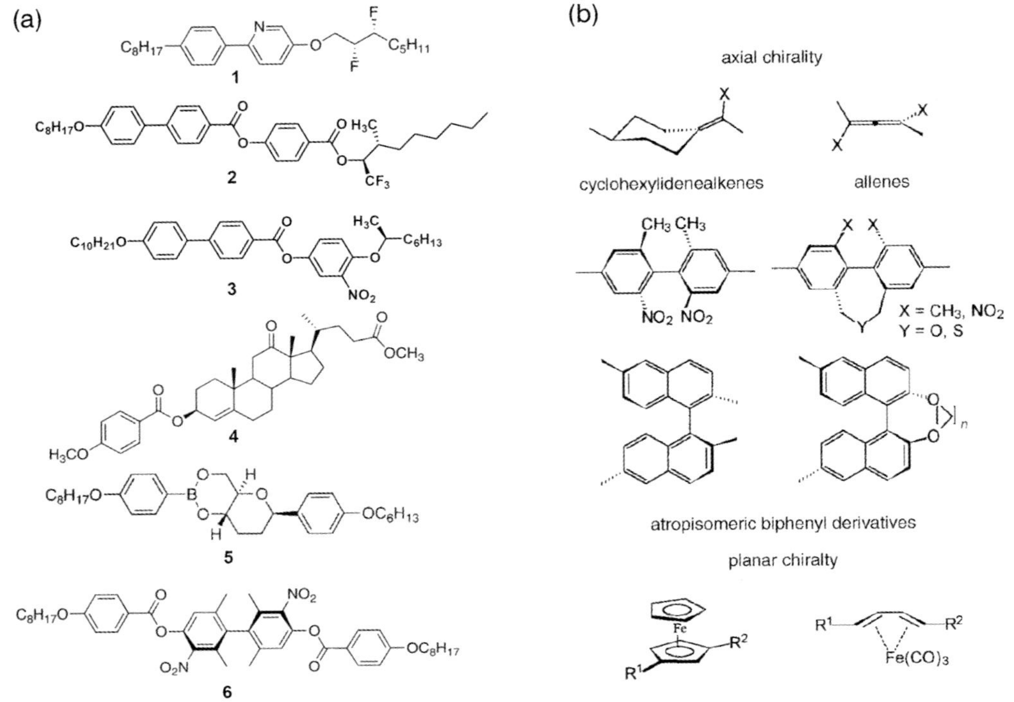

Figure 9.2 (a) Examples of chiral rod-like molecules [2, 10, 95]; (b) axial and planar chiral building blocks used for chiral LC [8a].

interaromatic bond that have a fixed chiral conformation under usual conditions. The second level of chirality arises from the organization of the molecules with formation of chiral superstructures by means of long-range positional and orientational order of molecules. Chirality at a molecular level is not always a prerequisite for chirality at the superstructural or macroscopic level, that is, it is not essential to have chiral molecules for chiral LC phases. Hence, chirality in LC systems can be either based on a chirality of molecules or dopants, as discussed in this section, or it can be formed spontaneously from achiral or racemic molecules as discussed in Sections 9.3–9.5 [7].

Molecular chirality in LC is mostly based on stereogenic centers (compounds **1–5**), but there are also examples of mesogens having a stereogenic axis and those having a plane of chirality, as shown in Figure 9.2b [8]. Chirality modifies the fundamental phase structures described in Section 9.1 [9, 10].

9.2.1
Chiral Nematic Phases and Blue Phases

As a general rule, chiral and rigid molecules cannot align perfectly parallel in soft-matter systems. The reason is that stereogenic center(s) couple with the chiral molecular conformations and this coupling provides a slight imbalance in the

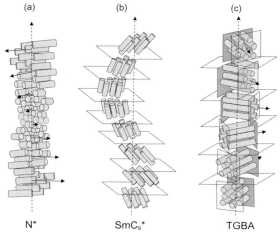

Figure 9.3 Helical superstructures in LC phases [2c].

distribution of enantiomorphic conformations, inducing a helicity of the molecules. As helical objects can be more densely packed if there is a slight deviation of their alignment in a uniform sense, this gives rise to helical superstructures (Figure 9.3), which changes the nematic phase (N) into a chiral nematic (cholesteric) phase (N*). The star (*) in the phase assignment is generally used to indicate chiral mesophases, formed by enantiomerically enriched or enantiomerically pure chiral mesogens or induced by chiral dopants in mesophases of achiral or racemic compounds. The length of the helical pitch depends on the molecular structure, the optical purity, the temperature and it is reversed for enantiomeric molecules with opposite chirality. If the pitch of this helix is of the order of the wavelength of light, then interference effects can be observed. Such materials selectively reflect light of a wavelength corresponding to the length of the helical pitch. Hence, these LCs have a photonic bandgap and can be used as photonic materials, for example in tuneable lasers [11]. The length of the pitch is inversely proportional to the helical twisting power (HTP) of the LC material and the HTP is often used as a measure of the "strength of chirality" [12, 13].

Material with a high strength of chirality and high optical purity can give rise to new phase structures, not achievable with achiral or racemic mesogens. For example, in nematic phases the helical organization can develop in all three dimensions (instead of only one as in the N* phases) and this is not possible without defects. These defects are organized on a cubic lattice, leading to the so-called blue phases (BP I, BPII) [4, 14, 15]. However, with increasing temperature the number of disordered defects increases and above a critical temperature this defect lattice becomes unstable and a transition to an isotropic mesophase without long-range order takes place [14a]. This defect-mediated melting is a quite general phenomenon in highly chiral LC systems and it is indicated by differential scanning calorimetry (DSC) where the sharp LC-to-isotropic phase transition peak is followed by an additional broad endotherm for the continuous transition from the defect-saturated isotropic

9.2.2
Chirality in Smectic Phases

In smectic phases the supramolecular helix usually develops perpendicular to the layer planes (parallel to the layer normal, Figure 9.3b). This has no effect on the structure of the nontilted SmA phases, but in tilted phases a helix is induced parallel to the layer normal (SmC* phases, Figure 9.3b). The most important chiral smectic phases are the synclinic tilted chiral SmC_s^* phase where the tilt direction is (nearly) identical in adjacent layers, and the anticlinic tilted SmC_a^* phase, where the tilt direction changes between each layer [17].

In high chirality strength smectic LC materials, between the SmC_s^* phase and the SmC_a^* phase, additional phases can arise, where the tilt direction changes in certain intervals different from 0° (SmC_s^*) and 180° (SmC_a^*), for example, by about 120° in the SmC_γ phase and about 90° in SmC_β phase [2, 9, 17, 18]. In such high chirality strength smectic materials the helix can also develop *parallel* to the layer planes. As a result the layers become deformed in a helical fashion, leading to the so-called twist grain-boundary phases (TGB phases, Figure 9.3c) [19]. This type of superstructure can be formed in tilted (TGBC phase types) and nontilted smectic phases (TGBA phase). Other chirality-induced phases are smectic blue phases [15, 20] and birefringent mesophases with 3D lattice (e.g., SmQ phases [21]) as well as cubic and disordered isotropic mesophases, among them there are also defect-melted isotropic mesophases (Figure 9.4) [15, 16, 22]. Usually, these high-chirality mesophases require "high-chirality materials," high enantiomeric purity (>90% ee) and exist only in small temperature ranges close to the transition to the isotropic liquid (Figure 9.5), but in molecules with stereogenic units at both sides, for mesogenic dimers or oligomers, and in mixtures containing dimers and oligomers or polymers, broader regions of these mesophases can be achieved [14b,22,23].

9.2.3
Polar Order and Switching in Chiral LC Phases

9.2.3.1 Ferroelectric and Antiferroelectric Switching

In tilted smectic phases, chirality of the molecules reduces the symmetry of the layers to C_2. In these layers a spontaneous polarization (P_S) arises, which is parallel to the layer planes and perpendicular to the tilt direction (i.e. along the C_2 axis, Figure 9.6). This polarization is an inherent consequence of the coupling of chirality and tilt, as predicted theoretically by Meyer *et al.* (Figure 9.6a) [24]. At the molecular level, the SmC* phase is considered to be a supramolecular host, behaving like a receptor or "binding site" for any of the involved LC molecules (Boulder model, Figure 9.6b [25]). This binding site is C_{2h} symmetric and has a zigzag shape corresponding to a preferred rotational state of the molecules in which the aromatic cores are more tilted

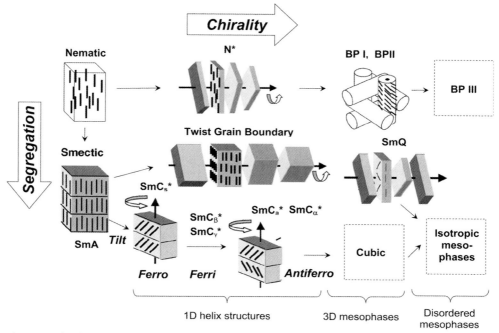

Figure 9.4 Chirality-induced mesophases of rod-like molecules, adopted from [22c].

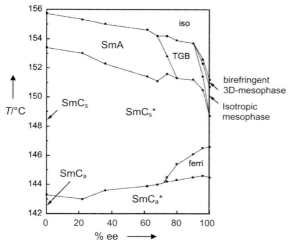

Figure 9.5 Typical phase diagram of a "high-chirality" smectic LC material depending on the enantiomeric purity. [22a] Reproduced by permission of The Royal Society of Chemistry.

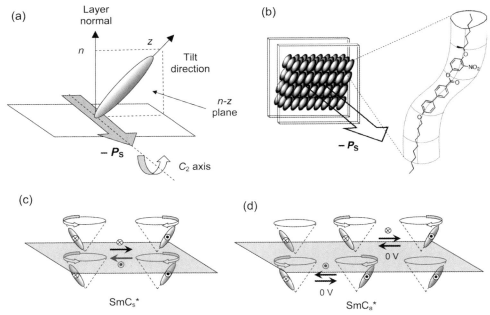

Figure 9.6 (a, b) Origin of spontaneous polarization in SmC* phases (a) coupling of chirality and tilt gives rise to a polarization (P_S) perpendicular to the tilt plane; (b) Boulder model explaining the origin of spontaneous polarization by a binding site [10b]; (c) bistable (ferroelectric) switching in the synclinic SmC$_s$* phases and (d) tristable (antiferroelectric) switching in the anticlinic SmC$_a$* phases as observed after removal of the helical superstructure (the direction of the applied field and the polarization of the molecules are indicated as dots and crosses).

than the paraffinic side chains. Steric coupling between a polar functional group and the stereogenic unit (e.g., compound **3** in Figure 9.2) results in an orientational bias of the corresponding dipole moment along the polar axis that contributes to the spontaneous polarization P_S occurring parallel to the layer planes.

This polarization is, however, cancelled on a macroscopic scale due to the helical superstructure, induced by chirality. If the helix is removed (unwound) either by surface alignment or by pitch compensation, in the resulting helix-free SmC$_s$* phases the polarization of the molecules is uniform and the polarization direction can be switched between two polar states by means of an electric field in a bistable switching process (ferroelectric switching, FE, see Figure 9.6c) [9]. In anticlinic SmC$_a$* phases, in contrast, polarity is cancelled in adjacent layers. Here the uniformly polar states can only be formed under an applied external electric field. In this case the switching does not take place directly between the two polar states, but instead there is a third apolar state at 0 V, which leads to an in total tristable (antiferroelectric, AF) switching process (see Figure 9.6d) [9, 17]. These (anti)ferroelectric soft materials provide interesting applications, for example, as fast switching light modulators or as switchable NLO materials. SmC$_a$* phases with 45° tilt (orthoconic SmC$_a$* phases) have special properties useful for applications in fast-switching displays [26].

9.2.3.2 Electroclinic Effect

An interesting effect is seen in nontilted SmA phases formed by enantiomerically enriched or enantiopure mesogens (SmA* phases) where an electric field applied parallel to the layers induces a tilt. This electroclinic effect is due to the fact that tilt induces a polarization parallel to the layer planes and the resulting polar structure (the ferroelectric SmC_s^* phase) is stabilized by the applied electric field [9, 27]. This effect can be used for fast electrooptical switching, as a sensitive tool for detection of chirality and for determination of enantiomeric purity in high-throughput parallel screening [28].

9.2.3.3 Electric-Field-Driven Deracemization

According to the arguments of Meyer *et al.* [24], tilt, polarization and chirality are coupled. For a SmC phase of achiral or racemic mesogens, the *n-z* plane (*n* = layer normal, *z* = tilt direction, Figure 9.7) is a mirror plane. If an electric field *E* is applied normal to this plane, the mirror plane is eliminated, reducing the symmetry of the system to C_2 and the phase becomes field-induced chiral, the handedness depending on the sign of the electric field, which renders the SmC phase unstable with respect to deracemization (diastereomeric relations between the field-induced chiral situation and molecular chirality). Deracemization (resolution) is accompanied by the appearance of a polarization normal to the *n–z* plane, that is the SmC phase, now chiral becomes ferroelectric (SmC*) and this polar structure is stabilized by the electric field [29]. Development of a conglomerate of chiral domains of opposite handedness was recently seen during prolonged switching of a SmC phase composed of a racemic mixture of rod-like LC material in an electric ac field [29]. The enantiomeric excess obtained in the distinct chiral domains was calculated to be about 1% ee. Strangely, in other reports only one sign of chirality was apparently observed in such experiments [30].

9.2.4
Chirality Transfer via Guest–Host Interactions

Chirality can also be induced by homogeneously chiral or enantiomerically enriched guest molecules (which do not necessarily need to be LC themselves) into an achiral

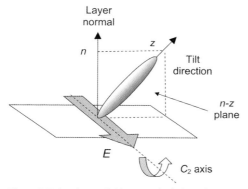

Figure 9.7 An electric field (*E*) applied along the C_2 axis removes the *n–z* mirror plane in the LC phase and leads to a chiral situation.

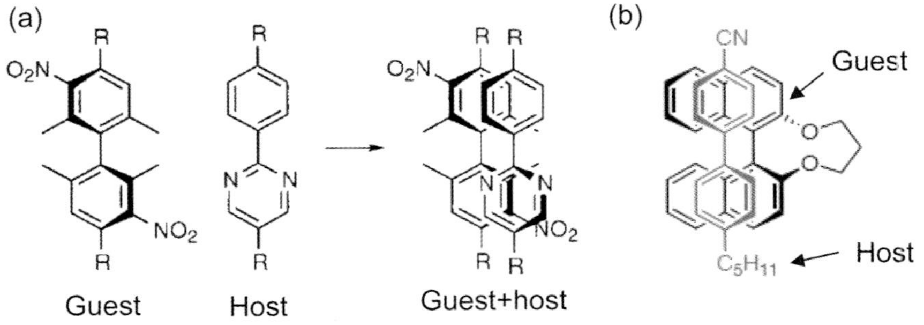

Figure 9.8 Chiral conformational interactions between axial chiral dopants and rod-like liquid-crystal hosts [10, 32].

or racemic LC host [9, 10]. In these mixtures the handedness of the dopant is transferred to the host via chiral conformational interactions (Figure 9.8). Chirality transfer is used to introduce helicity in N phases (and SmC phases, HTP) or ferroelectricity in SmC phases (polarization power). The HTP and polarization power of a dopant largely depend on the strength of chirality, which is a function of the deviation of the molecular shape from the next achiral structure [31].

Dopants having a stereogenic unit located in one of the flexible side-chains have a relatively small helical twisting as well as polarization power and these dopants are more or less invariant of the achiral host structure ("Type I" dopants, Figure 9.2, compounds **1–3**) [10]. This is due to the high degree of conformational disorder among side chains in the diffuse layer structure of the smectic phases. Dopants with stereogenic units located in the rigid core parts ("Type II" dopants', Figure 9.2 compounds **4–6**) usually have higher twisting power and polarization power which is also more sensitive to the structure of the achiral host, which may be viewed as a manifestation of molecular recognition via core–core interactions in the more ordered layers containing the rigid cores [10]. Axial chiral biphenyl and binaphthyl derivatives are especially efficient "Type II" dopants, because the chirality of these cores can directly couple with the fluxional conformational chirality of the host (twisted conformations of the biaryl units, see Figure 9.8) [9, 10, 32].

The chiral dopant itself causes a chiral perturbation of the surrounding host molecules that results in a chiral distortion of the binding site that, as a feedback effect, causes a shift in the conformational equilibrium of the chiral dopant ("chirality transfer feedback" effect) [10]. Hence, chirality transfer is a complex recognition process that can influence helicity and polarization to a different extent.

As a consequence of chirality transfer and feedback effects, configurational achiral molecules bearing highly chiral (helical) conformations as energetic minima of their conformational equilibrium can also behave as chiral dopants in the chiral environment of a chiral LC phase. For example, the helical twisting power of chiral nematic and SmC phases was increased, that is, the helical pitch became shorter, as the concentration of an achiral phenyl benzoate or bent-core mesogen (see Section 9.4) was increased [33, 34]. Moreover, blue phases that have a helical superstructure in all three dimensions (and therefore require high-chirality LC materials) have been induced with achiral bent-core molecules in relatively broad temperature ranges in chiral nematic phases [35]. Though the achiral dopant has chiral conformations with equal probability, in the chiral field of the LC phase conformers of one handedness are stabilized (diastereomeric relations) and the degeneracy of conformers is removed. If the induced chirality of the guest is stronger than the chirality provided by the host molecules themselves, an amplification of chirality is observed. In the absence of a chirality transfer from the chiral host to the dopant, or if the dopant has only weakly chiral conformations, addition of achiral molecules to a chiral host would only dilute the chirality of the system (helical pitch increases).

9.2.5
Induction of Phase Chirality by External Chiral Stimuli

In nematic and also SmC phases formed by achiral or racemic mesogens chirality can be induced by a director field between aligning surfaces of different orientations. In this configuration the director of the LC compound (i.e. the average direction of the molecular axis of the mesogenic molecules in a domain) rotates from one surface to the other, providing a helical superstructure (Figure 9.9a). This was reported as far back as 1911, when Mauguin had made several observations that liquid-crystal domains could be aligned by placing them in contact with crystal surfaces [36]. Nowadays, this effect is essential for LC display technology [37] and it must be considered as one of the possible artefacts if "spontaneous" formation of homogeneous chirality is observed in LC systems.

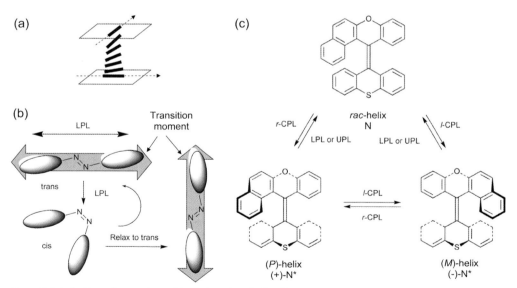

Figure 9.9 Induction of mesophase chirality by external stimuli: (a) Mauguin effect; (b) Weigert effect [38]; (c) CPL-induced deracemization of an overcrowded alkene guest that induces a positive or negative helix in a N host [40].

Irradiation of LC polymers, dimers and disc-like aggregates containing azobenzene units with circularly polarized light (CPL) was also used to induce chirality [38, 39]. This photoinduced circular anisotropy is based to the "Weigert effect" where azobenzene molecules tend to reorient under linearly polarized light irradiation to the direction perpendicular to the polarization direction of the electric field vector (Figure 9.9b). With CPL the angular momentum is transferred from the CPL to the azobenzene chromophores, inducing a helical orientation of the azobenzene chomophores, similar to that in cholesteric LC [38]. CPL can also be used to induce deracemization of dopants that then induce a helical superstructure in the host LC phase. Overcrowded alkenes, shown in Figure 9.9c, have been most efficiently used for this purpose [40].

9.2.6
Chirality in Columnar LC Phases

Chirality effects similar to those observed in nematic and SmC phases of rod-like molecules were also found in mesophases formed by chiral disc-like molecules (see Figure 9.10) [9, 5, 41]. Chiral nematic and also blue phases were observed

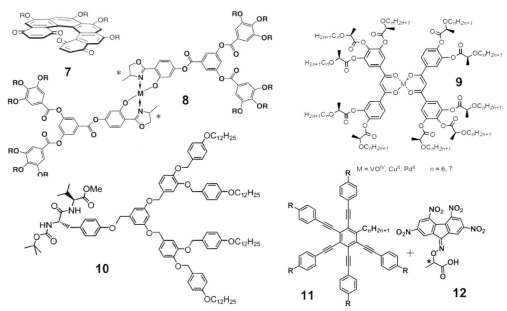

Figure 9.10 Examples of chiral molecules forming chiral columnar LC phases: **7** = helicene [45a], **8** = polycatenar metallomesogen (R = $C_{12}H_{25}$, M = Pd(II),Cu(II)) [41b], **9** = disc-like metallomesogen [41], **10** = dendritic molecule with chirality at the apex [51], **11** + **12** = donor–acceptor complex of an achiral disc-like mesogen with a chiral electron-deficient dopant [46].

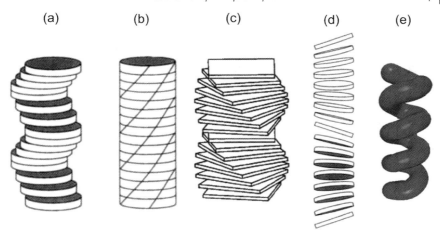

Figure 9.11 (a–d) Modes of helical arrangements in columns formed by disc-like mesogens as proposed in [43] and (e) helical filament formed by soft columns.

as chiral variants of nematic phases formed by orientational ordered disc-like molecules (N_D phases) or short columnar stacks (N_{Col} phases) [42].

In columnar phases helical structures can form as different types of intracolumnar helices along the column long axis as shown in Figures 9.11a–d [43]. All shown variants can be coupled and individual strands can form double helices, triple helices or ropes twisted around each other that leads to a wide variety of different helical superstructures. One can also imagine that several columns spiral around a straight column in the middle and an infinite number of these cords can pack into a hexagonal lattice as suggested by Kamien and Nelson ("Moiré" phase) [44]. Ferroelectric switching was found (after removal of the helical superstructure) for columnar phases, in which enantiomerically pure or enriched chiral disc-like molecules are organized in columns and where the disc-like cores are tilted with respect to the column long axis [42].

In most columnar phases the chirality is induced by stereogenic centers located in the flexible chains at the periphery of the disc-like cores, but also helicenes, where the chirality is located in the core, were investigated (Figure 9.10) [45]. Moreover, addition of chiral molecules, capable of forming electron donor–acceptor interactions [46] or hydrogen bonding with the cores were used to induce chirality [47].

As shown in Figure 9.10, formation of columnar phases is not restricted to disc-like molecules with flat aromatic cores [5]. Chiral polycatenar molecules [41] and molecules with a propeller-like [48, 49], taper-like or dendritic shape [50, 51] can also self-organize into columnar aggregates with helical superstructures. Chiral nematic (N_{Col}^* phases) and columnar LC phases can also be found for synthetic and naturally occurring chiral polymers [52] (see also Chapter 5).

9.3
Chirality Due to Molecular Self-Assembly of Achiral Molecules

9.3.1
Helix Formation in Columnar Phases

There are several reports about helix formation in columnar phases formed by stacking of achiral disc-like molecules where a close packing of the disc-like cores requires a small twist of the discs with respect to each other, allowing a staggering of the position of the peripheral substituents (Figure 9.11b and c) [53]. The helical organization within the columnar low-temperature phase of achiral discotic hexa-(hexylthio)triphenylenes belongs to the most intense investigated cases [54]. There are also reports about spontaneous helix formation in columnar mesophases of nondiscotic achiral molecules. For example, a helical columnar structure was found for a low-temperature phase of star-shaped mesogens (compound **13** in Figure 9.12) [55] and Percec et al. found X-ray diffraction patterns with the typical signature of a helical superstructure for hexagonal columnar phases formed by achiral dendritic molecules of type **14** (Figure 9.12) [56]. The helical superstructure seems to be determined by the dense packing of aromatics in a pine-tree-like (7/2 helix) or cone-like (1/5 helix) organization of the dendritic wedges around the column long axes (Figure 9.13). These columnar phases have additional intracolumnar order due to the stacking of the aromatics and some of them undergo at higher temperature a reversible first-order transition to a hexagonal columnar phase without intracolumnar order. Interestingly, at this transition the helix structure is also lost [56]. This indicates that a certain packing density is required for spontaneous formation of a chiral superstructure within the columns.

C_3-Symmetric molecules, like **16** and **17**, incorporating amide bonds can spontaneously adopt propeller-like conformations [48]. The chiral form is transferred to adjacent molecules in the columnar aggregates and provides a helical structure of the resulting columns, as shown in Figure 9.14. For the small molecule **16** the helicity mainly results from a mismatch between the π–π distance and the N–H$\cdots$O=C distance along the stacking direction [57], whereas for the foldamer **17** microsegregation and π-stacking seems to be the main forces stabilizing the helical columnar organization and the helix formation seems to be due to packing constraints at the periphery [48]. However, helical columns with opposite handedness are mixed in the bulk materials and there is no symmetry breaking on a macroscopic scale and therefore these mesophases are optically inactive. Proof of helical superstructures comes in these cases from X-ray diffraction (helix periodicity see Figure 9.13) [56] and indirectly from the study of related chiral compounds, showing optical activity and CD due to the helical organization [41] or from investigations of mixtures with chiral dopants that induce a uniform helicity (the sergeant-and-soldier effect).

Helical superstructure were also observed for 3 : 1 complexes of V-shaped benzoic acids with triaminotriazines, which was explained by a propeller-like overall shape of these complexes [49]. For complexes of this type with azobenzene-derived V-shaped

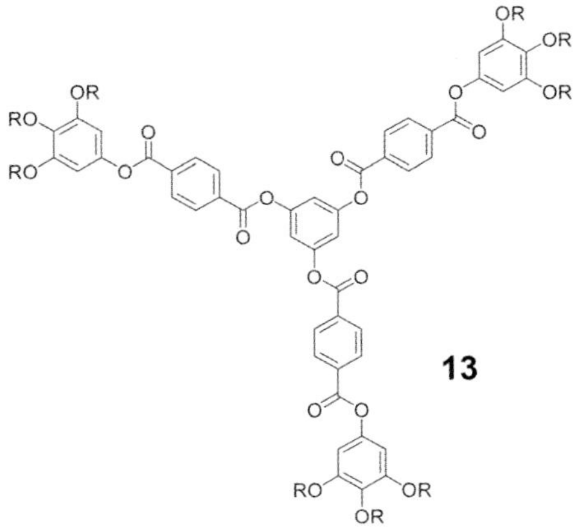

13

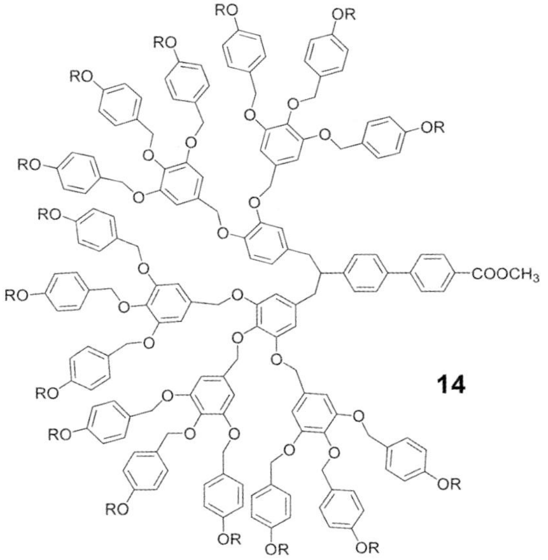

14

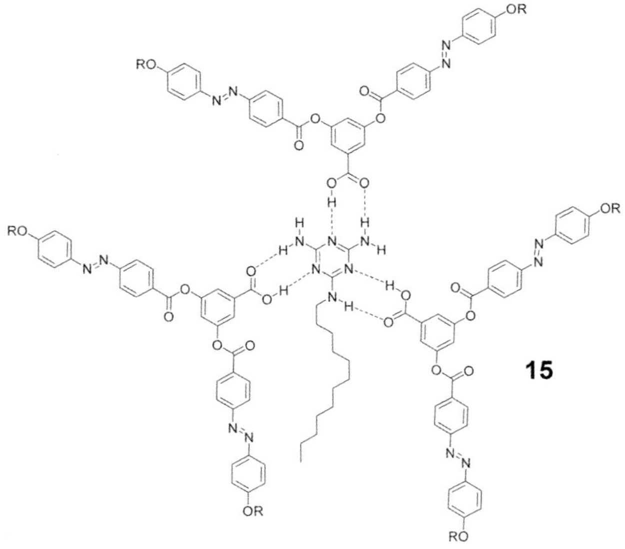

15

Figure 9.12 Examples of achiral molecules forming columnar LC phases with a helical superstructure; **13**: star-shaped molecule ($R = C_{12}H_{25}$) [55], **14**: dendritic molecules ($R = C_{12}H_{25}$) [56] and **15**: hydrogen-bonded complex formed by V-shaped molecules with a melamine derivative ($R = C_{14}H_{29}$) [39].

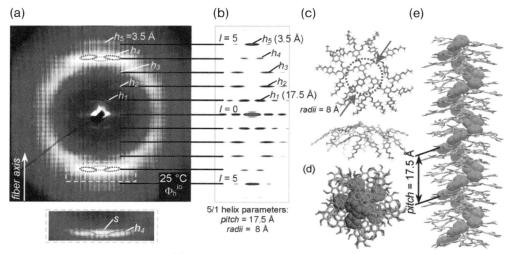

Figure 9.13 (a) X-ray diffraction pattern of an oriented fiber of the dendritic molecule **14** (see Figure 9.12) in the 1/5 helical ordered hexagonal columnar phase; (b) simulation of the fiber pattern from an ideal 5/1 helix; (c and d) views of the supramolecular helical column. Reprinted with permission from [56], copyright 2007, American Chemical Society.

molecules (complex **15**, Figure 9.12) a uniform chirality was induced by irradiation with CPL [39].

Soft columns formed by polymers or by amphiphilic molecules without rigid units can spontaneously form helical filaments (see Figure 9.11e), because in these soft columns helix formation is entropically favored as it reduces the excluded volume

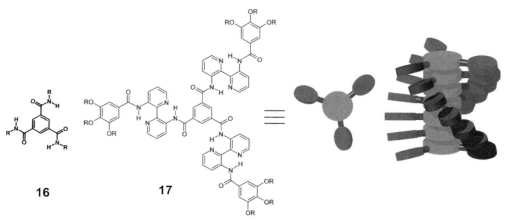

Figure 9.14 Examples of C_3 symmetric molecules and a cartoon showing their helical supramolecular stacking into columns. Reprinted with permission from [48a] copyright 2002, American Chemical Society.

around these columns [58, 59]. Block copolymers with hexagonal columnar morphologies spontaneously form single and double-helical geometries inside alumina nanochannels [60]. Hence, confinement of soft columns can induce and stabilize helical superstructures.

9.3.2
Helical Filaments in Lamellar Mesophases

There are different possible modes of helical superstructures [61] in smectic phases of achiral LC systems, as shown in Figures 9.15a–c and e. Helical ribbons (a–c) can result from deformation of relatively stiff layers or ribbons formed by densely packed crystalline lamellae and therefore these structures are typically found in gels if the packing in flat layers or ribbons is frustrated [62]. In fluid LC phases (SmA, SmC) the layers are flexible and have a certain tendency to organize into multilamellar cylinders (Figure 9.15d). These cylinders can wind up to form helical filaments (Figure 9.15e), which are entropically stabilized with respect to linear cylinders [58]. Therefore, formation of helical filaments can quite often be observed during the slow growth of smectic LC phases from the isotropic liquid state [63], especially in lyotropic systems [64] and for some thermotropic phases (e.g., B7-type phases) of bent-core molecules (see for example Figures 9.15f and g) [65–68]. A high tendency for helix formation was also reported for SmC phases of achiral biphenyl carboxylic acids, as compound **18** and related compound **18** and related compounds [69].

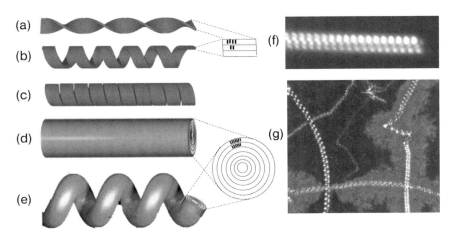

Figure 9.15 Helical superstructures; (a–c) helical ribbons [62]; (d) cylinder stack formed by rolling up flexible smectic layers; (e) helical superstructure obtained from (d); (f and g) helical filaments formed by bent-core molecules in their polar smectic phases, as seen under a polarizing microscope: (f) single helix and (g) helices coexisting with a dark conglomerate texture (slightly decrossed polarizers) [65].

18

In these achiral systems the helical filaments are generally formed with both handedness in a ratio 1 : 1 and represent an expression of macroscopic chirality that does not require chirality on a smaller length scale. Any layer instability can favor formation of cylinder stacks and if these cylinders have a sufficient flexibility, then helical filaments can develop.

9.4
Polar Order and Chirality in LC Phases Formed by Achiral Bent-Core Molecules

9.4.1
Phase Structures and Polar Order

Bent-core molecules (banana-shaped molecules) represent one of the most fascinating new directions in LC research [70]. These mesogens have a bent chevron-like aromatic core, mostly provided by a 1,3-substituted benzene ring in the middle of an aromatic core structure, incorporating in total at least five rings (e.g., compounds **19**, **21**, **22**, Figure 9.16), but also mesogenic dimers with short odd-numbered alkylene spacers (e.g., compound **20**) have been reported. The restricted rotation of such bent

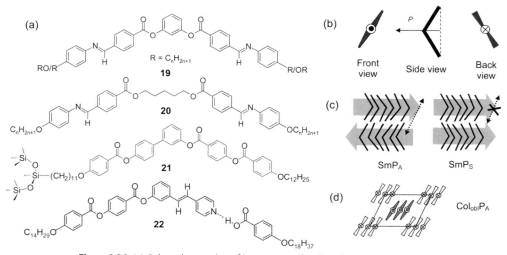

Figure 9.16 (a) Selected examples of bent-core molecules, (b) different views on the models used for the representation of bent-core molecules (c) polar smectic phases (gray arrows indicate the polar direction, dotted arrows indicate interlayer fluctuations) and (d) example of a modulated smectic = columnar phase [70, 73].

molecules around their long axes gives rise to smectic phases with a uniform orientation of the bent direction of the mesogens in the layers that leads to a polarization parallel to the layer planes (Figure 9.16c). These mesophases are assigned as polar smectic phases (SmAP, SmCP).

In these polar layers, there is a relatively dense packing of the bent aromatic cores whereas the chains are more disordered. This gives rise to an unequal area required by the aromatic and aliphatic segments at the interfaces between them and this

(a) **Tristable ("antiferroelectric") switching**

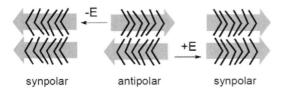

synpolar   antipolar   synpolar

(b) **Bistable ("ferroelectric") switching**

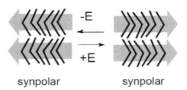

synpolar   synpolar

(c)
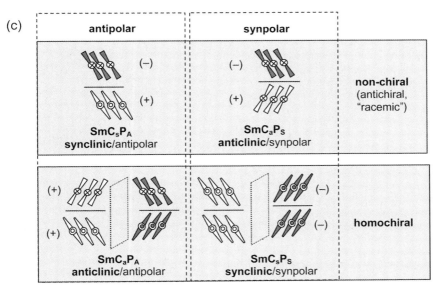

Figure 9.17 (a) AF and (b) FE switching of bent-core molecules and (c) the four distinct phase structures due to the combination of tilt direction and polar direction.

induces a tilt of the molecules. Therefore, in most of these polar smectic mesophases the molecules are strongly tilted (SmCP phases). In adjacent layers of the polar smectic phases tilt direction (synclinic = s, anticlinic = a) and polar direction (synpolar = P_S, antipolar = P_A) can be either identical (syn) or opposite (anti), leading to four different phase structures in total, as shown in Figure 9.17c [70b–e]. In two of them the polar directions in adjacent layers are identical (synpolar), providing macroscopically polar order (SmC_sP_S and SmC_aP_S phases)[1], the other two are macroscopically nonpolar, because the polar direction is opposite in adjacent layers (antipolar, SmC_aP_A and SmC_sP_A phases). For bent-core molecules the antipolar states (P_A) are entropically favored due to the possibility of interlayer fluctuations between adjacent layers (see dotted arrows in Figure 9.16c) [71] whereas the synpolar states (P_S) are stabilized by electric fields or by surface interactions. This leads to a tristable, that is, AF switching of these materials in most cases (Figure 9.17a). However, if the synpolar states are stable also at zero voltage, then a direct switching between these synpolar states without relaxation to an antipolar state is observed (bistable = FE switching, as shown in Figure 9.17b) [72, 73]. Hence, FE and AF switching can be observed for mesophases formed by achiral molecules and the spontaneous polarization is about 800 nC cm^{-2}, which is even higher than the values usually observed in the FE and AF switching SmC* phases of chiral rod-like molecules (5–200 nC cm^{-2}).

9.4.2
Superstructural Chirality and Diastereomerism

According to the predictions of Meyer *et al.* tilt, polar order and chirality are coupled [24]. If chiral molecules become tilted, the smectic layers become polar, which leads to the development of ferroelectric liquid crystals (see Section 9.2.3.1). The *electroclinic effect* (see Section 9.2.3.2) works in the opposite direction, that is, if nontilted chiral molecules become polar ordered in the electric field, a tilt appears. In achiral, but polar banana materials, the missing link between the triad of clincity, chirality and polarity was found, which means that tilted and polar ordered layer structures are chiral. Due to the orthogonal combination of polarization and tilt the smectic layers of bent-core molecules have C_2 symmetry, which lacks a mirror plane. Hence, as shown in Figure 9.18, layer normal, tilt-direction and polarization vector describe either a right-handed or a left-handed system, leading to an intrinsic chirality of the layers. Combination of individual layers leads to diastereomeric structures. In the SmC_sP_A and SmC_aP_S phases layers with opposite chirality sense (see Figure 9.17b, indicated by gray vs. white color) are combined, leading to an overall nonchiral (antichiral,[2] "racemic") structure, whereas in the SmC_aP_A and SmC_sP_S phases the layers have identical chirality sense and hence these LC structures are homochiral ("homogeneously chiral") [70b, 74]. This layer chirality represents a new kind of

---

1) Often, P_F where F stands for "ferroelectric" is used to indicate synpolar order.
2) As in these mesophases the chirality sense of adjacent layers is opposite, these non-chiral structures are also assigned as "antichiral" [74].

superstructural chirality, distinct from the usually observed helical organizations. It was proposed that mesophases formed by homochiral layers have a direction-dependent and tilt-dependent optical activity that can be detected if the birefringence of the LC phase itself is small [75]. Another interesting effect, which still requires experimental proof, is the so-called *chiroclinic effect*, which was predicted for nontilted and therefore nonchiral polar smectic phases (SmAP) of bent-core molecules, where addition of a chiral dopant is thought to induce a tilt of the molecules [74].

9.4.3
Switching of Superstructural Chirality

FE or AF switching of these polar smectic mesophases under an applied electric field usually takes place by a collective rotation of the molecules on a cone that changes both polar direction as well as tilt direction, and this maintains the chirality sense of the layers (Figure 9.18b). However, if the rotation on a cone is hindered for any reason, then switching can alternatively take place by rotation around the molecular long axes that changes only the polar direction and therefore reverses the chirality sense of the layers (Figure 9.18b). This means that in this case the superstructural layer chirality can be switched by means of external electric fields [76–78], either from an antichiral to a homochiral state (SmC_sP_A to SmC_sP_S, Figure 9.18d) and vice versa (SmC_aP_A to SmC_aP_S) or between homochiral states of opposite chirality sense (FE switching between enantiomeric SmC_sP_S states, Figure 9.18d).

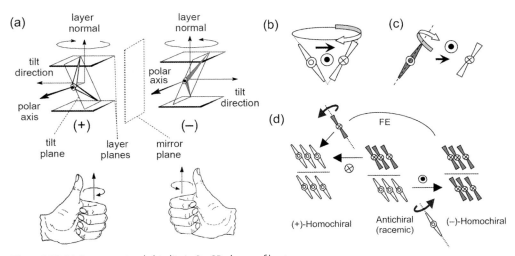

Figure 9.18 (a) Superstructural chirality in SmCP phases of bent-core molecules resulting from the combination of tilt and polar order; (b) switching on a cone retains layer chirality, whereas (c) switching around the long axis reverses layer chirality, (d) switching an antichiral (racemic) SmC_sP_A structure into homochiral SmC_sP_S structures by field-induced collective rotation around the long axes [70b–e].

Switching around the long axis takes place in smectic phases of only few bent-core mesogens at low switching frequencies if the tilt angle is small (<20°) and the molecules have a relatively large bending angle [76]. It can also be induced if the bent aromatic cores are separated from each other by crowded end groups [77] and it is the typical switching mechanism in polar ordered modulated smectic phases, where the layers are broken into ribbons (Figure 9.16d). In these ribbons the rotation of the molecules on a cone is hindered by the additional interfaces between the ribbons [78]

9.4.4
Macroscopic Chirality and Spontaneous Reflection Symmetry Breaking in "Banana Phases"

9.4.4.1 Layer Chirality

Under an electric field circular domains can be grown from SmCP phases that then can be investigated under the polarizing microscope (Figures 19.9a–c). In these circular domains the positions of the extinction crosses occur parallel and perpendicular to the main optical axis and therefore the position of these extinction brushes indicates the tilt direction of the molecules in the layers. For example, during the AF switching on a cone the reorganization of the molecules from the field-induced SmC_sP_s structure via the SmC_aP_A structure at zero voltage to the SmC_sP_s structure with opposite polar direction is associated with a change of the tilt direction. Therefore, there is a rotation of these extinction crosses and the direction of rotation upon flipping the electric field is opposite (clockwise or counterclockwise) for domains with (+)- and (−)-chirality [70b,72]. This indicates that there is a conglomerate of macroscopically chiral domains with opposite layer chirality.

9.4.4.2 Dark Conglomerate Phases

Another expression of reflection-symmetry breaking on a macroscopic scale is found in the "dark conglomerate" phases. Between crossed polarizers the textures of these mesophases appear uniformly dark, but by slightly decrossing of the polarizers, micrometer- to millimeter-sized domains with different brightness become visible (Figures 9.19d and e). The brightness of the domains is exchanged by decrossing the polarizers either in one or the other direction, indicating the optical activity of these domains. Different domains have opposite sense of chirality and opposite CD and the distribution of areas with (+)- and (−)-handedness is on average equal (conglomerate). The superscript "[*]" in the phase assignment is used to indicate this type of chiral mesophases formed by achiral molecules.

There are several subtypes of dark conglomerate phases that can be divided into two major groups. The first group represents nonswitchable mesophases with crystallized aromatic cores and fluid or conformationally disordered alkyl chains (soft crystals). The second one is derived from polar ordered and fluid smectic LC phases that can show either AF or FE switching under an applied external electric field ($SmCP_A^{[*]}$ and $SmCP_{FE}^{[*]}$ phases).

The $B4^{[*]}$ phase is the most prominent representative of the soft crystalline type of dark conglomerate phases [79, 80]. A complex structure composed of bundles of

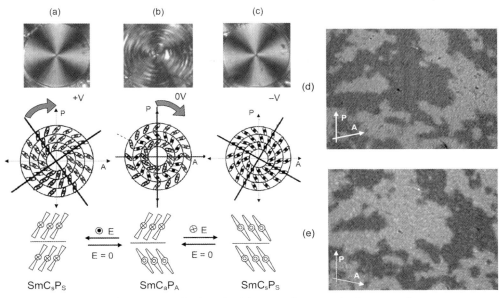

Figure 9.19 Expressions of macroscopic chirality in mesophases of bent-core molecules: (a–c) Rotation of extinction crosses by reversing the sign of the applied electric field as seen for an AF switching of a SmC$_a$P$_A$ phase [70d]; (d and e) chiral domains of opposite chirality, as seen for a dark conglomerate phase (SmCP$_{FE}$[*]) of bent-core molecules between slightly decrossed polarizers (A and P indicate the positions of polarizer and analyzer) [73].

crystalline helical ribbons (type a in Figure 9.15) formed by the bent aromatic cores between conformationally disordered alkyl chains and dominated by saddle splay defects was proposed for this mesophase [81]. According to second-harmonic generation CD investigations the helix axis is along the bent direction of the molecules [79c]. Therefore, it is likely that in these crystalline ribbons the molecules adopt a preferred chiral conformation that is coupled with the helix sense of the ribbons. It is remarkable that these B4[*] phases can be mixed with achiral nematic phases of appropriate rod-like molecules (cyanobiphenyls) and that in these mixtures the achiral nematic phase is removed and replaced by the chiral B4[*] phase with nearly unchanged rotation power and X-ray diffraction pattern [82]. This B4[*] phase covers a huge concentration range, between the pure B4[*] material to a concentration as low as only 2.5% (!) of the B4[*] material in the nematic host. It seems that the nematic material can be accommodated in the fluid regions between the crystalline helical ribbons and swells these regions. The helical ribbons might form a filament network and the helicity of the crystalline ribbons is transferred to the organization of the rod-like molecules in the space between the filaments.

In the fluid dark conglomerate SmCP$_{A/FE}$[*] phases [70d,e,73,83,84] the molecules are strongly tilted (∼35–45°). The characteristic feature of the X-ray diffraction pattern is a layer reflection with several higher-order reflections, indicating a well-defined layer structure with relatively sharp interfaces. On the other hand, the wide-angle scattering is diffuse, indicating a fluid smectic phase, and the small-angle diffraction peaks are not resolution limited, indicating a limited size of the uniformly aligned smectic domains. According to AFM investigations [73b] and electron microscopy of freeze-fractured samples [81, 85] the layers are strongly folded and deformed. As the size of the layer sections with uniform direction is smaller than the wavelength of light, these mesophases are not birefringent and appear dark between crossed polarizers. It was proposed by Hough and Clark that the optical activity in these dark conglomerate phases arises from the layer chirality in homochiral phase structures (SmC_sP_S or SmC_aP_A) [75]. Chirality in these $SmCP_{A/FE}^{[*]}$ phases develops at the transition from the isotropic liquid state to the LC state and it is fixed once it is formed. This means that the chiral domains are usually stable and do not racemize or change their shape as long as the LC state is maintained and that chirality develops as long as the liquid crystal and the isotropic liquid are in a thermodynamic equilibrium. As suggested for development of homogeneous chirality during crystallization of solids, the loss of chirality in the liquid state ("chiral amnesia") could probably also be a driving force for development of large areas with uniform chirality during "crystallization" of LCs [86].

Several experiments have been performed to break the inevitable degeneracy of the chiral domains in the dark conglomerate phases. For example, addition of only small amounts of a chiral dopant shifts the ratio of right and left domains and can even eliminate domains of one-handedness, indicating a "sergeant-and-soldier" behavior [83]. Similarly, a predominance of only one chirality sense was found if the mesophase was grown between chiral-modified surfaces (chiral polymer coating) [87]. The ratio of the domains was also changed under the influence of CPL as demonstrated for compounds incorporating azo groups (see Section 9.3) [88]. The most interesting case is reported for a compound with an additional nematic phase above a dark conglomerate phase. In this nematic phase a uniform helical pitch can be induced by surface alignment (Section 9.2.5). Upon cooling, from this helical nematic organization to the dark conglomerate phase predominantly only one sign of handedness was found for the chiral domains. Heterogeneity was observed starting with a twist angle of 40° and almost complete symmetry breaking was observed between 60 and 80° twist configuration. Reversed twist sense between the surfaces gave reversed chirality [89].

In most $SmCP_{A/FE}^{[*]}$-type dark conglomerate phases optical isotropy and domain structure can be removed and replaced by birefringent textures after application of a sufficiently strong electric field [73]. This indicates that the layer distortion is removed and flat layers are formed and it seems that the chirality of the domains is retained during this field-induced transformation [90]. However, in some cases the dark conglomerate textures could not be removed by means of an electric field. In one of these cases a FE switching was recorded and the optical activity of the chiral domains was reversed by reversing the field direction [66]. A bistable switching between two enantiomeric SmC_sP_S structures by collective rotation around the molecular long axes could be a possible explanation for this observation.

Also, a temperature-dependent and reversible inversion of chirality was observed at the transition between two different types of dark conglomerate phases, a fluid one at high temperature and a semicrystalline version at lower temperature [73b]. As this phase transition is associated with a significant change of the tilt angle from

43° at high temperature to 53° at low temperature it was proposed that the chiral superstructure itself is retained, but the optical activity is reversed due to the tilt dependence of the layer optical activity [73b].

Beside the optically active dark conglomerate phases there are also examples of optically isotropic polar smectic phases without visible optical activity [91] and in some cases optical active and inactive domains coexist [92]. A very small size of the domains, a nonchiral structure, either SmC_sP_A or SmC_aP_S or the presence of antichiral defects between homochiral SmC_sP_S layer stacks were proposed as possible reasons for optical inactivity [91, 92]. Remarkably, there are also cases where a dark conglomerate texture with chiral domains only becomes visible after application of an electric field to an apparently optically inactive isotropic mesophase or to a birefringent smectic phase [93]. Overall, these mesophases are still mysterious and not yet understood.

9.5
Spontaneous Reflection-Symmetry Breaking in Other LC Phases

The observation of reflection-symmetry breaking in smectic mesophases of achiral bent-core molecules stimulated the search for related phenomena in other LC systems. It also produced the question of whether spontaneous separation of stable configurational and fluxional conformational enantiomers could be possible in LC systems.

9.5.1
Chirality in Nematic Phases of Achiral Bent-Core Molecules

There are several reports about nematic phases of bent-core molecules showing spontaneous formation of areas with different chirality sense [94]. Also, stripe patterns similar to those known from chiral nematic phases were observed occasionally [94d]. As indicated by X-ray diffraction, most of these nematic phases represent a kind of cybotactic nematic phases composed of small smectic clusters [94b]. In principle, transfer of layer chirality between adjacent clusters could lead to the observed effects. In another model Goodby *et al.* proposed a helical stacking of the molecules in clusters where conformers of one handedness pack together to give a helical structure that in turn stabilizes one handedness of the chiral conformer [94d].

9.5.2
Spontaneous Resolution of Racemates in LC Phases of Rod-Like Mesogens

Spontaneous resolution of a racemate in a fluid LC phase was reported for the SmC phase of a racemic mixture of the rod-like compound **2** (Figure 9.2a) incorporating two stereocenters. Molecules with opposite chirality segregate and like enantiomers partially self-assembled to form a conglomerate of chiral domains, identified by ferroelectric switching (SmC* phase) as well as by texture observations and CD [95]. Deracemization was also observed under an applied electric field, as already discussed in Section 9.2.3.3).

9.5.3
Deracemization of Fluxional Conformers via Diastereomeric Interactions

The importance of molecular conformational chirality and diastereomeric relations for the transfer of chirality between LC molecules and from guests to LC hosts was already discussed in Section 9.2.4. For fluid smectic phases of bent-core molecules it was predicted theoretically that the diastereomeric relations between layer chirality (SmC_sP_S and SmC_aP_A) and molecular conformational chirality can shift the conformational equilibrium towards one of the enantiomeric conformers [96]. This induced deracemization was calculated to be of the order of only one to a few per cent in the fluid SmCP-type phases. However, in smectic phases with additional order between the aromatic cores this effect could be significantly larger.

On the other hand, there are reports that enantiomerically pure chiral bent-core mesogens form a SmC_aP_S phase structure that is nonchiral, that is, the layer chirality alternates from layer to layer despite being composed of uniformly chiral mesogens [97]. This would mean that in this case the factors selecting the layer chirality are sufficiently strong that they are not affected by the molecular chirality.

9.5.4
Chirality in Nematic, Smectic and Cubic Phases of Achiral Rod-Like Molecules

Strigazzi et al. observed chiral domains in nematic phases of achiral 4-alkylcyclohexane carboxylic acids and 4-alkyloxybenzoic acids and proposed twisted H-bonding dimers as the source of a helical structure [98].

The formation of chiral domains was recently also reported for simple SmC phases of nonchiral rod-like phenyl benzoates and it was proposed that this should be due to a spontaneous separation of chiral conformational isomers (Figure 9.20a [99]) [100], but reinvestigations have shown that no polar switching can be observed and chirality most probably derives from surface interactions [101].

Moreover, a conglomerate of chiral domains was found for an optically isotropic cubic phase of the achiral amide **23** (Figure 9.20) [102]. Addition of a chiral dopant led to an unequal distribution of the chiral areas with opposite handedness. It was assumed that the intermolecular hydrogen bonding between the amide groups of adjacent molecules generates a twisted conformation around the N–Ar bond and conformers with the same twist sense pack preferably side-by-side, as shown in Figure 9.20b [102].

9.5.5
Segregation of Chiral Conformers in Fluids, Fact or Fiction?

As shown in Sections 9.5.1–9.5.4 there are several examples of reflection-symmetry breaking events in different types of LC phases that are not yet understood. This led to a controversial discussion of whether segregation of fluxional conformers could in principle be possible in the fluid but ordered LC state.

The *conformational chirality hypothesis* holds that chiral conformations with barriers separating enantiomeric conformations can be "frozen" in the LC state by intermolecular interactions in the local chiral environment [70d,e,79,94d,99]. This is well

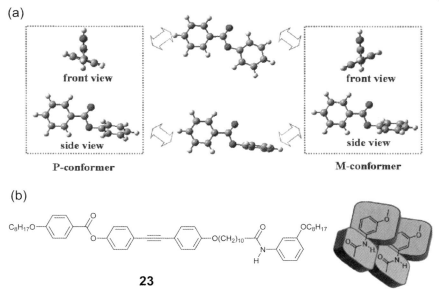

Figure 9.20 Conformational chirality (a) conformers of phenyl benzoates ([99] Reproduced by permission of The Chemical Society of Japan); (b) mesogen forming a cubic phase and model proposed for the origin of chirality in this mesophase [102].

known for crystalline solids (see Chapter 10), but it was not found in any isotropic liquid. The challenging question arises, if spontaneous chiral segregation were possible in LC phases, which represent intermediate states between the crystalline solids and disordered liquids. This might be possible at least in LC phases with positional order (Sm, Col, Cub) and rather well-defined (sharp) interfaces, which restrict the position and orientation of the molecules in space. Moreover, it should be more likely for mesophases with densely packed mesogens, which allows a larger enthalpic contribution to chiral segregation. There should also be an impact of the size of the molecules, because for larger molecules a smaller number of molecules per volume unit would reduce the unfavorable influence of the entropy of mixing.

Chiral conformers are involved in the formation of helical superstructures in columnar aggregates, but formation of larger domains with uniform chirality was not observed for columnar LC phases of achiral or racemic compounds. It seems that despite columns providing relatively strong restrictions for the organization of the molecules, giving rise to chirality in the columns (see Section 9.3.1), there is no sufficiently strong coupling between adjacent columns, due to the decoupling of the individual columns by the fluid peripheral chains.

In smectic layers the position of the molecules is less well defined, especially in layers without polar order, but nevertheless, macroscopic domains with uniform chirality were observed in several cases. It seems that the chiral coupling between layers is stronger than that between columns because the contact area between adjacent layers is larger than between columns. Moreover, chirality can be transferred between the layers by the parallel organization of the terminal chains, by interlayer fluctuations and by partial intercalation of these chains. Therefore, it could be expected that in smectic phases there is a significant coupling of chirality in adjacent layers, which is advantageous for formation of macroscopic domains with uniform chirality (cooperative effect). However, the layers also provide a strong coupling with surfaces and therefore small chirality effects provided by these surfaces can easily induce macroscopic chirality in the smectic phases of mesogens having highly chiral helical conformations as energetic minima [e.g., [100, 101]].

In bicontinuous cubic phases (see Figure 9.1) the situation is again different. There are quasi-infinite 3D networks of branched and interconnected cylinders and infinite layers located at the minimal surfaces [4]. Hence, interaggregate chirality transfer, as required for macroscopic reflection-symmetry breaking in lamellar and columnar phases, is not required in these cubic phases.

Interestingly, saddle splay, as is typical for cubic phases [4], was also observed for the dark conglomerate phases of bent-core molecules [85] and probably the stability and large size of the chiral domains in these mesophases can be explained by regarding them as (disordered) 3D networks of layers or ribbons [73b]. In the soft-crystalline B4$^{[*]}$ type of "dark conglomerate" phases the aromatic cores are crystallized and segregation of conformers is very likely in this case [79–81]. In the fluid dark conglomerate phases ($SmCP_{A/FE}^{[*]}$) chiral segregation of conformers could be stabilized by the coupling with layer chirality and the coupling to surfaces.

Besides the conformational chirality hypothesis, there is a second hypothesis, which suggests that spontaneous reflection-symmetry breaking based on segregation and freezing of chiral fluxional conformers is generally not possible in fluid phases and that simpler alternative explanations should exist for these cases [80, 101]. Surface constraints are discussed as the main origin of macroscopic chirality. Moreover, it is suggested that freezing of enantiomeric conformations is a characteristic feature defining a solid crystal, while conformational mobility defines the fluid state. Accordingly, a mesophase in which segregation of chiral conformers can be proven should be classified as a crystalline mesophase.

Hence, investigation of spontaneous reflection-symmetry breaking and enantiomer resolution in LC systems remains a challenge. If it should turn out that in LC soft-matter structures there are mechanism for spontaneous deracemization of stable or rapidly racemizing enantiomers, this would certainly have a significant impact on the general discussion around the problem of evolution of homochirality in biological systems. At least, the LC state provides efficient ways for chirality transfer and chiral amplification of weak external chirality sources. In particular, the highly efficient transfer of chirality from surfaces to bulk soft matter structures with formation of highly chiral superstructures with macroscopic dimensions is of significant importance.

9.6
Liquid Crystals as Chiral Templates

There are numerous applications of chiral LC phases in electro-optics, photonics, nonlinear optics, and so on. It should be mentioned here, that chiral nematic phases are also useful as highly twisted reaction medium to produce chiral conjugated polymers and helical fibers [103] and possibly also other chiral structures. Polymerization of chiral amphiphilic LC in their lyotropic columnar phases lead to chiral porous solid-state catalysts, useful for enantioselective catalysis [104]. Chiral lyotropic systems could be also of interest as templates for the production of chiral porous materials in sol-gel processes.

9.7
Perspective

As described above many interesting phenomena, new phase structures, specific physical properties and technological applications are due to chirality in LC systems. The recent research in bent-core mesogens led to the question of whether fluxional molecular chirality can lead to spontaneous symmetry breaking in the liquid-crystalline soft-matter states. The understanding of the distinct modes of chiral self-assembly at the nanoscale and its transfer to a macroscopic scale are still at a very initial stage and represents a challenge for future research in this fast-developing area.

References

1 (a) Kirsch, P. and Bremer, M. (2000) *Angewandte Chemie*, **39**, 4216–4235; (b) Kato, T., Mizoshita, N. and Kishimoto, K. (2006) *Angewandte Chemie*, **45**, 38–68; (c) Woltman, S.J., Jay, G.D. and Crawford, G.P. (2007) *Nature Materials*, **6**, 929–938.

2 (a) Demus, D., Goodby, J., Gray, G.W., Spiess, H.-W. and Vill, V. (1998) *Handbook of Liquid Crystals*, Vol. 1–3, Wiley-VCH, Weinheim; (b) Collings, P.J. and Hird, M. (1997) *Introduction to Liquid Crystals*, Taylor & Francis, London; (c) Tschierske, C. (2006) *Encyclopedia of Supramolecular Chemistry* (eds. Steed, J.W. and Atwood, J.L.), Marcel Dekker, New York, DOI: 10.1081/E-ESMC-120012801.

3 (a) Tschierske, C. (1998) *Journal of Materials Chemistry*, **8**, 1485–1508; (b) Tschierske, C. (2001) *Journal of Materials Chemistry*, **11**, 2647–2671; (c) Tschierske, C. (2001) *Annual Reports on the Progress of Chemistry, Section C: Physical Chemistry*, **97**, 191–267; (d) Saez, I.M. and Goodby, J.W. (2005) *Journal of Materials Chemistry*, **15**, 26–40; (e) Ungar, G. and Zeng, X. (2005) *Soft Matter*, **1**, 95–106; (f) Tschierske, C. (2007) *Chemical Society Reviews*, **36**, 1930–1970.

4 Seddon, J.M. and Templer, R.H. (1995) *Handbook of Biological Physics*, Vol. 1 (eds. Lipowsky, R. and Sackmann, E.), Elsevier, Amsterdam, pp. 97–160.

5 Laschat, S., Baro, A., Steinke, N., Giesselmann, F., Hägele, C., Scalia, G., Judele, R., Kapatsina, E., Sauer, S., Schreivogel, A. and Tosoni, M. (2007) *Angewandte Chemie-International Edition*, **46**, 4832–4887.

6 (a) Borisch, K., Diele, S., Göring, P., Kresse, H. and Tschierske, C. (1998) *Journal of Materials Chemistry*, **8**, 529–543; (b) Cheng, X.H., Diele, S. and Tschierske, C. (2000) *Angewandte Chemie-International Edition*, **39**, 592–595; (c) Fuchs, P., Tschierske, C., Raith, K., Das, K. and Diele, S. (2002) *Angewandte*

Chemie-International Edition, **41**, 628–631.

7 Chirality at the level of atoms, caused by the parity breaking of weak interactions is not considered here, because these effects are extremely small and there is presently no proven example of a measurable chirality transfer from this level of chirality to LC systems.

8 (a) Lunkwitz, R., Tschierske, C., Langhoff, A., Giesselmann, F. and Zugenmaier, P. (1997) *Journal of Materials Chemistry*, **7**, 1713–1721, and references 4–11 cited in this manuscript. (b) Lunkwitz, R., Zab, K. and Tschierske, C. (1998) *Journal fur Praktische Chemie*, **340**, 662–668.

9 (a) Kitzerow, H.-S. and Bahr, C. (2001) *Chirality in Liquid Crystals*, Springer, New York; (b) Lagerwall, S.T. (1999) *Ferroelectric and Antiferroelectric Liquid Crystals*, Wiley-VCH, Weinheim; (c) Goodby, J.W. (1991) *Journal of Materials Chemistry*, **1**, 307–318; (d) Lagerwall, S.T. (2004) *Ferroelectrics*, **301**, 15–45.

10 (a) Lemieux, R. (2001) *Accounts of Chemical Research*, **34**, 845–853; (b) Lemieux, R. (2007) *Chemical Society Reviews*, **36**, 2033–2045.

11 (a) Kitzerow, H. (2002) *Liquid Crystal Today*, **11**, 3–7; (b) Ford, A.D., Morris, S.M. and Coles, H.J. (2006) *Materials Today*, **9**, 36–42.

12 Harris, A.B., Kamien, R.D. and Lubensky, T.C. (1999) *Reviews of Modern Physics*, **71**, 1745–1757.

13 Osipov, M.A. and Kuball, H.-G. (2001) *European Physical Journal E*, **5**, 589–598.

14 (a) Stegemeyer, H., Blumel, T., Hiltrop, K., Onusseit, H. and Porsch, F. (1986) *Liquid Crystals*, **1**, 3–28; (b) Kitzerow, H.-S. (2006) *ChemPhysChem*. **7**, 63–66.

15 Pansu, B. (1999) *Modern Physics Letters B*, **13**, 769–782.

16 Goodby, J.W., Dunmur, D.A. and Collings, P.J. (1995) *Liquid Crystals*, **19**, 703–709.

17 Fukuda, A., Takanishi, Y., Isozaki, T., Ishikawa, K. and Takezoe, H. (1994) *Journal of Materials Chemistry*, **4**, 997–1016.

18 Gleeson, H.F. and Hirst, L.S. (2006) *ChemPhysChem*, **7**, 321–328.

19 Goodby, J.W. (2002) *Current Opinion in Colloid & Interface Science*, **7**, 326–352.

20 Grelet, E. (2003) *Liquid Crystal Today*, **12**, 1–5.

21 Levelut, A.M., Hallouin, E., Bennemann, D., Heppke, G. and Loetzsch, D. (1997) *Journal de Physique II*, **7**, 981–1000.

22 (a) Nishiyama, I., Yamamoto, J., Goodby, J.W. and Yokoyama, H. (2002) *Journal of Materials Chemistry*, **12**, 1709–1716; (b) Takanishi, Y., Ogasawara, T., Yoshizawa, A., Umezawa, J., Kusumoto, T., Hiyama, T., Ishikawaa, K. and Takezoe, H. (2002) *Journal of Materials Chemistry*, **12**, 1325–1330; (c) Nishiyama, I., Goodby, J.W. and Yokoyama, H. (2005) *Molecular Crystals and Liquid Crystals*, **443**, 25–41; (d) Nguyen, H.T., Ismaili, M., Isaert, N. and Achard, M.F. (2004) *Journal of Materials Chemistry*, **14**, 1560–1566.

23 (a) Yamamoto, J., Nishiyama, I., Inuoe, M. and Yokoyama, H. (2005) *Nature*, **437**, 525–528; (b) Coles, H.J. and Pivnenko, M.N. (2005) *Nature*, **436**, 997–1000.

24 Meyer, R.B., Liebert, L., Strzelecki, L. and Keller, P. (1975) *Journal de Physique (France)*, **36**, L69–L71.

25 Walba, D.M., Razavi, H.A., Horiuchi, A., Eidman, K.F., Otterholm, B., Bengt, C., Haltiwanger, R., Clark, N.A., Shao, R., Parmar, D.S., Wand, M.D. and Vohra, R.T. (1991) *Ferroelectrics*, **113**, 21–36.

26 Lagerwall, S., Dahlgren, A., Jägemalm, P., Rudquist, P., D'havė, K., Pauwels, H., Dabrowski, R. and Drzewinski, W. (2001) *Advanced Functional Materials*, **11**, 87–94.

27 Garoff, S. and Meyer, R.B. (1977) *Physical Review Letters*, **38**, 848–851.

28 Walba, D.M., Eshdat, L., Korblova, E., Shao, R. and Clark, N.A. (2007) *Angewandte Chemie-International Edition.*, **46**, 1473–1475.

29 Kane, A., Shao, R.-F., Maclennan, J.E., Wang, L., Walba, D.M. and Clark, N.A. (2007) *ChemPhysChem*, **8**, 170–174.

30 (a) Cowling, S.J., Hall, A.W. and Goodby, J.W. (2005) *Chemical Communications*, 1546–1548; (b) Cowling, S.J., Hall, A.W. and Goodby, J.W. (2005) *Advanced Materials*, **17**, 1077–1080.

31 Harris, A.B., Kamien, R.D. and Lubensky, T.C. (1999) *Reviews of Modern Physics*, **71**, 1745–1757.

32 Gottarelli, G., Hilbert, M., Samori, B., Solladié, G., Spada, G.P. and Zimmermann, R. (1983) *Journal of the American Chemical Society*, **105**, 7318–7321.

33 (a) Thisayukta, J., Niwano, H., Takezoe, H. and Watanabe, J. (2002) *Journal of the American Chemical Society*, **124**, 3354–3358; (b) Choi, S.-W., Fukuda, K., Nakahara, S., Kishikawa, K., Takanishi, Y., Ishikawa, K., Watanabe, J. and Takezoe, H. (2006) *Chemistry Letters*, **35**, 896–897; (c) Kawauchi, S., Choi, S.-W., Fukuda, K., Kishikawa, K., Watanabe, J. and Takezoe, H. (2007) *Chemistry Letters*, **36**, 750–751.

34 Gorecka, E., Cepic, M., Mieczkowski, J., Nakata, M., Tajezoe, H. and Zeks, B. (2003) *Physical Review E*, **67**, 061704.

35 Nakata, M., Takanishi, Y., Watanabe, J. and Takezoe, H. (2003) *Physical Review E*, **68**, 041710.

36 Mauguin, C. (1911) *Bulletin de la Société Française de Minéralogie*, **34**, 71–117.

37 Hoogboom, J., Rasing, T., Rowan, A.E. and Nolte, R.J.M. (2006) *Journal of Materials Chemistry*, **16**, 1305–1314.

38 Choi, S.W., Kawauchi, S., Ha, N.Y. and Takezoe, H. (2007) *Physical Chemistry Chemical Physics*, **9**, 3671–3681.

39 Vera, F., Tejedor, R.M., Romero, P., Barbera, J., Ros, M.B., Serrano, J.L. and Sierra, T. (2007) *Angewandte Chemie-International Edition*, **46**, 1873–1877.

40 Eelkema, R. and Feringa, B.L. (2006) *Organic and Biomolecular Chemistry*, **4**, 3729–3745.

41 (a) Serrano, J.L. and Sierra, T. (2000) *Chemistry – A European Journal*, **6**, 759–766; (b) Serrano, J.L. and Sierra, T. (2003) *Coordination Chemistry Reviews*, **242**, 73–85.

42 (a) Heppke, G., Krüerke, D., Löhning, C., Lötzsch, D., Moro, D., Müller, M. and Sawade, H. (2000) *Journal of Materials Chemistry*, **10**, 2657–2661; (b) Krüerke, D., Rudquist, P., Lagerwall, S.T., Sawade, H. and Heppke, G. (2000) *Ferroelectrics*, **243**, 207–220.

43 (a) van Nostrum, C.F., Bosman, A.W., Gelinck, G.H., Schouten, P.G., Warman, J.M., Kentgens, A.P.M., Devillers, M.A.C., Meijerink, A., Picken, S.J., Sohling, U., Schouten, A.-J. and Nolte, R.J.M. (1995) *Chemistry – A European Journal*, **1**, 171–181; (b) Engelkamp, H., van Nostrum, C.F., Picken, S.J. and Nolte, R.J.M. (1998) *Chemical Communications*, 979–980.

44 Kamien, R.D. and Nelson, D.R. (1996) *Physical Review E*, **53**, 650–666.

45 (a) Vyklicky, L., Eichhorn, S.H. and Katz, T.J. (2003) *Chemistry of Materials*, **15**, 3594–3601; (b) Praefcke, K., Eckert, A. and Blunk, D. (1997) *Liquid Crystals*, **22**, 113–119.

46 Praefcke, K. and Holbrey, J.D. (1996) *Journal of Inclusion Phenomena and Macrocyclic Chemistry*, **24**, 19–41.

47 Barbera, J., Puig, L., Romero, P., Serrano, J.L. and Sierra, T. (2005) *Journal of the American Chemical Society*, **127**, 458–464.

48 (a) van Gorp, J.J. Vekemans, J.A.J.M. and Meijer, E.W. (2002) *Journal of the American Chemical Society*, **124**, 14759–14769; (b) Palmans, A.R.A. and Meijer, E.W. (2007) *Angewandte Chemie-International Edition*, **46**, 8948–8968.

49 Barbera, J., Puig, L., Romero, P., Serrano, J.L. and Sierra, T. (2006) *Journal of the American Chemical Society*, **128**, 4487–4492.

50 (a) Davis, J.T. and Spada, G.P. (2007) *Chemical Society Reviews*, **36**, 296–313; (b) Kato, T., Matsuoka, T., Nishii, M., Kamikawa, Y., Kanie, K., Nishimura, T., Yashima, E. and Ujiie, S. (2004) *Angewandte Chemie-International Edition*, **43**, 1969–1972.

51 Percec, V., Dulcey, A.E., Peterca, M., Adelman, P., Samant, R.,

Balagurusamy, V.S.K. and Heiney, P.A. (2007) *Journal of the American Chemical Society*, **129**, 5992–6002.

52 Livolant, F. and Leforestier, A. (1996) *Progress in Polymer Science*, **21**, 1115–1164.

53 Pisula, W., Tomovic, Z., Watson, M.D., Müllen, K., Kussmann, J., Ochsenfeld, C., Metzroth, T. and Gauss, J. (2007) *The Journal of Physical Chemistry. B*, **111**, 7481–7487.

54 Fontes, E., Heiney, P.A. and De Jeu, W.H. (1988) *Physical Review Letters*, **61**, 1202–1205.

55 Gearba, R.I., Anokhin, D.V., Bondar, A.I., Bras, W., Jahr, M., Lehmann, M. and Ivanov, D.A. (2007) *Advanced Materials*, **19**, 815–820.

56 Percec, V., Won, B.C., Peterca, M. and Heiney, P.A. (2007) *Journal of the American Chemical Society*, **129**, 11265–11278.

57 (a) Bushey, M.L., Hwang, A., Stephens, P.W. and Nuckolls, C. (2002) *Angewandte Chemie-International Edition*, **41**, 2828–2831; (b) Lightfoot, M.P., Mair, F.S., Pritchard, R.G. and Warren, J.E. (1999) *Chemical Communications*, 1945–1946.

58 Snir, Y. and Kamien, R.D. (2005) *Science*, **307**, 1067.

59 Zhong, S., Cui, H., Chen, Z., Wooley, K.L. and Pochan, D.J. (2008) *Soft Matter*, **4**, 90–93.

60 Wu, Y., Cheng, G., Katsov, K., Sides, S.W., Wang, J., Tang, J., Fredrickson, G.H., Moskovits, M. and Stucky, G.D. (2004) *Nature Materials*, **3**, 816–822.

61 Maeda, K. and Yashima, E. (2006) *Topics in Current Chemistry*, **265**, 47–88.

62 (a) Shimizu, T., Masuda, M. and Minamikawa, H. (2005) *Chemical Reviews*, **105**, 1401–1443; (b) Hill, J.P., Jin, W., Kosaka, A., Fukushima, T., Ichihara, H., Shimomura, T., Ito, K., Hashizume, T., Ishii, N. and Aida, T. (2004) *Science*, **304**, 1481–1483.

63 Todorokihara, M., Fujihara, K. and Naito, H. (2004) *Molecular Crystals and Liquid Crystals*, **412**, 77–83.

64 Sakurai, I., Kawamura, Y., Sakurai, T., Ikegami, A. and Seto, T. (1985) *Molecular Crystals and Liquid Crystals*, **130**, 203–222.

65 Dantlgraber, G., Keith, C., Baumeister, U. and Tschierske, C. (2007) *Journal of Materials Chemistry*, **17**, 3419–3426.

66 (a) Eremin, A., Diele, S., Pelzl, G. and Weissflog, W. (2003) *Physical Review E*, **67**, 020702; (b) Pelzl, M.W., Schröder, A., Eremin, S., Diele, B., Das, S., Grande, H., Kresse, W. and Weissflog (2006) *European Physical Journal E*, **21**, 293–303.

67 (a) Pelzl, G., Diele, S., Jakli, A., Lischka, C., Wirth, I. and Weissflog, W. (1999) *Liquid Crystals*, **26**, 135–139; (b) Lee, C.-K. and Chien, L.-C. (1999) *Liquid Crystals*, **26**, 609–612; (c) Jákli, A., Lischka, C., Weissflog, W., Pelzl, G. and Saupe, A. (2000) *Liquid Crystals*, **27**, 1405–1409; (d) Achard, M.-F., Kleman, M., Nastishin, Y.A. and Nguyen, H.-T. (2005) *European Physical Journal E*, **16**, 37–47; (e) Bailey, C., Cartland, E.C., Jr and Jakli, A. (2007) *Physical Review E*, **75**, 031701.

68 Coleman, D.A., Fernsler, J., Chattham, N., Nakata, M., Takanishi, Y., Korblova, E., Link, D.R., Shao, R.-F., Jang, W.G., Maclennan, J.E., Mondainn-Monval, O., Boyer, C., Weissflog, W., Pelzl, G., Chien, L.-C., Zasadzinski, J., Watanabe, J., Walba, D.M., Takezoe, H. and Clark, N.A. (2003) *Science*, **301**, 1204.

69 Jeong, K.-U., Knapp, B.S., Ge, J.J., Jin, S., Graham, M.J., Harris, F.W. and Cheng, S.Z.D. (2006) *Chemistry of Materials*, **18**, 680–690.

70 (a) Niori, T., Sekine, F., Watanabe, J., Furukawa, T. and Takezoe, H. (1996) *Journal of Materials Chemistry*, **6**, 1231–1233; (b) Link, D.R., Natale, G., Shao, R., Maclennan, J.E., Clark, N.A., Körblova, E. and Walba, D.M. (1997) *Science*, **278**, 1924–1927; (c) Pelzl, G., Diele, S. and Weissflog, W. (1999) *Advanced Materials*, **11**, 707–724; (d) Reddy, R.A. and Tschierske, C. (2006) *Journal of Materials Chemistry*, **16**, 907–961; (e) Takezoe, H. and Takanishi, Y.

(2006) *Japanese Journal of Applied Physics*, **45**, 597–625.

71 (a) Glaser, M.A. and Clark, N.A. (2002) *Physical Review E*, **66**, 021711; (b) Nishida, K., Cepic, M., Kim, W.J., Lee, S.K., Heo, S., Lee, J.G., Takanishi, Y., Ishikawa, K., Kang, K.-T., Watanabe, J. and Takezoe, H. (2006) *Physical Review E*, **74**, 021704.

72 Walba, D.M., Körblova, E., Shao, R., Maclennan, J.E., Link, D.R., Glaser, M.A. and Clark, N.A. (2000) *Science*, **288**, 2181–2184.

73 (a) Dantlgraber, G., Eremin, A., Diele, S., Hauser, A., Kresse, H., Pelzl, G. and Tschierske, C. (2002) *Angewandte Chemie-International Edition*, **41**, 2408–2412; (b) Keith, C., Reddy, R.A., Hauser, A., Baumeister, U. and Tschierske, C. (2006) *Journal of the American Chemical Society*, **16**, 3051–3066.

74 Cepic, M. (2006) *Europhysics Letters*, **75**, 771–777.

75 (a) Hough, L.E. and Clark, N.A. (2005) *Physical Review Letters*, **95**, 107802; (b) Hough, L.E., Zhu, C., Nakata, M., Chattham, N., Dantlgraber, G., Tschierske, C. and Clark, N.A. (2007) *Physical Review Letters*, **98**, 037802.

76 (a) Heppke, G., Jakli, A., Rauch, S. and Sawade, H. (1999) *Physical Review E*, **60**, 5575–5579; (b) Schröder, M.W., Diele, S., Pelzl, G. and Weissflog, W. (2004) *ChemPhysChem*, **5**, 99–103; (c) Bedel, J.P., Rouillon, J.C., Marcerou, J.P., Nguyen, H.T. and Achard, M.F. (2004) *Physical Review E*, **69**, 061702; (d) Nakata, M., Shao, R.-F., Maclennan, J.E., Weissflog, W. and Clark, N.A. (2006) *Physical Review Letters*, **96**, 067802; (e) Weissflog, W., Dunemann, U., Schröder, M.W., Diele, S., Pelzl, G., Kresse, H. and Grande, S. (2005) *Journal of Materials Chemistry*, **15**, 939–946.

77 (a) Keith, C., Reddy, R.A., Baumeister, U. and Tschierske, C. (2004) *Journal of the American Chemical Society*, **126**, 14312–14313; (b) Keith, C., Reddy, R.A., Baumeister, U., Kresse, H., Lorenzo Chao, J., Hahn, H., Lang, H. and Tschierske, C. (2007) *Chemistry – A European Journal*, **13**, 2556–2577.

78 Szydlowska, J., Mieczkowski, J., Maraszek, J., Bruce, D.W., Gorecka, E., Pociecha, D. and Guillon, D. (2003) *Physical Review E*, **67**, 031702.

79 (a) Niwano, H., Nakata, M., Thisayukta, J., Link, D.R., Takezoe, H. and Watanabe, J. (2004) *The Journal of Physical Chemistry. B*, **108**, 14889–14896; (b) Kurosu, H., Kawasaki, M., Hirose, M., Yamada, M., Kang, S., Thisayukta, J., Sone, M., Takezoe, H. and Watanabe, J. (2004) *Journal of Physical Chemistry A*, **108**, 4674–4678; (c) Araoka, F., Ha, N.Y., Kinoshita, Y., Park, B., Wu, J.W. and Takezoe, H. (2005) *Physical Review Letters*, **94**, 137801.

80 Walba, D.M., Eshdat, L., Körblova, E. and Shoemaker, R.K. (2005) *Crystal Growth & Design*, **5**, 2091–2099.

81 Clark, N.A., Hough, L.E., Heberling, M.-S., Spannuth, M., Nakata, M., Glaser, M., Krüerke, D., Jones, C.D., Heppke, G., Jung, H., Zasadzinski, J., Rabe, J., Stocker, W., Körblova, E. and Walba, D. (2007) Presented at the 11th International Conference on Ferroelectric Liquid Crystals FLC07, Sapporo, Japan, I–09.

82 Takanishi, Y., Shin, G.J., Jung, J.C., Choi, S.-W., Ishikawa, K., Watanabe, J., Takezoe, H. and Toledano, P. (2005) *Journal of Materials Chemistry*, **15**, 4020–4024.

83 Thisayukta, J., Nakayama, Y., Kawauchi, S., Takezoe, H. and Watanabe, J. (2000) *Journal of the American Chemical Society*, **122**, 7441–7448.

84 Alonso, I., Martinez-Perdiguero, J., Ortega, J., Folcia, C.L. and Etxebarria, J. (2007) *Liquid Crystals*, **34**, 655–658.

85 Spannuth, M., Hough, L., Coleman, D., Jones, C., Nakata, M., Takanishi, Y., Takezoe, H., Watanabe, J., Tschierske, C., Korblova, E., Walba, D., Clark, N.A. and Poster, (2004) Presented at the 9 th International Liquid Crystal Conference, Book of Abstracts, Ljubljana, July 4–9, SYN-P026.

86 Blackmond, D.G. (2007) *Chemistry – A European Journal*, **13**, 3290–3295.

87 Shiromo, K., Sahade, D.A., Oda, T., Nihira, T., Takanishi, Y., Ishikawa, K. and Takezoe, H. (2005) *Angewandte Chemie-International Edition*, **44**, 1948–1951.

88 Choi, S.-W., Izumi, T., Hoshino, Y., Takanishi, Y., Ishikawa, K., Watanabe, J. and Takezoe, H. (2006) *Angewandte Chemie-International Edition*, **45**, 1382–1385.

89 (a) Choi, S.-W., Kang, S., Takanishi, Y., Ishikawa, K., Watanabe, J. and Takezoe, H. (2006) *Angewandte Chemie-International Edition*, **45**, 6503–6506; (b) Choi, S.-W., Kang, S., Takanishi, Y., Ishikawa, K., Watanabe, J. and Takezoe, H. (2007) *Chirality*, **19**, 250–254.

90 Lee, S.K., Shi, L., Tokita, M., Takezoe, H. and Watanabe, J. (2007) *The Journal of Physical Chemistry. B*, **111**, 86998–8701.

91 Liao, G., Stojadinovic, S., Pelzl, G., Weissflog, W., Sprunt, S. and Jakli, A. (2005) *Physical Review E*, **72**, 021710.

92 (a) Hahn, H., Keith, C., Lang, H., Reddy, R.A. and Tschierske, C. (2007) *Advanced Materials*, **18**, 2629–2633; (b) Keith, C., Dantlgraber, G., Reddy, R.A., Baumeister, U., Prehm, M., Hahn, H., Lang, H. and Tschierske, C. (2007) *Journal of Materials Chemistry*, **17**, 3796–3805.

93 Martínez-Perdiguero, J., Alonso, I., Folcia, C.L. and Etxebarria, J. (2006) *Physical Review E*, **74**, 031701.

94 (a) Pelzl, G., Eremin, A., Diele, S., Kresse, H. and Weissflog, W. (2002) *Journal of Materials Chemistry*, **12**, 2591–2593; (b) Weissflog, W., Sokolowski, S., Dehne, H., Das, B., Grande, S., Schröder, M.W., Eremin, A., Diele, S., Pelzl, G. and Kresse, H. (2004) *Liquid Crystals*, **31**, 923–933;

(c) Niori, T., Yamamoto, J. and Yokoyama, H. (2004) *Molecular Crystals and Liquid Crystals*, **409**, 475–482; (d) Görtz, V. and Goodby, J.W. (2005) *Chemical Communications*, 3262–3264.

95 Takanishi, Y., Takezoe, H., Suzuki, Y., Kobayashi, I., Yajima, T., Terada, M. and Mikami, K. (1999) *Angewandte Chemie-International Edition*, **38**, 2353–2357.

96 Earl, D.J., Osipov, M.A., Takezoe, H., Takanishi, Y. and Wilson, M.R. (2005) *Physical Review Letters*, **71**, 021706.

97 Nakata, M., Link, D.R., Araoka, F., Thisayukta, J., Takanishi, Y., Ishikawa, K., Watanabe, J. and Takezoe, H. (2001) *Liquid Crystals*, **28**, 1301–1308.

98 Torgova, S.I., Petrov, M.P. and Strigazzi, A. (2001) *Liquid Crystals*, **28**, 1439–1449.

99 Kawauchi, S., Choi, S.-W., Fukuda, K., Kishikawa, K., Watanabe, J. and Takezoe, H. (2007) *Chemistry Letters*, **36**, 750–751.

100 Kajitani, T., Masu, H., Kohmoto, S., Yamamoto, M., Yamaguchi, K. and Kishikawa, K. (2005) *Journal of the American Chemical Society*, **127**, 1124–1125.

101 Walba, D.M., Körblova, E., Huang, C.-C., Shao, R., Nakata, M. and Clark, N.A. (2006) *Journal of the American Chemical Society*, **128**, 5318–5319.

102 Kajitani, T., Kohmoto, S., Yamamoto, M. and Kishikawa, K. (2005) *Chemistry of Materials*, **17**, 3812–3819.

103 (a) Yoneyama, H., Tsujimoto, A. and Goto, H. (2007) *Macromolecules*, **40**, 5279–5283; (b) Goh, M., Kyotani, M. and Akagi, K. (2007) *Journal of the American Chemical Society*, **129**, 5519–8527.

104 Pecinovsky, C.S., Nicodemus, G.D. and Gin, D.L. (2005) *Chemistry of Materials*, **17**, 4889–4891.

10
The Nanoscale Aspects of Chirality in Crystal Growth: Structure and Heterogeneous Equilibria

Gérard Coquerel and David B. Amabilino

10.1
An introduction to Crystal Symmetry and Growth for Chiral Systems. Messages for Nanoscience

The growth of a crystal is a hierarchical process: Starting from a nanoscopic nucleation point molecules join up in a kinetically controlled but thermodynamically influenced way to generate strong one-, two-, or three-dimensional connectivities in a periodic macroscopic object [1]. Nanoscience frequently takes inspiration from crystal structures because they allow easy identification of effective patterns of noncovalent bonding interactions [2]. To take a very simple example, trimesic acid forms an interpenetrating hexagonal network in the solid state [3], which has been used to generate layers incorporating spaces that might be used for binding of other molecules [4]. Furthermore, a derivative of this compound, the tris-amide **1**, forms a chiral one-dimensional hydrogen-bonded column in the solid state (Figure 10.1) [5], a fact that has motivated the preparation of several systems that form chiral nanofibers [6].

The growth of a crystal can be envisioned from different viewpoints, including the noncovalent interactions between the components and the thermodynamic and kinetic aspects of their formation, but the packing of molecules and ions in three dimensions is characterized to a large extent by the symmetry of the space groups that they crystallize [7]. When racemic or achiral source materials are crystallized, the inversion center, glide plane, and two-fold screw symmetry are the most frequent symmetry elements in the solids (Figure 10.2). Centrosymmetric pairs of molecules are especially favored [8]. These symmetry principles are important for systems in the nanometer-size regime, because in certain instances, such as two-dimensional systems where some of the elements are evaded [9], but they will be important in all systems involving three-dimensional packing.

Apart from the relatively trivial case of chiral molecules that are forced to crystallize in chiral space groups, the racemic modification of a molecule or achiral molecules could crystallize, leading to a conglomerate (with enantiomers in different crystals), a

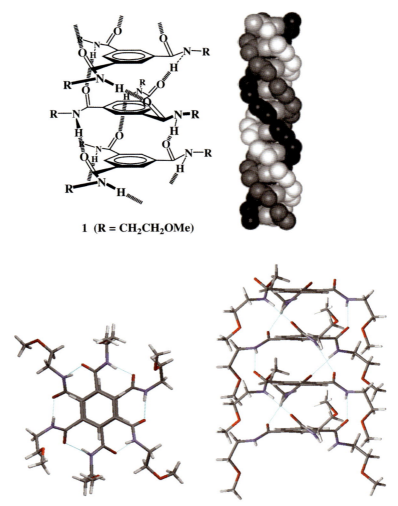

Figure 10.1 Chemical structure of stacks of **1** in its crystals, showing the amide hydrogen bonds, and a partial view of the helices that the hydrogen bonds form.

racemic compound, or a pseudoracemate [10]. The most wide ranging analysis of the propensities of racemic organic compounds to crystallize in different chiral space groups was published many years ago [11], and there has not been such a comprehensive reappraisal since. That data indicates that over 90% of molecules devoid of metal ions crystallize in centrosymmetric space groups: These are usually $P2_1/c$, $P\text{-}1$, $C2/c$, Pbca, etc... that is to say the usual list of most populated space groups for all organic compounds. Some controversial arguments have generated hot debates on the reasons why these racemic compounds are so favored [12]. The remaining

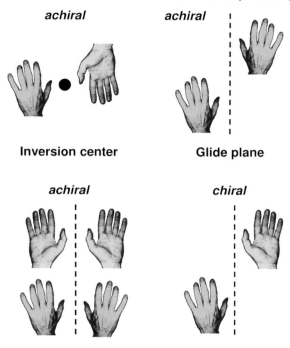

Figure 10.2 Common symmetry elements leading to achiral and chiral superstructures.

sub-10% fall in the 65 chiral space groups.[1] Over all compound types, the vast majority of structures are in the $P2_12_12_1$ and $P2_1$ space groups, followed by (an order of magnitude lower) $P1$ and $C2$ [8]. Why these preferences exist is a matter of intense debate. Thermodynamic factors are clearly important in determining the packing of molecules in crystals (particularly the efficiency of packing, more dense packing being favored, in which entropy ($T\Delta S$) and the enthalpy (ΔH) offset one another [13]) and kinetics of crystallization also play a role, because the molecules are kinetically restrained in one conformation (more often than not) and relative orientation. An authoritative analysis of the division of chiral and nonchiral space groups that cover coordination compounds has yet to be revealed.

Under usual circumstances, crystallization of a racemic material that resolves spontaneously will lead to the formation of crystals with opposite optical rotation, and

---

1) There are 65 chiral space groups, and 165 nonchiral ones. The 65 are: Cubic $P23$, $P2_13$, $I23$, $I2_13$, $F23$, $P432$, ($P4_132$, $P4_332$), $P4_232$, $I432$, $I4_132$, $F432$, $F4_132$, tetragonal $P4$, ($P4_1$, $P4_3$), $P4_2$, $I4$, $I4_1$, $P422$, $P42_12$, ($P4_122$, $P4_322$), ($P4_12_12$, $P4_32_12$), $P4_222$, $P4_22_12$, $I422$, $I4_122$, monoclinic $P2$, $P2_1$, $C2$, orthorhombic $P222$, $P222_1$, $P2_12_12$, $P2_12_12_1$, $C222$, $C222_1$, $I222$, $I2_12_12_1$, $F222$, triclinic $P1$, trigonal $P3$, ($P3_1$, $P3_2$), $P312$, $P321$, ($P3_112$, $P3_212$), ($P3_121$, $P3_221$), $R3$, $R32$, hexagonal $P6$, ($P6_1$, $P6_5$), ($P6_2$, $P6_4$), $P6_3$, $P622$, ($P6_122$, $P6_522$), $P6_322$, ($P6_222$, $P6_422$). The space groups between brackets are enantiomorphous pairs. All chiral space groups are noncentrosymmetric, but not all noncentrosymmetric space groups are chiral.

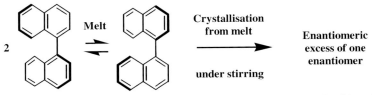

Figure 10.3 A schematic representation of the spontaneous resolution of 1,1'-binaphthyl.

even different form – enantiomorphism – as in Pasteur's classic discovery of the enantiomorphic crystals of sodium ammonium tartrate [14]. However, from an achiral or racemizing compound, spontaneous symmetry breaking can occur where an excess of just one enantiomer is crystallized, for one of a number of reasons. In a stirred solution, an initial nucleation point can be split by the mechanical contact of the stirrer to generate a secondary nucleation, and so on [15]. If the molecule in question can racemize, an example being 1,1'-binaphthyl (**2**, Figure 10.3) [16], then a racemic material can be converted into an optically active product through so-called spontaneous resolution under racemizing conditions [17]. Rarely, crystals can be grown from a nonstirred, apparently homogeneous solution to generate just one enantiomer [18]. This situation may be due to the presence of very low concentrations of chiral impurities that are amplified, or can arise when a single nucleus grows [19].

Up to now, we have treated samples whose mixture of enantiomers is split spontaneously upon crystal formation. On the other hand, symmetry can be forcibly broken by introducing a chiral agent that forms crystalline diastereomeric complexes with the two enantiomers, usually salts [11, 20]. This method is used successfully on an industrial scale to separate enantiomers because one of the salts is less soluble than the other, a point that will be dealt with in more detail in the sections that follow.

From this brief introduction to the crystallization of chiral compounds, it is clear that there are a variety of factors that can lead to the formation of chiral structures and to the separation of enantiomers. These aspects will be addressed in the coming sections, which deal first with the most relevant aspects of noncovalent interactions in crystals, followed by coverage of the different types of chiral crystalline solid-state structures that have been described.

10.2
Supramolecular Interactions in Crystals

The control of intermolecular noncovalent forces lies at the heart of most (if not all) rational attempts to influence chirality in the solid state, and the control of structure is at the scale of nanometers. It is in the spirit of any scientist to believe that it is possible to understand things using the scientific process and design logically a way of overcoming problems to resolve, in this case, compounds. Yet, logic and reason have yet to overcome in a general way the intrinsic problem of how to predict crystal structures. Although one- and even two-dimensional packing motifs can be relied upon, control in the third dimension, which is necessary for the separation of enantiomers, has not been conquered. Still, valiant efforts have resulted in the implication of intermolecular forces in the separation of enantiomers. It has been shown that in many systems, noncovalent interactions can help compounds to crystallize in chiral space groups, and the conformations of the molecules influence the chirality transfer in the solid state. It is an oversimplification to believe that any one interaction can lead to a system that resolves, but it is useful to spell them out one by one, in the knowledge that more often than not the combination of two or more of them is necessary to lead to a successful system.

10.2.1
Hydrogen Bonds

The hydrogen bond is the interaction *par excellence* in the set of noncovalent forces in crystals [21]. Its strength and directionality [22] make it a reliable ally of the "crystal engineer," and one-, two- and three-dimensional hydrogen-bonding networks can be achieved with a relative degree of control in cases where conformational possibilities are limited. Chiral hydrogen-bonded networks are reliably formed from a number of different molecules. For example, the helical tubuland hosts are a family of aliphatic chiral dialcohols (e.g., **3**, Figure 10.4) that exhibit an almost predictable behavior when cocrystallizing with achiral guests of a compatible size, forming conglomerates [23]. The chiral crystals contain parallel helical tubes of the host molecules linked through hydrogen bonds with guest molecules located in the cavities. The common "supramolecular synthon" is the [H···O−H···O−] hydrogen-bonded chain, manifesting itself either as an infinite helix or cyclic tetramer (Figure 10.4), for example. Depending on the length of the spacer between the two hydroxyl groups, one or other of these packing arrangements is achieved. Boronic acids are another example of compounds in which interhydroxyl bonds determine the packing and leads to spontaneous resolution [24]. The kind of hydrogen-bonding pattern in chiral crystals was at the center of a study by Leiserowitz and Weinstein on the occurrence of noncovalent arrangements in crystals [25].

The *trans* amide hydrogen bond is another strong interaction that leads to chains of molecules linked by the N–H to carbonyl connection [26]. As seen in the example in the introduction, this bond can lead to quite elegant structures, and the type of

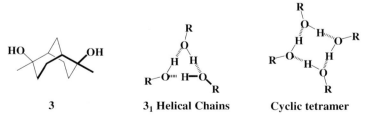

Figure 10.4 A representative tubuland-forming molecule and two hydrogen-bonding patterns that occur in tubuland crystals.

association in the crystalline state can be used to infer the mechanisms for chiral recognition in chromatographic separations [27].

10.2.2
Interaromatic Interactions

While the forces that hold aromatic systems together are weak and poorly directional [28], in comparison with classical hydrogen bonds, π–π stacking [29] and the [C−H···π] interaction [30] are able to influence greatly the chiral packing of molecules in crystals, as we shall see later in this chapter.

An example of π–π stacking in resolution of simple complexes, with no coordination bonds between the units, is seen for the hexafluorophosphate salt of the ruthenium (II) complex with two 2,2′-bipyridine molecules and one phenenthroline ligand [31]. A stack of phenanthroline ligands is generated in the crystals of the complex (π–π separation 3.52 Å) giving a homochiral pillar, which is surrounded by like ones (Figure 10.5). The tetrafluoroborate salt of the same complex, on the other hand, is a racemic compound even though π-stacking interactions are present (though the stacking distance is longer than in the hexafluorophosphate salt). The Ru(II) tris(2,2′-bipyridine) complex as its hexafluorophosphate salt is also a racemic compound [32]. Very subtle differences in packing tip the balance between conglomerate and racemate in these systems.

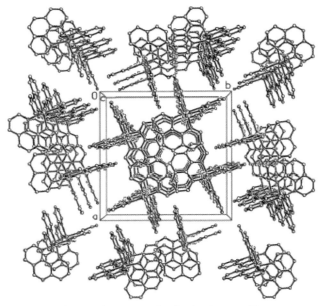

Figure 10.5 The crystal structure of a chiral ruthenium (II) complex showing the π overlap of the phenanthroline ligands.

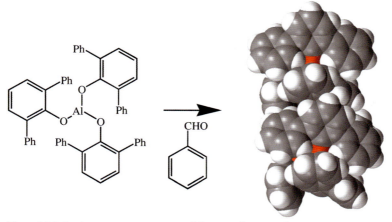

Figure 10.6 A schematic representation of the crystallization of the complex between benzaldehyde and aluminum tris(2,6-diphenylphenoxide), with the asymmetric unit of the crystal structure showing two complex molecules where the close edge-to-face contacts can be seen.

A quite remarkable preference to form crystalline conglomerates is shown when benzaldehyde derivatives coordinate through their carbonyl oxygen atom to the metal atom in aluminum tris(2,6-diphenylphenoxide) [33]. The relative orientations of the phenoxy ligands create helical-type conformations, and these chiral arrangements are spread in the solid state thanks to several C–H···π interactions (Figure 10.6) in three dimensions. High yields of one or the other enantiomer can be obtained by slow crystallization because the enantiomers interconvert rapidly in solution.

10.2.3
Electrostatic Interactions

The strongest single noncovalent interaction is that between oppositely charged ions [34], but at the same time it is the least specific in the absence of any other interaction. In addition, the presence of charge in hydrogen-bonded [35] or π-organized systems [36] makes the interactions much stronger in comparison with the equivalent bonds in the absence of mutually compatible ions. It is no coincidence that the vast majority of resolutions involving crystalline materials through diastereomers take advantage of salts.

The stereoselectivity can be at the level of the crystal formation, or in individual complexes. For example, electrostatic interactions between ligands attached to a copper (II) ion results in the optical resolution of α-amino acids [37]. Specifically, DL-aspartic and glutamic acid were resolved through the formation of ternary complexes composed of the acid and enantiopure arginine, lysine, or ornithine. The complementary electrostatic interactions between the side chains of the pure compounds with the racemate lead to the specific formation of one of the possible diastereomers preferentially.

10.2.4
Modulation of Noncovalent Interactions with Solvent

This section on intermolecular forces would not be complete without mentioning the vital role that the solvent has on molecular recognition processes, a well-documented phenomenon [38]. The binding strength between molecules in a liquid depends on the physicochemical characteristics of that medium, such as its hydrogen-bonding accepting and/or donating characteristics, etc. [39], but there is also the possibility that the solvent molecules are incorporated in the eventual structure and can be determining in the formation of racemic compound or conglomerate [40].

The composition of the solvent has also been shown to play a vital influence over the chiral selection of certain resolving agents [41], and can even invert the stereoselectivity of resolving agents from one enantiomer of the target racemate to the other [42]. The solvate of a diastereomeric salt can result in optical resolution, while the pure salt does not, an example being the incorporation of methanol in a salt of *trans*-chrysanthemic acid [43]. In this case, the solvent was not incorporated stoichiometrically, and was hypothesized to play a role in crystal nucleation and growth.

10.2.5
Polymorphism

No discussion on the crystal forms of chiral compounds would be complete without a mention of polymorphism – the existence of distinct crystal forms of the same composition of a compound [44]. Polymorphs arise because of the similarity in energy of distinct arrangements of molecules in crystals, a phenomenon that can be caused because of competition between the different noncovalent interactions mentioned above. For this reason, the topic is particularly relevant for chiral systems, because the polymorphs may result in a process where resolution can be achieved or not. For example, the spontaneous resolution of silver complexes from achiral ligands can take place, while a polymorph is a racemic compound [45]. The chiral polymorph resulting from differences in [C—H···O] hydrogen bonding of an achiral molecule has also been described [46].

10.3
Symmetry Breaking in Crystal Formation

The spontaneous separation of enantiomers – either configurational or conformational – upon crystallization is not a frequent occurrence, as we have pointed out (between 5 and 10% of organic compounds), but, nonetheless, it is an interesting way of generating chiral systems, provided it can be controlled. Here, we will describe some pertinent examples of the phenomenon. How do we know if a compound will break symmetry upon crystallization? This question remains unanswerable today because of the unpredictability of crystal structures. While some 1D or even 2D aspects of crystals can be designed, in most instances the 3D structure cannot because of the weak forces acting between chains or sheets of molecules.

10.3.1
Spontaneous Resolution of Chiral Compounds

A plethora of odd examples of conglomerates are known, from which it is difficult to learn anything in isolation, but certain families of compounds apparently tend to crystallize in chiral space groups [11]. An example is a series of 4-arenesulfonyliminocyclohex-2-en-1-one derivatives (**4** and **5**, Figure 10.7) that produced eleven new conglomerates [47]. One of the compounds crystallizes as a 2:1 inclusion compound with CCl_4 (space group $P1$). This achiral solvent molecule is considered to act as a "conglomerator," and preferential spontaneous resolution can be achieved by its use in substoichiometric amounts. In contrast, the structurally related O-benzenesulfonyl-oximes **6** are racemic compounds, demonstrating how a small variation in the molecular structure can influence the nature of the racemic mixture.

Achiral compounds can be used to generate conglomerates from otherwise racemic compounds, because the supramolecular structure of the compound is changed by its interaction with the achiral guest, which presumably reduces symmetry possibilities of the host. Again, supramolecular structure of the host–guest complex directs conglomerate formation. As an example, the axially chiral 1,1'-binaphthyl-2,2'-dicarboxylic acid (**7**, Figure 10.8) is a racemic compound and forms racemic complexes with a variety of guests. However, the salt-type complex of **7** with

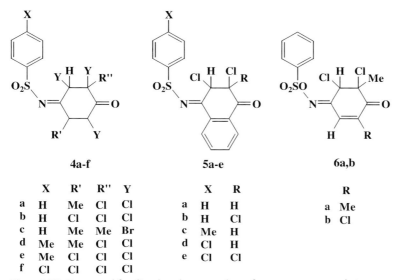

Figure 10.7 Compound families that show a tendency for spontaneous resolution.

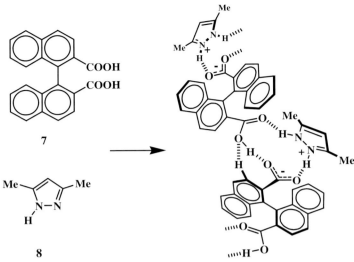

Figure 10.8 Schematic representation of the chains of the monoanion of **7** and protonated **8** that are present in the conglomerate crystals [48].

3,5-dimethylpyrazole (**8**) spontaneously resolves upon crystallization [48]. The 1:1 complex is formed by a carboxylate anion and a pyrazolium cation linked by an [$^+$N–H···O$^-$], [$^+$N–H···O] and [O–H···O$^-$] hydrogen bonds as well as [C–H···O] interactions in a cyclic arrangement, leading to the formation of chains of the same configuration, i.e. each crystal is formed by only on enantiomer of **7** (Figure 10.6). The crystals showed hemihedrism and they were separated and their physical properties evaluated. The spontaneous resolution of the 1:1 complex of **7** with **8** into a conglomerate constitutes a unique example amongst many complexes reported for the bis-acid. It should be noted, though, that the frequency of conglomerate formation in salts appears to be much greater than in neutral compounds [49].

The spontaneous resolution of chiral coordination complexes (which may or may not interconvert in solution) is a common process also [50], although a quantification of the tendency by analysis of structures has not been undertaken. Spontaneous resolution has led to the isolation of a series of mononuclear seven-coordinate complexes with the general formula [Ln(dbm)$_3$H$_2$O] [51]. Three lanthanide complexes of dibenzoylmethane (dbm) have a molecular structure most easily envisaged as a capped octahedron and crystallize as conglomerates (for Pr, Sm and Er) because of the formation of a "quadruple helix" and a number of C–H···π interactions. The spontaneous resolution is under racemizing conditions (where nucleation is slower than crystal growth) because the complexes are labile in solution, and essentially enantiopure samples can be recovered.

A particularly nice example of the spontaneous resolution of a coordination complex is that of a silver(I) complex laced with aromatic groups, which as well as resolving shows a predominance of one handedness of crystals in each batch [52].

Face-to-face and edge-to-face interactions aid in the transmission of chirality in the crystals.

10.3.2
Spontaneous Resolution of Achiral Compounds

Achiral substances can pack in chiral ways to give materials that are optically active, quartz being the classic example. The "chiral crystallization" – as it is referred – of achiral organic compounds that pack in a chiral way was most recently reviewed by Matsuura and Koshima for organic compounds [53]. It was realized early on that (photo)chemical reactions in chiral crystals comprised of molecules with no stereogenic centers would be a way of performing absolute asymmetric synthesis [54].

A structural database screening of purely organic achiral compounds that crystallize in chiral space groups suggests that the flexibility of the molecules is influential in their spontaneous resolution, or not [55]. One of the most important indications from this work was that chiral conformationally flexible molecules are 7.9 times more likely to crystallize in $P2_1/c$ rather than $P2_12_12_1$, while the same preference is shown by more rigid molecules but "only" 5.9 times. In addition, if the rigid molecule has mirror symmetry, the preference is cut to 4.3 times. Apparently, when an achiral molecule crystallizes in a chiral space group, some of its features make it prefer packing with the formation of a screw axis rather than an inversion centre and glide plane. This study – a preliminary search that is hoped to lead further – indicates a role of flexibility in determining separation and packing of organic molecules.

A nice example of the formation of chiral crystals is that of tertiary amides (no hydrogen-bond-forming capacity) that are a source of conformational chirality. For instance, 1-2-bis(N,N'-benzoyl-N,N'-methylamino)benzene (**9**, Figure 10.9) is a rather intriguing molecule in the sense that crystallization conditions can modulate whether it crystallizes as a racemic compound (space group $P2_1/n$) or as a conglomerate (space group $P2_12_12_1$) [56]. In various dry solvents amide **9** spontaneously resolves under racemizing conditions, giving enantiomorphic crystals that were morphologically distinguishable, exhibited different physical properties, and were characterized by their solid-state CD spectra. Slow crystallization, as well as seeding,

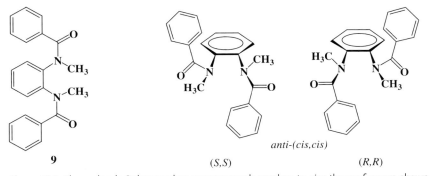

Figure 10.9 The molecule **9** that resolves spontaneously resolves to give the conformers shown.

induces preferential formation of one of the enantiomers, but water of crystallization results in a racemic compound.

Curiously, crystal structures of both racemic and enantiomorphic crystals of **9** are similar. There are two possible enantiomeric *anti-(cis,cis)* conformers. In the chiral crystals, four molecules of the same chirality are present in the unit cell, whereas racemic crystals include two of each enantiomer and four molecules of water. Therefore, water molecules act as a "racemizer." This fact is even more interesting considering that chirality induction takes place under racemizing conditions, and that conformational conversion is fast in solution (about 16 kcal/mol at 200–300 K), and therefore chirality in solution is only retained at low temperatures (173 K).

Cocrystals of achiral compounds can also form conglomerates [57]. As an example, we cite the crystallization of a variety of benzoic acid derivatives with organic bases (Figure 10.10), in which the 2_1 screw axes is present in their structures, and three of them are chiral [58]. Hydrogen bonds between the acid and pyridine-type basic heterocycles are present in all the complexes, although beyond this level the connectivities are quite dissimilar. The complex of 4-nitrobenzoic acid with 2-aminopyridine reveals a structure with heterodimers between the components, then N−H···O hydrogen bonds between the "free" amine hydrogen atom and one of the carboxyl oxygen atoms generates the 2_1 chain. The 2:1 complex of 3,5-dinitrobenzoic acid and 2,2′-bipyridine contains spiraling supramolecular chains of the acid pillared

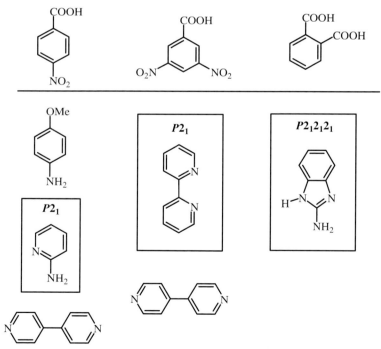

Figure 10.10 Achiral pyridine-type bases and carboxylic acids that form crystals with the conglomerates boxed.

together by the *trans*-conformer of bipyridine, forming voids that are filled by another network. In contrast, the complex of phthalic acid and 2-aminobenzimidazole contains sheets of supramolecular rings thanks to the strong hydrogen bonds between the components. Given the diverse structures, it is unclear if there is a relation between them that explains the occurrence of chirality. They do illustrate the ease with which chiral cocrystals can be generated from achiral components.

Many coordination compounds comprised of metal centers and achiral ligands exhibit spontaneous resolution. Symmetry breaking in this kind of system can be achieved by adding molecules to the solutions that retard crystal growth and encourage nucleation from a limited number of points. A beautiful example of this effect is the addition of ammonia to the mixture from which the conglomerate [{Cu(succinate)(4,4′-bipyridine)}$_n$]4(H$_2$O)$_n$ crystallizes [59]. The ammonia complexes the metal ion in solution, and the nucleation events are reduced, as seen visibly by the number and size of the crystals being formed. In addition, the CD spectra of the samples with few and large crystals (with more ammonia) showed optical activity (implying symmetry breaking) while those of the small crystals formed without ammonia or with small quantities showed no bulk optical activity form the mixture of the crystals, the behavior of a conglomerate.

10.4
Resolutions of Organic Compounds

The "classic" resolution of an organic compound requires its functionalization as an acid or amine and subsequent crystallization with the pure enantiomer of a resolving agent (RA) with a basic or acidic chiral intermediate, respectively (Figure 10.11). It constitutes the industrially most important resolution method for racemates [60]. In the ideal case, one of the diastereomeric salts is favored and crystallizes from the mixture. But, how starting from the developed formulae of the molecule to resolve, is

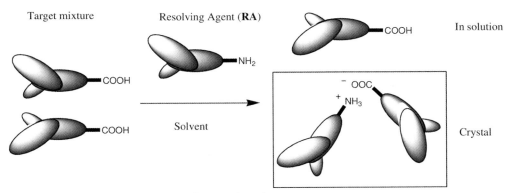

Figure 10.11 A schematic representation of "classical" resolution of enantiomers – in this case resolution of a mixture of acids by an amine.

it possible to predict some efficient RA? This long lasting question has received several fragmented answers, but to date no general satisfying answer has been validated. The different issues (detailed below) in the discussion of phase diagrams show that it is a multisided problem, which is difficult to tackle from every angle. When dealing with nonsolvated diastereomers, most of the time the crystal structures are not so different and this is why the lattice energies are similar.

In many examples of diastereomeric acid-amine complexes there is a kind of nanophase segregation in which a hydrogen-bonded layer, often incorporating a 2_1 axis, is formed with a planar boundary surface formed by the substituents that interact through van der Waals interactions [61]. Among the rules that have been proposed in view to cover a large number of successful Pasteurian resolution [62], one can cite:

1. The principle of maximum similarity between the molecule to resolve and the RA [63].
2. The relative rigidity of the RA.

In the first rule there is some kind of consistency with the principle of formation of centrosymmetric-like pairs. However, no strict rule exists for a clear definition of maximum similarity! The second rule stipulates that rigid RA would increase the difference in the solid state between the two diastereomers: i.e. the rigidity renders the RA less adaptive to the two enantiomers. Nevertheless, when the crystal structures of both diastereomers are known, there is not a significant proportion of very different RA conformations [64].

The "space-filler concept" has also been invoked [65]; it stipulates that the molecular lengths of base and acid should be similar for optimum results. If the base and the acid, laid side by side, leave an empty space a protic additive – such has a water molecule – should fill the empty space.

A further complication arises when considering solvates, and more particularly hydrates things can differ more deeply, and for instance different diastereomers can display a bidimensional hydrogen-bond-connected net or columnar hydrogen-bonded network [66]. Therefore, for a given racemic mixture to resolve, a real prediction of a good RA would suppose that the organic solid state can be predicted. The current state of the art is far from that level of knowledge, although notable progress is being made [67]. In a similar way as solvents molecules can be active partners for a high chiral recognition at the solid state; it is likely that the so-called "cocrystals" could also help in finding a nice Pasteurian process. An interesting recent case of cocrystallization resulted in the successive resolution of 1,1′-binaphthyl-2,2′-dicarboxylic acid and secondary aliphatic alcohols using enantiopure (1R,2R)-1,2-diphenylethylenediamine [68]. The resolution of the alcohol in the second step is achieved thanks to the cavities formed between the 2_1 chains of the acid and amine components.

Complexity can be taken advantage of in other ways: A promising advance in terms of speeding up the screening of resolving agents is the family resolution strategy, or "Dutch resolution" because of the origin of its inventors [69]. Here, several (usually three) resolving agents of the same family of derivatives and of the same handedness are used simultaneously [70]. The major benefits of this methodology are fourfold:

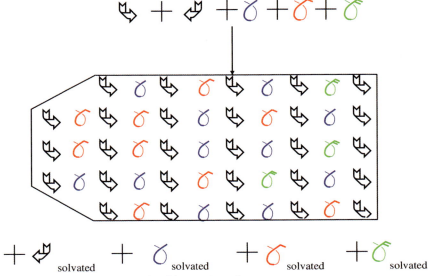

Figure 10.12 A schematic example of the Dutch resolution.

- The less soluble diastereomers are easily crystallized,
- Good crystallinity of the less soluble salts,
- The solute to be resolved shows a very poor domain of solid solution if any,
- A strong inhibition effect on spontaneous nucleation of the opposite diastereomer is usually observed.

High dissymmetry in the phase diagram (*vide infra*) means sharp contrast in the molecular interactions (a high degree of discrimination in the solid state) (Figure 10.12).

Some bifunctional analogs of a given resolving agent can also induce selective effect on the nucleation of a couple of diastereomers. The same additive can also be a habit modifier of the less soluble salt and then improve the filterability of the slurry [71].

Besides the screening of an efficient resolving agent, it can be convenient to find a crystallizing inhibitor of the more soluble salt. As already evoked, a family resolution (i.e. Dutch resolution) can involve an inhibiting effect on the crystallization of the most soluble salt [72]. A successful nucleation inhibitor of the most soluble salt seems to form a solid solution with the more soluble salt and barely affects the less soluble salt. Unfortunately, nucleation in itself is a critical but poorly understood step in the resolution process [73]. Despite this, "tailor-made" additives can be used for resolution of enantiomers of amino acids that form conglomerates [74]. Monte Carlo simulations of growth sites on crystals may prove to be helpful in the identification of these tailor-made crystal-growth inhibitors [75].

The crystallization of achiral organic compounds in a chiral lattice, while being more a case of chiral induction than resolution of the conformations of the achiral system, does lead specifically to chiral structures with supramolecular tilt chirality [76]. For example, the cocrystallization of cholic acid with different benzene derivatives in what is an inclusion complex [77]. These systems are dealt with in more detail in Section 10.8.

10.5
Resolutions of Coordination Compounds with Chiral Counterions

Coordination compounds show all the behavior that organic compounds do. This was demonstrated for the first time by the resolution of a coordination compound carried out in Werner's group at the end of the 19th century or early 20th century (the first results remained unpublished for many years) [78]. The most widely known resolution is that of hexol (a hexacation), which was accomplished by treating the hexol chloride salt with silver (+)-bromocamphorsulphonate in dilute acetic acid.

"Hexol"

$$\left[\text{Co}\{(\mu\text{-OH})_2\text{Co}(NH_3)_4\}_3 \right]^{6+}$$

The metal ions are normally positively charged, and the chiral anions afforded by nature – tartrate, mandelate, camphorsulphonate, and their derivatives – are more often than not the resolving agents. The area has witnessed a positive boom in recent years, because of the interest in the optical [79], magnetic [80] and chemical (catalysis [81]) properties of chiral metal ion complexes. The separation of the enantiomers of the metal complexes in research laboratories is often achieved by chromatographic separation of diastereomers because it is in general faster and more reliable than crystallization.

A relatively recent addition to the family of resolving anions for cationic complexes is the so-called TRISPHAT phosphorous-based anion [82], which can be prepared on a large scale in very high optical purity by resolution using a cinchonidinium derivative that solubilizes one of the enantiomers preferentially [83]. Among other coordination compounds, this anion makes possible the resolution of the ruthenium (II) complex Ru(PhenMe$_2$MeCN)$_2$, which was shown to be a new type of enantioselective catalyst [84]. The highly solvated single crystals of the complex implies that the interaction between the π-face of one of the anion's rings with the terminal ring of the phenanthroline unit aids in promoting diastereoselectivity. The dication has both phenanthroline rings interacting with different chiral anions. Furthermore, if the stereoselective precipitation is performed under irradiation (resulting in equilibration at the metal center), only one of the two enantiomers of the metal complex is present at the end of the reaction [85].

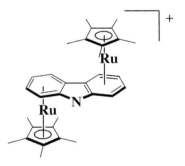

Δ TRISPHAT   Λ Ru(PhenMe₂MeCN)₂

Sp,Sp Carb(Cp*Ru)₂

The π-stacking of phenanthroline residues in another ruthenium complex with aromatic rings of (+)-O,O'-di-4-tolyl–D-tartrate is present in the crystal structure of the hydrated salt that allows separation of enantiomers, in this case by chromatography [86], as it is in the hydrated salt of the dibenzoyl derivative of tartrate with a ruthenium complex incorporating pyridine [87].

Resolution of enantiomers using the TRISPHAT anion is not restricted to "chiral at metal" stereoisomers. The interaction between the aromatic rings of the TRISPHAT anion and pentamethylcyclopentadienyl rings of the planar chiral organometallic cation Carb(Cp*Ru)₂ is responsible for the formation of homochiral columns of the ions (Figure 10.13). In the crystals, which belong to the $P2_1$ space group, this stacking leaves the other two rings of the anion at the sides of the stack, presumably aiding the stereodifferentiation. It is a nice example of resolution under supramolecular control [88].

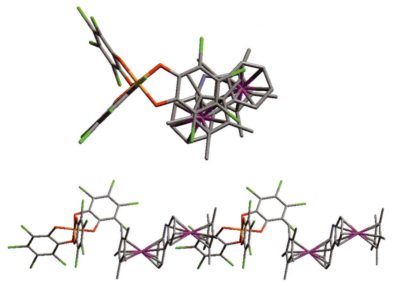

Figure 10.13 Top and side views of the stack of the TRISPHAT anion and the cation Carb(Cp*Ru)$_2$ in the crystals of the salt (hydrogen atoms and anions omitted for clarity).

It is important to point out that in resolution with resolving agents, a selective *one-dimensional aggregation* (surrounded by solvent, as is the case of the structures described in this section) *is sufficient to separate enantiomers*, while in spontaneous resolution *a three-dimensional extension of stereoselectivity is necessary*.

10.6
Thermodynamic Considerations in the Formation of Chiral Crystals

10.6.1
What is the Order of a System Composed of Two Enantiomers?

If *R* and *S* enantiomers of a compound interconvert in the vapor state and the liquid state very easily, *R* and *S* are simply dynamic forms and this is a unary system. Let us consider for example 3a,10-dihydro-5,5-dioxo-4H-(*S*)pyrrolidino-[1,2c][1,2,4]benzothiadiazine) (**10**). After enantioselective synthesis and purification, the enantiomers (ee > 99%) can be obtained. DSC and HPLC analyses reveal that as soon as the enantiomer reaches its "melting point," it quasi-instantaneously converts into the racemic mixture. Obviously the same applies for any composition different from ee = 0%.

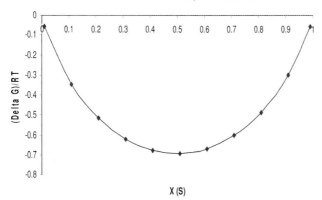

Figure 10.14 Gibbs free energy/RT of an interconvertible system of enantiomer vs. composition.

The R ⇔ S equilibrium has some particularities: it is a completely athermal reaction:

$$\Delta H_{\text{reaction}} = \Delta H_f(R) - \Delta H_f(S) = 0 \text{ and } (\forall T \ C_p(R) = C_p(S)) \quad (10.1)$$

The equilibrium constant is thus 1 whatever the temperature:

$$\frac{\partial \ln K}{\partial T} = \frac{\Delta H}{RT^2} \Rightarrow K = 1 \quad (10.2)$$

The interconversion equilibrium is therefore an exclusively entropy-driven process whose ultimate state corresponds to the racemic mixture (the most probable state, Figure 10.14).

$$\Delta G = -RT(x_R.\ln x_R + x_S.\ln x_S) \quad (10.3)$$

This equilibrium might be associated to a fast kinetics, by contrast, the racemization can be a very long process (of the order of thousands years or even more) and it can be used for dating purposes [89].

When there is no interconversion between R and S, the enantiomers are genuine independent components. The R–S system is then a binary system. Nevertheless because $G(R) = G(S)$, whatever T and P, *the phase diagram must be symmetrical*. Scott was the first to detect that there was a possible conflict between the Gibbs phase rule and systems containing pairs of enantiomers, which is why he proposed a derived expression that accounts for the additional equality between the chemical potentials of enantiomers [90]

$$v = \frac{n_2}{2} + n_1 + 2 - \frac{\varphi_2}{2} - \varphi_1 \quad (10.4)$$

n_1 and φ_1: stand, respectively, for the number of independent components and number of phases that are not symmetrical. n_2 and φ_2: stand respectively for the number of independent components and number of phases that are symmetrical. It is worth noting that this relation is applicable only if $\varphi_2 \geq 2$ [91].

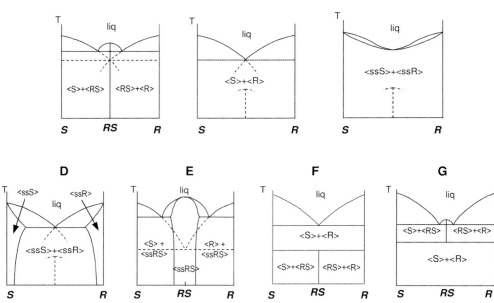

Figure 10.15 Diagram A: a stable racemic compound exists; Diagram B stable conglomerate: no stable racemic compound exists whatever the temperature; Diagram C a miscibility at the solid state exists whatever the composition and the temperature. D: *idem* as case B with partial solid solutions. E: *idem* as case A but the racemic compound is not stoichiometric, Diagram F: mixed case between A (at low temperature) and B (at high temperature); G mixed case between A (at high temperature) and B (at low temperature).

When a system, composed of two nonracemizable enantiomers, is examined in terms of nature of the crystallized phases versus composition, several cases can be encountered:

1. The binary system contains the two enantiomers and a (1:1) stoichiometric compound named racemic compound (Figure 10.15A)

2. The binary system contains the two enantiomers without any stable intermediate compound, this case is termed a conglomerate, but is simply a eutectic (Figure 10.15B).

3. The binary system contains a single solid phase: the two enantiomers form a complete solid solution. (Figure 10.15C). The solid–liquid equilibria can also lead to a maximum at the racemic composition or almost a flat curve i.e. there is hardly any modification of the melting point when the enantiomeric composition is changed (not represented).

On the average, cases (1), (2) and (3) account respectively for ca. 90–95% – 5–10% – ε% ($\varepsilon \ll 1$%) of the total number of cases encountered. Nevertheless, there are large fluctuations from one molecular series to another. In addition, rather rare additional cases might exist:

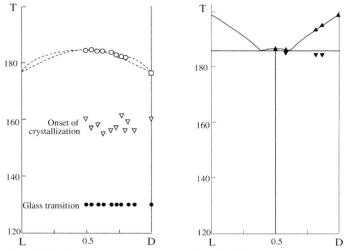

Figure 10.16 Binary system between the (−) and (+) limonene. Left binary system: metastable equilibria with a complete solid solution similar to Figure 10.15C. Right binary system corresponding to stable equilibria (Figure 10.15A) without any detectable solid solution (adapted from Ref. [96]).

4. Stable partial solid solutions between enantiomers (Figure 10.15D) [92].

5. Stable partial solid solution associated with the racemic compound (Figure 10.15E) [93]

6. "Abnormal" racemic compound [94] (with stoichiometries $(n-m)$ and $(m-n)$ with $n \neq m$)

7. Various stable equilibria at different temperatures, for example those corresponding to the anilide of 2-(1-nitro-2-naphthoxy)-propionic acid (Figure 10.15F) and 2,2′-diamino-1,1′-binaphtyl (Figure 10.15G) [11].

8. Mixed situations such as that depicted above but corresponding to a different state of equilibrium (stable and metastable equilibria, Figure 10.16) [95].

The most striking feature about crystallization of enantiomers is the overwhelming proportion of the racemic compound, as mentioned in the introduction to this chapter. It is worth noting that starting from a crystal structure of an enantiomer it is always possible to construct several dozen racemic compounds by using simple molecular modeling tools [96]. Among this series of virtual racemic compounds some of them possess competitive lattice energies: this is why Figures 10.15A–C show metastable racemic compounds.

In some cases, the racemic compounds display a large advantage in terms of thermal stability compared to the enantiomers. Figure 10.17 illustrates this very large difference in melting point – a situation that is sometimes referred to as an "anticonglomerate" [97]. This feature can be used to design an efficient enantiopurifying process. Starting from a mixture of enantiomers with a poor enantiomeric excess, a stable racemic

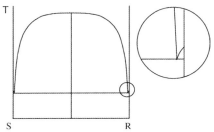

Figure 10.17 Binary system between enantiomers displaying a great thermal stability of the racemic compound compared to that of the enantiomers.

compound can be formed so that the aqueous mother liquor could exhibit an enantiomeric excess as high as 99% (great stability means very low solubility compared to that of the enantiomer in excess). Of course the amount of solvent is also an important issue when one wants to optimize the ee of the mother liquor. It is worth noting that only very few examples are known where stable thermodynamic equilibria are involved in the evolution towards homochirality in solution [98].

The trend towards centrosymmetric species is so strong that racemate-like packing of a pure enantiomer can be observed. The enantiomers in the crystal structure have different conformations so that they can mimic a "pseudoracemic compound." Long ago, Petterson and others used this trend to correlate the absolute configuration of related molecules (Figure 10.18) [99].

One of the common misconceptions is to consider that the solid phases involved in solid–liquid equilibria remain unchanged when temperature is dropped (or pressure is modified to a large extent). For instance, eutectoid, peritectoid invariants correspond to an inversion of relative stability between the conglomerate and the racemic compound. At the exact temperatures of these invariants the systems are triphasic and the following equilibria ($\Delta G = 0$) can be depicted (see Figures 10.15f and g):

$$<RS> \Leftrightarrow <R> + <S> (\text{eutectoid } \Delta H < 0; \text{ peritectoid } \Delta H > 0)$$

The sign of the heat transfer (ΔH) is negative for the eutectoid and positive for the peritectoid.

Other heterogeneous equilibria can also interfere in the chiral discrimination in the solid state. Indeed, solvent can be an active partner in the solid state. In other words the crystallization of solvates (or cocrystals) below a certain temperature can completely overturn the relative stability of a conglomerate versus a racemic compound. The case of Pasteur salts is a well-known illustration of this fact. Below 27 °C the mixture of homochiral crystals of tetrahydrated sodium-ammonium tartrate is stable. But above this temperature a monohydrated racemic compound (Sacchi's salt) becomes more stable. Therefore, a ternary peritectic invariant of the first kind leads to the stability of the conglomerate below 27 °C and the stability of the racemic compound above that temperature [100].

In a recent study [101], the triethanolamonium modafinate led to an interesting case of contrasted situation between the binary system of enantiomers and the

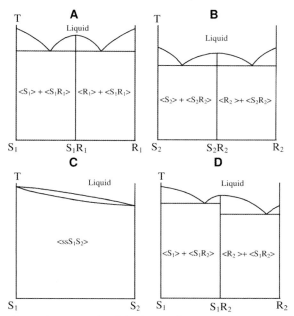

Figure 10.18 Principle of absolute configuration assignment by using pseudoracemic compound formation. 1 and 2 are chemically similar compounds. S1 and S2 lead to a complete solid solution by means of substitution: they have the same chirality. S1 and R2 have opposite chirality and form a pseudoracemic compound.

ternary system with water. The binary system corresponds to a complete solid solution between the two enantiomers, whereas the R and S monohydrates crystallize as a conglomerate without any detectable partial solid solution. A schematic isotherm in the temperature domain of stability of the monohydrates is presented in Figure 10.19. The single water molecule entrapped in the crystal lattice is sufficient to switch from a complete absence of chiral recognition in the solid state to the best possible chiral discrimination i.e. the conglomerate. The crystal structure justifies the important role of the water molecule. Although this phase is a salt, there is no direct electrostatic link between the carboxylic group and the core of the cation (i.e. the hydrogen connected to the nitrogen of the ammonium group). There is a dual relay as illustrated by Figure 10.20. The desolvation is therefore of the destructive–reconstructive type and there is no real surprise in observing the big contrast in the chiral recognition in the solid state, with and without a water molecule.

The research group in Rouen has long been intrigued by the difference between conglomerates in terms of performances in preferential crystallization. Briefly we will recall here the basics of this method (also called resolution by entrainment) that is applicable at the laboratory as well as the industrial scale. Starting from a homogeneous solution slightly enriched in one enantiomer (represented by point E in Figure 10.21), the homogeneous solution is rapidly cooled to the lowest possible

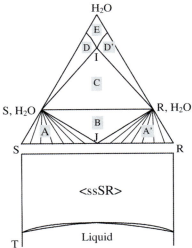

A	Solid solution + S, $1H_2O$
A'	Solid solution + R, $1H_2O$
B	Solid solution J + S, $1H_2O$ + R, $1H_2O$
C	S, $1H_2O$ + R, $1H_2O$ + liquid I
D	S, $1H_2O$ + saturated solution
D'	R, $1H_2O$ + saturated solution
E	Unsaturated solution

Figure 10.19 Triethanolamonium modafinate: complete solid solution in the binary system; conglomerate without partial miscibility in the solid state for the hydrate in the ternary isotherm at 20 °C Adapted from Ref. [101].

temperature without uncontrolled nucleation of any solid (T_F). At this temperature the supersaturated solution is seeded with very pure small crystals of the enantiomer that is initially in excess. This inoculation triggers a stereoselective secondary nucleation of crystal of the same handedness and of course crystal growth of this enantiomer. The process keeps going for a certain time, which depends on numerous factors (supersaturation, interactions between enantiomers at the crystal mother liquor interface, stirring mode and stirring speed, viscosity of the slurry, etc.). When the maximum enantiomeric excess in the mother liquor (point F) is attained a swift filtration is implemented. The next operation consists of adding the same mass of racemic mixture as that of the enantiomer collected by filtration and heating the system again at T_{Homo} (point E'). The system is therefore in a symmetrical situation as that previously described. The operator will simply drop the temperature of the

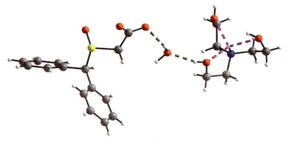

Figure 10.20 Role of the water molecule in the crystal structure of the triethanolamonium modafinate monohydrate. Adapted from Ref [101].

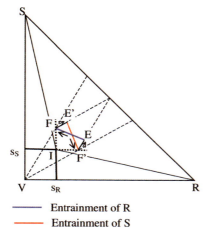

——— Entrainment of R
——— Entrainment of S

Figure 10.21 Projection of the mother liquor pathway during preferential crystallization (seeded and isothermal variant). Points s_R and s_S stand for the solubility of R and S enantiomers respectively in solvent V at T_F. I is the ternary liquid invariant. E, E' are compositions of the system at the beginning of the entrainments (respectively, R and S). F, F' are compositions of the system at the end of the entrainments (respectively, R and S). Segment EF represents the crystallization of the R enantiomer. E'F' represents the crystallization of the S enantiomer. From F to E' and F' to E show the addition of the racemic mixture after every entrainment.

system down to T_F and inoculate very pure and fine S crystals while stirring the suspension so that the mother liquor will reach F' point due to the entrainment effect. At this stage the pure solid S is collected by filtration and the mother liquor is once again added with the same mass of racemic mixture as that of the crude crops. The system is, therefore, back to square one and the process can be repeated indefinitely up to the exhaustion of the amount of racemic mixture to resolve.

Several variants of this process have been proposed and the reader is invited to look at the literature for details [102]. The yield of every batch operation depends on the concentration and how large are the wings of the "butterfly" i.e. how long is the projection of the crystallization pathway in the ternary system, i.e. the length of FE segment compared to FR. This means that for a given concentration the yield is proportional to the enantiomeric excess reached at the end of the entrainment ee_f. By definition, ee_{fmax} is the maximum enantiomeric excess reachable at the end of the entrainment. It is related to:

- The metastable solubility of the crystallizing enantiomer at T_F (a thermodynamic limit),
- The interactions of the counter enantiomers at the crystal – mother liquor interface $E_{hkl}^{p/n}$ this parameter is specific to every orientation (*hkl*) of the growing crystal [103].

The behaviors of conglomerates with regards to preferential crystallization can be roughly divided into three groups:

- Group 1: ee_{fmax} is situated between 0 and 5%

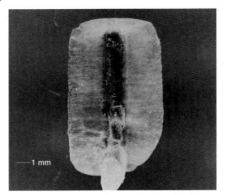

Figure 10.22 Photo of a crystal of norfenfluramine dichloroacetate showing the heteronucleation of the counter enantiomer on the (10–1) face of the growing crystal. Starting from a pure enantiomer – central single crystal – immersed in slightly supersaturated racemic solution, the added material – the translucent peripheral part – is perfectly racemic. Adapted from Ref. [104].

- Group 2: ee_{fmax} is between 6 and 9%
- Group 3: ee_{fmax} is above 10%.

In Group 1, if the external parameters have been well tuned (solvent, supersaturation, stirring mode, stirring speed, mode of filtration, etc.), there is a tough competition between hetero- and homochiral interactions at the mother liquor–crystal interface. At least one of the (*hkl*) orientations of the growing crystal can be used by the counter enantiomer for heteronucleation. A quasi-irreversible docking of the wrong enantiomer can also be observed (Figure 10.22) [105]. There is hardly any possible application of the preferential crystallization.

In the intermediate Group 2, there are still some basic problems such as epitaxy between enantiomers or an easy heteronucleation of the counter enantiomer but homochiral interactions are strong enough so that a large-scale application is conceivable [105].

In Group 3, the heterochiral interactions are definitively weaker than their homochiral counterparts. The ee_{fmax} value can go beyond 10%: 15–20% is sometimes reachable with a very good reproducibility. The stereoselective crystallization can proceed up to or close to the metastable solubility of the crystallizing enantiomer, even if this point is situated in the biphasic domain of the counter enantiomer and its saturated solution [106]. A very robust process can be tuned and large-scale production is feasible. In a recent article these different performances were analyzed not only in terms of interactions between the enantiomers at the interface crystal/mother liquor but also in terms of local supersaturation [103]. Conversely, when the preferential crystallization has been optimized, ee_{fmax} gives a rough approximation of the balance between the best heterochiral interaction at a given *hkl* interface and the corresponding homochiral interaction.

10.6.2
Resolution by Diastereomeric Associations

The addition of the so-called resolving agent (RA) is designed to introduce as much dissymmetry as possible in the phase diagram. We will consider hereafter that a racemic mixture of basic (or acidic) components has to be resolved. In a first approach the resulting couple of diastereomeric salts D1 and D2 will be formed by adding a stoichiometric amount of RA. Therefore the binary system to be considered is D1–D2. Most of the time the resolution is not performed in the molten state but rather in a medium, that is to say in a solvent (V) or in a mixture of solvents (V1, V2). The system to consider is therefore at least ternary: D1 – D2 – V for a single solvent or D1 – D2 – V1 – V2 i.e. quaternary system when two cosolvents are used. These solvents are not simple physical diluents, but can be active partners in the solvation and can modify the interactions in the solid state and/or the interactions between the (poly)saturated solution and the solid phase(s). Figure 10.23 shows how dissymmetrical the ternary system D1–D2–V can be at a given temperature. The more point I departs from the 50–50 composition the better. Indeed the yield of the resolution is strictly dependent on the IK/ID1 ratio.

Starting from a 50–50 composition of diastereomers and considering systems in thermodynamics equilibrium, it can de deduced from diagrams A, B and C in

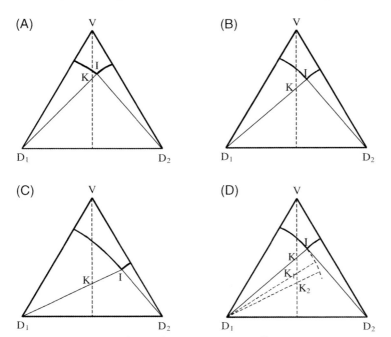

Figure 10.23 Various isotherms showing an increasing efficiency in the resolution from A to B and to C; note that the solubility of D1 is kept constant. D is identical to B but metastable equilibria involving the less soluble salt are considered.

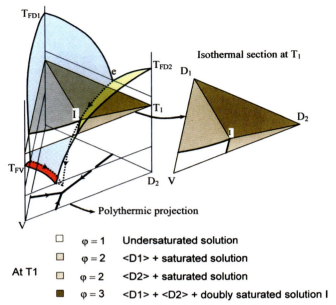

Figure 10.24 A perspective view of the phase diagram of ternary system D1–D2–V. The plane corresponding to the ternary eutectic has been omitted for clarity.

Figure 10.23 that the best composition for an optimum resolution correspond to point K. A simple thermal technique has been devised to precisely spot this fundamental point [20b]. When metastable equilibria are robust enough to be considered, the yield of a Pasteurian resolution can be enhanced. Figure 10.23d details these more favorable resolving processes that can be tuned if the overall synthetic mixture is adjusted in K1 or even K2 without the crystallization of D2.

Figure 10.24 shows the context in which a ternary isotherm arises. This schematic illustration is designed to represent the simplest case where there are:

- Thermal stabilities of D1 and D2 up to their melting points.
- No deviation from ideality.
- No intermediate compound (double salt).
- No solid solution (partial or complete).
- No solvate.
- No polymorphism.

In this ideal case, the binary system D1–D2–T contains all the relevant information for prediction of the overall yield of the resolution. The nature of the solvent will simply change the volumic yield. The temperature of the ideal eutectic T_e can be calculated by solving the following equation (where ΔH_{D1} and ΔH_{D2} stand for the enthalpy of fusion of D1 and D2, respectively, and T_{D1} and T_{D2} stand, respectively, for the melting temperature of D1 and D2):

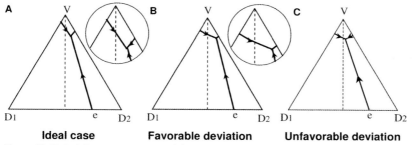

Figure 10.25 Polythermic projection of the invariant point versus temperature.

$$\exp((\Delta H_{D1}/R)(1/T_{D1}-1/T_e))+\exp((\Delta H_{D2}/R)(1/T_{D2}-1/T_e))=1 \quad (10.5)$$

From this temperature it is easy to reckon the ideal composition X_e of the eutectic mixture by using the simplified Schröder–Van Laar relation:

$$X_e = \exp((\Delta H_{D1}/R)(1/T_{D1}-1/T_e)) = 1-\exp((\Delta H_{D2}/R)(1/T_{D2}-1/T_e)) \quad (10.6)$$

The projection of the monovariant curves on the horizontal plane at T_ε (i.e. locus of point I versus temperature) is a straight segment: proj_e – ε it shows that starting from point e at T_e belonging to the binary system D1–D2–T there is no deviation in composition of the polysaturated solution versus temperature (Figure 10.25A. The solvent here is a neutral diluter that favors neither D1 nor D2. In other words the solvent does not promote any difference in the interactions between the D1 and D2.

The polythermic projections in Figures 10.25B and C show, respectively, how the interactions between the diastereomeric components could evolve with regards to temperature. The deviations from ideality result from a difference in variations of ΔC_p vs. temperature for D1 and D2 and/or from difference in solvation effects. For the former effect, a switch from one solvent to another would not introduce a change in the sign and the magnitude of the deviation. By contrast, for the latter effect a change in the nature of the solvent is likely to change at least the magnitude of the deviation from ideality.

The presence of an intermediate compound (double salt) is not exceptional. Two industrial resolution processes that worked quite well for months (illustrated by Figure 10.23c for instance) were abandoned because an intermediate compound crystallized suddenly. In a similar way as new polymorphs might appear [44], this (1-1) double salt never appeared before because of kinetic reason(s) or because a modification in the impurity profile of the sample [107]. The phase diagram of a double salt that has a metastable character is shown in Figure 10.26A. In accordance to the Ostwald rule of stages [108] this intermediate phase can be observed if the system is left far from equilibrium. In Figure 10.26B the (1:1) intermediate compound has a noncongruent solubility and it has a domain of stability that limits the yield of the resolution. The optimum conditions are represented by point K'.

$$Y = MT \times I'K'/I'D1 < MT \times IK/ID1 \quad (10.7)$$

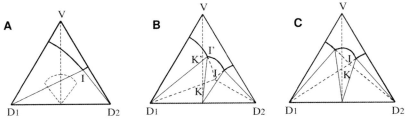

Figure 10.26 Effect of an increasing stability (from A to B to C) of an intermediate compound in a Pasteurian resolution process.

The kinetics (nucleation and/or crystal growth) associated with a double salt might be low enough to rely on metastable equilibria (dashes lines) for an optimized resolution (point K′) [107].

In Figure 10.26C the double salt has a congruent solubility, making the resolution impossible by using stable equilibria. Nevertheless, metastable equilibria, if they are attainable and robust enough, maintain the possibility to resolve the 50–50 mixture of diastereomers.

Polymorphism is a phenomenon with a high occurrence of any type of organic solid phases. Diastereomers as any other organic crystals can therefore be affected by monotropism or enantiotropism [109]. Figure 10.27 shows the phase diagram for a simple case of mono-enantiomorphism of the less soluble salt (D1) how *a priori* it may have a positive effect on the yield of the resolution.

The formation of a solid solution between diastereomeric salts can be an issue in the resolution process. An evolution from a dissymmetrical ternary isotherm to the existence of a continuous solid solution is shown in Figures 10.28A–D. In 28A, the domain of miscibility in the solid state does not affect the less-soluble diastereomer (D1); there is in practice almost no impact of the solid solution on the resolution and the purification process. By contrast, in situations B and C the less soluble salt leads to a partial solid solution. High diastereomeric excess is reachable only after several recrystallizations of the solid. Nevertheless, simultaneously the yield will drop depending on the orientation of the tie lines that connect the composition of the solid and the composition of the saturated mother liquor in equilibrium with this

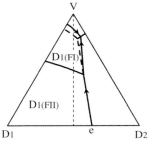

Figure 10.27 Impact of polymorphism of the less soluble salt on the heterogeneous equilibria between D1 and D2; D1(FII) high-temperature form, D1(FI) low-temperature form.

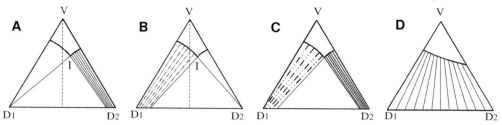

Figure 10.28 Pasteurian resolution under the influence of solid solution. A no impact, B and C detrimental impact, D hardly feasible.

solid of particular composition. Tie-line orientations in Figure 10.28C are more favorable to the purification of D1 compared to the case depicted in Figure 10.28B. In Figure 10.28D a complete solid solution spans from pure D1 to pure D2. This means that a single-crystal lattice can accommodate both components (resolution can hardly be achieved at that temperature in this solvent).

These cases should not be confused with the "Dutch resolution" where several (usually three) resolving agents of the same family of derivatives and of the same handedness are used simultaneously (see Section 10.4). The order of these systems is at least 6 (RA1, RA2, RA3, (−)amine, (+) amine and solvent V1).

10.7
Influencing the Crystallization of Enantiomers

10.7.1
Solvent

When a solvate (or a heterosolvate) crystallizes, it definitely means that the solvent(s) is (are) not a passive actor(s) in the chiral discrimination in the solid state but rather an active partner. To exemplify such a case at a temperature below T_p (corresponding to the binary peritectic invariant), consider the phase diagram in Figure 10.29, in which the resolution is feasible under more favorable conditions because the

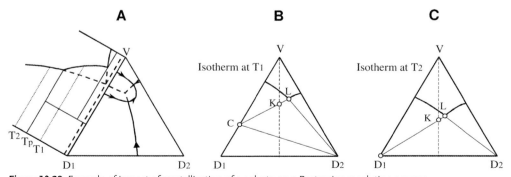

Figure 10.29 Example of impact of crystallization of a solvate on a Pasteurian resolution process.

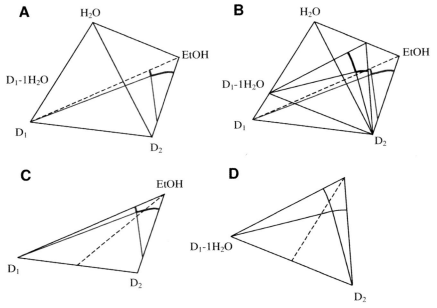

Figure 10.30 Possible inversion of the nature of the less soluble salt when a solvate crystallizes. In ethanol-rich composition D2 is obtained. In the water-rich region D1–1H$_2$O is obtained.

monovariant curve shifts towards the more soluble salt (D2). Two isothermal sections detail the benefit provided by the crystallization of the solvate (KL/CL > KL/D1L). This shows that when screening RA it is then highly recommended to simultaneously screen solvents and even mixtures of solvents.

Figure 10.30 illustrates in a quaternary system the fact (D1–D2–H$_2$O–ethanol) [110] that the less-soluble salt might be inverted according to the ratio R = H$_2$O/ ethanol. In pure ethanol and low concentration in water, D2 is obtained. Conversely, in pure water or high R ratio the D1-hydrate is obtained.

10.7.2
Preferential Nucleation and Inhibition

The title method is a general kinetic resolving method that is based on the difference in docking energies of a chiral additive on both enantiomers of a conglomerate. The principle is schematized in Figure 10.31. As for any physical property the kinetics of nucleation and crystal growth of two enantiomers are identical (Figure 10.31a). When an active chiral additive is dissolved in the medium, the symmetries of kinetic nucleation and crystal growth are broken (Figures 10.31b and c). The energy of adsorption on prenuclei, nuclei and crystal interfaces of the enantiomer having the same absolute configuration is greater than that on the opposite enantiomer. Therefore, the net result of the presence of this active chiral additive is to slow down the primary nucleation kinetic of the solute with the same absolute configuration as that of the chiral additive, whereas the opposite enantiomer is almost

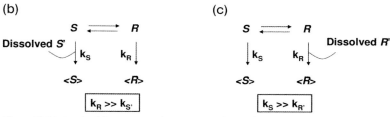

Figure 10.31 Break of the symmetry between R and S nucleation rates and crystal growth rates under the influence of chiral chemically related impurity.

unaffected. This is why the application of this method is submitted to the so-called "rule of reversal" [111]. Lahav's group made extensive studies on this elegant method and proposed three conditions for its application.

1. The solute to be resolved must crystallize as stable conglomerate (no racemic compound or solid solution between enantiomers).
2. Some chemical and crystallographic analogies must exist between the resolving additives and the solute.
3. The solubility of the racemic solute higher than that of the resolving additive in the used solvent.

The second condition raises some questions about the concepts of "chemical and crystallographic analogies." If, for instance, we consider the racemic mixture (±) **11**, it is quite obvious that: (S)-(+)-**12**, (S)-(+)-**13** and (S)-(+)-**14** share important structural features and could therefore be used as active chiral additives to trigger stereoselectively the primary nucleation of (R)-**11** [112].

(R)-11   (S)-12   (S)-13   (S)-14

Actually, (S)-**11** is selectively delayed in its primary nucleation. What is less obvious is why the authors needed 10 times more (S)-**12** than (S)-**13** or (S)-**14** for equivalent results in terms of optical purity and yield.

In addition to the three afforded mentioned conditions, tests carried out without stirring led to very poor results, showing that mechanical stirring is the necessary fourth condition for the success of the method. The collisions between the stirrer and the inoculated seeds induce a large number of secondary nuclei, as seen in the spontaneous resolution of different systems [15, 16].

10.8
Chiral Host–Guest Complexes

Crystalline host–guest complexes are a family of entities that form an ordered solid-state structure with neither covalent nor ionic bonds between the different components. Every member of these associations, like cocrystals or solvates possesses a minimum Gibbs free energy and appears because of favorable kinetic parameters. The relative size and characteristics of the components together with the stoichiometry between the different components make the solid-state chemist deciding about the right label of the phase, but there is no fundamental difference between host–guest complexes, cocrystals and solvates. On top of that, some kinds of fashions and drifts in terminologies are responsible for evolutions in naming these associations. For instance, clathrates were first used exclusively for gas-hydrate (Cl_2, methane, propane, Xe, etc. form crystals in which hydrogen-bonded water molecules built cages around the guest molecule or atom) but this restricted usage has turned into a more flexible one.

The host–guest associations represent a vast field of research and therefore an abundant literature exists on this issue [113]. These crystallized supramolecular complexes have been known for decades and include: urea, thiourea, perhydrotriphenylene, 9-9′-Biantryl, cyclophosphazene, tri-o-thymotide, anthracene bisresorcinol, hydroquinone and cholic-acid derivatives. When the host is a chiral molecule, some expectations of resolution are naturally raised. On steroidal bile-acid derivatives only, Miyata's group has been working for more than 20 years and resolved several hundred crystal structures [114]. The striking diversity of these molecular assemblies results from the unique inclusion behavior for every host molecule. As a rule more than as an exception, a minor change in the host molecular structure often induces large modifications in the host–guest behavior. The fit between the host molecule and the guest molecule can be analyzed at different levels. For instance, (i) the three axial chirality of the guest molecule, (ii) the helical assembly around a screw axis, (iii) the packing of this monodimensional supramolecular assembly compatible with symmetries of a crystal lattice (usually $P2_12_12_1$ or $P2_1$ or $C2$ or $P1$).

The "three-axial chirality" of a molecule is defined by assigning: head-tail, right-left, belly and back to any 3D object. The "tilt chirality" is defined likewise axial chirality; it can even be used to define the handedness of a screw axis including a 2_1 (Figure 10.32). These concepts are necessary to describe the supramolecular chirality. Some nonchiral objects can build chiral cavities in which one enantiomer of the guest can fit in while the other one barely accommodates the empty space.

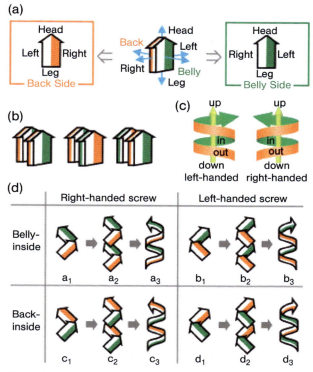

Figure 10.32 Schematic representation of a chiral molecule with three axes (a), three bimolecular association modes (b), an asymmetric helix with three axes (c), and four kinds of bimolecular and 2_1 helical assemblies by combination of three-axial and tilt chiralities (d) Reproduction by permission of the American Chemical Society.

Nevertheless, when considering chiral discrimination through supramolecular assemblies, experience gained in the past decades shows that it is difficult to reach ee values greater than 90% at the first crystallization. Therefore, molecular associations of a new type can be envisaged. Host–guest associations can be different in nature if one of the components (the host) is actually a cavity of sufficient size so that the other component (the guest molecule) can partially or completely be engulfed. If the macrocyclic cavity is chiral, then it means that conceptually couples of diastereomeric complexes can be formed and can be in competition both in solution [115] and in the solid state. Because of steric hindrance one can expect a good chiral selectivity. In addition to the formation of the solvated diastereomeric complexes, the construction of a 3D periodic crystal lattice could be another level of molecular recognition providing an improved chiral discrimination. These diastereomeric associations can be used for the classical problem of absolute configuration assignment of chiral organic molecules deprived of "heavy atoms" i.e. atoms leading to a sufficient anomalous scattering effect in single-crystal X-ray diffraction [116].

Up to now the following macrocyclic hosts have appeared attractive: cyclodextrins [117], cyclocholates, cyclophanes, calixarenes, cyclopeptides, and crown-ethers [118]. By far, cyclodextrins and derivatives have been submitted to the most extensive studies. It is worth noting that very successful applications of cyclodextrins in chiral discrimination via chromatography are of everyday use [119]. Nevertheless, for a given racemic mixture (guest molecules) and a given chiral macrocyclic cavity, a successful chiral selectivity in a chromatographic application [120] does not mean a high stereodifferentiation in the solid state [121].

A close match between the chiral host and a single enantiomer only means some flexibility in the molecular macrocyclic cavity. This is why α, β, and γ permethylated CDs (TRIMEA CD, TRIMEB CD and TRIMEG CD, respectively) have been recommended in order to ensure the famous "induced fit" between the host and the guest molecules [122]. Resolution of the 1,7-dioxaspiro[5,5]undecane with TRIMEB CD is one of the best examples found in the literature [123].

In a recent study on *para*-halogenated phenylethanol [124], various situations have been found including: disordered and partial solid solutions (fluorine derivative), odd stoichiometry (iodine derivative), very different inclusion geometries for the two enantiomers (bromine derivative). Although, the 4-bromophenyl ethanol gives the

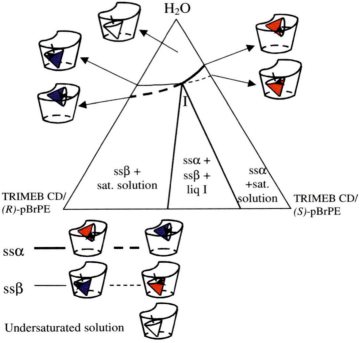

Figure 10.33 Schematic representation of the host–guest geometries in the ternary system: H_2O- TRIMEB CD (R)- 4-bromophenyl ethanol – TRIMEB CD (S)- 4-bromophenyl ethanol. Note that a single geometry has been observed by NMR in solution. Adapted from Ref. [124].

best resolution, a careful analysis of the heterogeneous equilibria has shown that actually, two solid solutions are in competition throughout the entire range of composition. Figure 10.33 schematizes the two modes of inclusion and the corresponding solid solutions. A simple molecular modeling approach led to consistent results and gave an overall picture of competition between the two molecular complexes and their crystal lattices $P2_12_12_1$ with the (S)-4-bromophenylethanol or $P2_1$ with the (R)-4-bromophenylethanol. Because native and permethylated CD have no specific endoreceptor (a specific binding site inside the cavity according to Lehn's definition [125]) a certain lack of stereoselectivity can be understood. The remedy of this drawback is in the hands of the organic chemists who have to design the appropriate endoreceptor for a given guest molecule so as to ensure the complete stereoselectivity.

10.9
Perspectives

The tracking of chiral structure transmission within crystals is achieved with more precision than in any other medium, at the atomic level, albeit in a nondynamic system. Important messages for nanoscience can be drawn from solid-state structures. Also, in particular, the precision with which phase diagrams can be determined in crystalline systems provides a benchmark for low-dimensional nanosystems.

Possible ways that nanoscience could influence the field of chiral crystals include the preparation of inhibitors and promoters of crystal growth in different media. In addition, the identification of the nuclei of crystals and the development of crystals can be greatly aided by the tools that have been developed in the nanoscience area. The symmetry principles that guide crystallization are well defined, but at this stage it is not clear at what size these factors become important, or exactly at what nucleus size long-range order is instigated. The apparatus used in nanoscience is sure to help identifying the effects in operation in this regime.

In all of these areas, molecular modeling is becoming a feasible method for investigating packing of chiral molecules. On some occasions, the outcome of crystallization can be mirrored in determination of the lattice energies of possible candidates as racemic compounds or conglomerates [96]. The modeling of nanometer-scale clusters of molecules is a particularly interesting area [126]. Hand in hand with experimental evidence from the bulk and nanometer scale, insight will surely be attained concerning the way in which chiral systems behave, in structural and thermodynamic ways.

References

1 (a) Hulliger, J. (1994) *Angewandte Chemie (International Edition in English)*, **33**, 143–162. (b) Desiraju, G.R. (2003) *Journal of Molecular Structure*, **656**, 5–15. (c) Melucci, M., Gazzano, M., Barbarella, G., Cavallini, M., Biscarini, F., Maccagnani, P. and Ostoja, P. (2003) *Journal of the American Chemical Society*,

125, 10266–10274; (d) Kato, K., Inoue, K., Tohnai, N. and Miyata, M. (2004) *Journal of Inclusion Phenomena and Macrocyclic Chemistry*, **48**, 61–67; (e) Du, M., Li, C.P. and Zhao, X.J. (2006) *Crystal Engineering Community*, **8**, 552–562; (f) Blatov, V.A. (2006) *Acta Crystal A*, **62**, 356–364; Patra, A., Hebalkar, N., Sreedhar, B., Sarkar, M., Samanta, A. and Radhakrishnan, T.R. (2006) *Small*, **2**, 650–659; Oaki, Y., Hayashi, S. and Imai, H. (2007) *Chemical Communications*, 2841–2843.

2 (a) De Feyter, S., and De Schryver, F.C. (2003) *Chemical Society Reviews*, **32**, 139–150; (b) Barth, J.V., Weckesser, J., Lin, N., Dmitriev, A. and Kern, K. (2003) *Applied Physics A-Materials Science & Processing*, **76**, 645–652; (c) Theobald, J.A., Oxtoby, N.S., Phillips, M.A., Champness, N.R. and Beton, P.H. (2003) *Nature*, **424**, 1029–1031; (d) Puigmartí-Luis, J., Laukhin, V., Pérez del Pino, A., Vidal-Gancedo, J., Rovira, C., Laukhina, E. and Amabilino, D.B. (2007) *Angewandte Chemie-International Edition*, **46**, 238–241; (e) Lena, S., Brancolini, G., Gottarelli, G., Mariani, P., Masiero, S., Venturini, A., Palermo, V., Pandoli, O., Pieraccini, S., Samori, P. and Spada, G.P. (2007) *Chemistry – A European Journal*, **13**, 3757–3764; (f) Nath, K.G., Ivasenko, O., MacLeod, J.M., Miwa, J.A., Wuest, J.D., Nanci, A., Perepichka, D.F. and Rosei, F. (2007) *Journal of Physical Chemistry C*, **111**, 16996–17007.

3 Duchamp, D.J., and Marsh, R.E. (1969) *Acta Crystallographica. Section B, Structural Science*, **25**, 5–19.

4 (a) Spillmann, H., Dmitriev, A., Lin, N., Messina, P., Barth, J.V., and Kern, K. (2003) *Journal of the American Chemical Society*, **125**, 10725–10728; (b) Lackinger, M., Griessl, S., Heckl, W.M., Hietschold, M. and Flynn, G.W. (2005) *Langmuir*, **21**, 4984–4988.

5 Lightfoot, M.P., Mair, F.S., Pritchard, R.G., and Warren, J.E. (1999) *Chemical Communications*, **19**, 1945–1946.

6 (a) Nguyen, T.-Q., Bushey, M.L., Brus, L.E., and Nuckolls, C. (2002) *Journal of the American Chemical Society*, **124**, 15051–15054; (b) van Gestel, J., Palmans, A.R.A., Titulaer, B., Vekemans, J.A.J.M. and Meijer, E.W. (2005) *Journal of the American Chemical Society*, **127**, 5490–5494; (c) van Hameren, R., Schon, P., van Buul, A.M., Hoogboom, J., Lazarenko, S.V., Gerritsen, J.W., Engelkamp, H., Christianen, P.C.M., Heus, H.A., Maan, J.C., Rasing, T., Speller, S., Rowan, A.E., Elemans, J.A.A.W. and Nolte, R.J.M. (2006) *Science*, **314**, 1433–1436; (d) Sakamoto, A., Ogata, D., Shikata, T., Urakawa, O. and Hanabusa, K. (2006) *Polymer*, **47**, 956–960; (e) Nguyen, T.Q., Martel, R., Bushey, M., Avouris, P., Carlsen, A., Nuckolls, C. and Brus, L. (2007) *Physical Chemistry Chemical Physics*, **9**, 1515–1532.

7 (a) Kitaigorodsky, A.I. (1973) *Molecular Crystals and Molecules*, Academic Press, New York; (b) Kitaigorodsky, A.I. (1978) *Chemical Society Reviews*, **7**, 133–163; (c) Wilson, A.J.C. (1993) *Acta Crystallographica. Section A, Crystal Physics, Diffraction, Theoretical and General Crystallography*, **49**, 210–212; (d) Wolff, J.J. (1996) *Angewandte Chemie (International Edition in English)*, **35**, 2195–2197.

8 Steiner, T. (2000) *Acta Crystallographica. Section B, Structural Science*, **56**, 673–676.

9 (a) Stewart, M.V., and Arnett, E.M. (1982) *Topics in Stereochemistry*, **13**, 195–262. (b) Kuzmenko, I., Weissbuch, I., Gurovich, E., Leiserowitz, L. and Lahav, M. (1998) *Chirality*, **10**, 415–424. (c) Ernst, K.-H. (2006) *Topics in Current Chemistry*, **265**, 209–252.

10 (a) Pérez-García, L., and Amabilino, D.B. (2002) *Chemical Society Reviews*, **31**, 342–356; (b) Kostyanovsky, R.G. (2003) *Mendeleev Communications*, 85–90; (c) Pérez-García, L. and Amabilino, D.B. (2007) *Chemical Society Reviews*, **36**, 941–967.

11 Jacques, J., Collet, A., and Wilen, S.H. (1994) *Enantiomers, Racemates, and*

Resolutions, Krieger Publishing Company, Malabar, Florida.
12. Pratt-Brock, C., Schweizer, W.B., and Dunitz, J.D. (1991) *Journal of the American Chemical Society*, **113** (26), 9811–9820.
13. Pratt-Brock, C., and Dunitz, J.D. (1994) *Chemistry of Materials*, **6**, 1118–1127.
14. (a) Pasteur, L. (1848) *Annales de Chimie et de Physique*, **24**, 442–459, 1848, 26, 535–539; (b) Tobe, Y. (2003) *Mendeleev Communications*, 93–94.
15. (a) McBride, J.M., and Carter, R.L. (1991) *Angewandte Chemie (International Edition in English)*, **30**, 293–295; (b) Kondepudi, D.K., Digits, J. and Bullock, K. (1995) *Chirality*, **7**, 62–68; (c) Saito, Y. and Hyuga, H. (2005) *Journal of the Physical Society of Japan*, **74**, 535–537; (d) Veintemillas-Verdaguer, S., Esteban, S.O. and Herrero, M.A. (2007) *Journal of Crystal Growth*, **303**, 562–567.
16. (a) Kondepudi, D.K., Kaufman, R.J., and Singh, N. (1990) *Science*, **250**, 975–977; (b) Kondepudi, D.K., Laudadio, J. and Asakura, K. (1999) *Journal of the American Chemical Society*, **121**, 1448–1451; (c) Kondepudi, D.K. and Asakura, K. (2001) *Accounts of Chemical Research*, **34**, 946–954; (d) Asakura, K., Soga, T., Uchida, T., Osanai, S. and Kondepudi, D.K. (2002) *Chirality*, **14**, 85–89; (e) Sainz-Diaz, C.I., Martin-Islan, A.P. and Cartwright, J.H.E. (2005) *The Journal of Physical Chemistry. B*, **109**, 18758–18764.
17. Bonner, W.A. (1999) *Origins of Life and Evolution of the Biosphere*, **29**, 317–328, and references therein.
18. Vestergren, M., Johansson, A., Lennartson, A., and Håkansson, M. (2004) *Mendeleev Communications*, **14**, 258–259.
19. Liu, M., Qiu, C., Guo, Z., Qi, L., Xie, M., and Chen, Y. (2007) *The Journal of Physical Chemistry. B*, **111**, 11346–11349.
20. (a) Kozma, D. (2002) *CRC Handndbook of Optical Resolutions via Diastereomeric Salt Formation*, CRC Press, New York; (b) Marchand, P., Lefèbvre, L., Querniard, F., Cardinaaeël, P., Perez, G., Counioux, J.-J. and Coquerel, G. (2004) *Tetrahedron Asymmetry*, **15**, 2455–2465.
21. (a) Steiner, T. (2002) *Angewandte Chemie (International Edition in English)*, **41**, 48–76; (b) Aakeroy, C.B. and Seddon, K.R. (1993) *Chemical Society Reviews*, **22**, 397–407; (c) Bernstein, J., Davis, R.E., Shimoni, L. and Chang, N.L. (1995) *Angewandte Chemie (International Edition in English)*, **34**, 1555–1573.
22. (a) Etter, M.C. (1990) *Accounts of Chemical Research*, **23**, 120–126; (b) Etter, M.C. (1991) *The Journal of Physical Chemistry*, **95**, 4601–4610.
23. (a) Dance, I.G., Bishop, R., and Scudder, M.L. (1986) *Journal of the Chemical Society-Perkin Transactions 2*, 1309–1318; (b) Bishop, R. and Dance, I.G. (1988) *Topics in Current Chemistry*, **149**, 137–188; (c) Ung, A.T., Bishop, R., Craig, D.C., Dance, I.G. and Scudder, M.L. (1993) *Journal of the Chemical Society. Chemical Communications*, 322–323; (d) Yue, W.M., Bishop, R., Craig, D.C. and Scudder, M.L. (2000) *Tetrahedron*, **56**, 6667–6673.
24. Filthaus, M., Oppel, I.M., and Bettinger, H.F. (2008) *Organic and Biomolecular Chemistry*, **6**, 1201–1207.
25. Leiserowitz, L., and Weinstein, M. (1975) *Acta Crystallographica. Section B, Structural Science*, **31**, 1463–1466.
26. Weinstein, S., and Leiserowitz, L. (1980) *Acta Crystallographica. Section B, Structural Science*, **36**, 1406–1418.
27. Weinstein, S., Leiserowitz, L., and Gil-Av, E. (1980) *Journal of the American Chemical Society*, **102**, 2768–2772.
28. (a) Hunter, C.A., Lawson, K.R., Perkins, J., and Urch, C.J. (2001) *Journal of the Chemical Society-Perkin Transactions 2*, 651–669; (b) Sinnokrot, M.O. and Sherrill, C.D. (2006) *Journal of Physical Chemistry A*, **110**, 10656–10668; (c) Lee, E.C., Kim, D., Jurecka, P., Tarakeshwar, P., Hobza, P. and Kim, K.S. (2007) *Journal of Physical Chemistry A*, **111**, 3446–3457.
29. (a) Goto, T., and Kondo, T. (1991) *Angewandte Chemie (International Edition in English)*, **30**, 17–33; (b) Janiak, C. (2000)

Journal of The Chemical Society-Dalton Transactions, 3885–3896; (c) Waters, M.L. (2002) Current Opinion in Chemical Biology, **6**, 736–741; (d) Ishi-i, T. and Shinkai, S. (2005) Topics in Current Chemistry, **258**, 119–160.

30 (a) Hishio, M., and Hirota, M. (1989) Tetrahedron, **45**, 7201–7245; (b) Ringer, A.L., Figgs, M.S., Sinnokrot, M.O. and Sherrill, C.D. (2006) Journal of Physical Chemistry A, **110**, 10822–10828.

31 Huang, W., and Ogawa, T. (2006) Polyhedron, **25**, 1379.

32 Breu, J., Domel, H. and Stoll, A. (2000) European Journal of Inorganic Chemistry, **11**, 2401–2408, and references cited therein.

33 Johansson, A., and Håkansson, M. (2005) Chemistry – A European Journal, **11**, 5238–5248.

34 (a) Weiss, E. (1993) Angewandte Chemie (International Edition in English), **32**, 1501–1523; (b) Honig, B. and Nicholls, A. (1995) Science, **268**, 1144–1149; (c) Dougherty, D.A. (1996) Science, **271**, 163–168; (d) Drain, C.M., Goldberg, I., Sylvain, I. and Falber, A. (2005) Topics in Current Chemistry, **245**, 55–88; (e) Angeloni, A., Crawford, P.C., Orpen, A.G., Podesta, T.J. and Shore, B.J. (2004) Chemistry – A European Journal, **10**, 3783–3791; (f) Cantuel, M., Bernardinelli, G., Muller, G., Riehl, J.P. and Piguet, C. (2004) Inorganic Chemistry, **43**, 1840–1849; (g) Sharma, R.P., Bala, R., Sharma, R., Kariuki, B.M., Rychlewska, U. and Warzajtis, B. (2005) Journal of Molecular Structure, **748**, 143–151.

35 (a) Gilli, P., Bertolasti, V., Ferretti, V., and Gilli, G. (1994) Journal of the American Chemical Society, **116**, 909–915; (b) Braga, D. and Grepioni, F. (1997) Comments on Inorganic Chemistry, **19**, 185–207; (c) Desiraju, G.R. (2002) Accounts of Chemical Research, **35**, 565–573; (d) Lee, S.O., Shacklady, D.M., Horner, M.J., Ferlay, S., Hosseini, M.W. and Ward, M.D. (2005) Cryst Growth Design, **5**, 995–1003; (e) Amendola, V., Boiocchi, M., Fabbrizzi, L. and Palchetti, A. (2005) Chemistry – A European Journal, **11**, 5648–5660; (f) Pakiari, A.H. and Eskandari, K. (2006) THEOCHEM Journal of Molecular Structure, **759**, 51–60; (g) Dalrymple, S.A. and Shimizu, G.K.H. (2007) Journal of the American Chemical Society, **129**, 12114–12116.

36 (a) Amabilino, D.B., Stoddart, J.F., and Williams, D.J. (1994) Chemistry of Materials, **6**, 1159–1167; (b) Alcalde, E., Pérez-García, L., Ramos, S., Stoddart, J.F., White, A.J.P. and Williams, D.J. (2004) Mendeleev Communications, 233; (c) Alcalde, E., Pérez-García, L., Ramos, S., Stoddart, J.F., Vignon, S., White, A.J.P. and Williams, D.J. (2003) Mendeleev Communications, 100.

37 (a) Sakurai, T., Yamauchi, O., and Nakahara, A. (1976) Journal of the Chemical Society. Chemical Communications, 553–554; (b) Yamauchi, O., Sakurai, T. and Nakahara, A. (1977) Bulletin of the Chemical Society of Japan, **50**, 1776–1779.

38 (a) Whitlock, B.J., and Whitlock, H.W. (1994) Journal of the American Chemical Society, **116**, 2301–2311; (b) Gilson, M.K., Given, J.A., Bush, B.L. and McCammon, J.A. (1997) Biophysical Journal, **72**, 1047–1069; (c) Beer, P.D., Graydon, A.R., Johnson, A.O.M. and Smith, D.K. (1997) Inorganic Chemistry, **36**, 2112–2118; (d) Klein, E., Ferrand, Y., Barwell, N.R. and Davis, A.P. (2008) Angewandte Chemie-International Edition, **47**, 2693–2696.

39 (a) Taft, R.W., Abboud, J.L.M., Kamlet, M.J., and Abraham, M.H. (1985) Journal of Solution Chemistry, **14**, 153–186; (b) Marcus, Y. (1993) Chemical Society Reviews, **22**, 409–416.40.

40 Solvent inclusion to give conglomerate(a) Tunyogi, T., Deak, A., Tarkanyl, G., Kiraly, P., and Palinkas, G. (2008) Inorganic Chemistry, **47**, 2049–2055; (b) Imai, Y., Kawaguchi, K., Tajima, N., Sato, T., Kuroda, R. and Matsubara, Y. (2008) Chemical Communications, 362–364.

41 Sakai, K., Sakurai, R., and Hirayama, N. (2007) Topics in Current Chemistry, **269**, 233–271.

42 Sakai, K., Sakurai, R., Nohira, H., Tanaka, R., and Hirayama, N. (2004) *Tetrahedron Asymmetry*, **15**, 3495–3500.

43 Kozsda-Kovács, E., Keser,ü, G.M., Böcskei, Z., Szilágyi, I., Simon, K., Bertók, B., and Fogassy, E. (2000) *Journal of the Chemical Society-Perkin Transactions 2*, 149–153.

44 Bernstein, J. (2002) *Polymorphism in Molecular Crystals*, Clarendon Press, Oxford.

45 Chen, X.-D., Du, M., and Mak, T.C.W. (2005) *Chemical Communications*, **35**, 4417–4419.

46 Gonnade, R.G., Bhadbhade, M.M., and Shashidhar, M.S. (2004) *Chemical Communications*, **22**, 2530–2531.

47 Kostyanovsky, R.G., Avdeenko, A.P., Konovalova, S.A., Kadorkina, G.K., and Prosyanik, A.V. (2000) *Mendeleev Communications*, **10**, 16–18.

48 Hager, O., Llamas-Saiz, A.L., Foces-Foces, C., Claramunt, R.M., López, C-. and Elguero, J. (1999) *Helvetica Chimica Acta*, **82**, 2213–2230.

49 Janiak, C. (2003) *Dalton Transactions*, **14**, 2781–2804.

50 (a) Jacques, J., Leclercq, M., and Brienne, M.J. (1981) *Tetrahedron*, **37**, 1727–1733; (b) Amouri, H. and Gruselle, M. (2006) *Actualite Chimique*, (Suppl 303), 26–33.

51 Lennartson, A., Vestergren, M., and Håkansson, M. (2005) *Chemistry – A European Journal*, **11**, 1757–1762.

52 Sun, Q., Bai, Y., He, G., Duan, C., Lin, Z., and Meng, Q. (2006) *Chemical Communications*, **26**, 2777–2779.

53 Matsuura, T., and Koshima, H. (2005) *Journal of Photochemistry and Photobiology C: Photochemistry Reviews*, **6**, 7–24.

54 Addadi, L., and Lahav, M. (1979) *Pure and Applied Chemistry*, **51**, 1269–1284; Green, B.S., Lahav, M. and Rabinovich, D. (1979) *Accounts of Chemical Research*, **12**, 191–197.

55 Pidcock, E. (2005) *Chemical Communications*, **27**, 3457–3459.

56 Azumaya, I., Okamoto, I., Nakayama, S., Tanatani, A., Yamaguchi, K., Shudo, K., and Kagechika, H. (1999) *Tetrahedron*, **55**, 11237–11246.

57 (a) Koshima, H. (2000) *Journal of Molecular Structure*, **552**, 111–116; (b) Casnati, A., Liantonio, R., Metrangolo, P., Resnati, G., Ungaro, R. and Ugozzoli, F. (2006) *Angewandte Chemie-International Edition*, **45** (12), 1915–1918.

58 Tan, T.-F., Han, J., Pang, M.-L., Song, H.-B., Ma, Y.-X., and Meng, J.-B. (2006) *Crystal Growth & Design*, **6**, 1186–1193.

59 Wu, S.-T., Wu, Y.-R., Kang, Q.-Q., Zhang, H., Long, L.-S., Zheng, Z., Huang, R.-B., and Zheng, L.-S. (2007) *Angewandte Chemie-International Edition*, **46**, 8475–8479.

60 (a) Murakami, H. (2007) *Topics in Current Chemistry*, **269**, 273–299; (b) Fogassy, E., Nogradi, M., Kozma, D., Egri, G., Palovics, E. and Kiss, V. (2006) *Organic and Biomolecular Chemistry*, **4**, 3011–3030.

61 Kinbara, K., Sakai, K., Hashimoto, Y., Nohira, H., and Saigo, K. (1996) *Journal of the Chemical Society-Perkin Transactions 2*, 2615–2622.

62 Newman, P. (1983) Optical resolution procedures for chemical compounds. Optical resolution information center, New York, Vol. 1–3; Faigl, F. and Kozma, D. (2004) *Enantiomer Separation, Fundamentals and Practical Methods*, Kluwer, Dordrecht.

63 Kinbara, K., Sakai, K., Hashimoto, Y., Nohira, H., and Saigo, K. (1996) *Tetrahedron Asymmetry*, **7** (6), 1539–1542.

64 Sakai, K., Sakurai, R., and Nohira, H. (2007) New Resolution Technologies Controlled by Chiral Discrimination Mechanisms, in *Novel Optical Resolution Technologies*, Vol. 269, Springer, Berlin, pp. 160–99.

65 Sakai, K., Sakueai, R., and Nohira, H. (2007) New Resolution Technologies Controlled by Chiral Discrimination Mechanisms, in *Novel Optical Resolution Technologies*, Vol. 269, Springer, Berlin, pp. 199–231.

66 (a) Coquerel, G. (2007) *Preferential Crystallization in Novel Optical Resolution Technologies*, Vol. 269 Springer, Berlin, pp. 1–51; (b) Coquerel, G., Catroux, L. and Combret, Y. (1997) PCT FR97/02158.

67 (a) Leusen, F.J.J. (2003) *Crystal Growth & Design*, **3**, 189–192; (b) Karamertzanis, P.G., Anandamanoharan, P.R., Fernandes, P., Cains, P.W., Vikers, M., Tocher, D.A., Florence, A.J. and Price, S.L. (2007) *The Journal of Physical Chemistry. B*, **111**, 5326–5336.

68 Imai, Y., Takeshita, M., Sato, T., and Kuroda, R. (2006) *Chemical Communications*, **10**, 1070–1072.

69 Vries, T.R., Wynberg, H., van Echten, E., Koek, J., ten Hoeve, W., Kellogg, R.M., Broxterman, Q.B., Minnaard, A., Kaptein, B., van der Sluis, S., Hulshof, L., and Kooistra, J. (1998) *Angewandte Chemie (International Edition in English)*, **37**, 2349–2354.

70 Kellogg, R.M., Nieuwenhuijzen, J.W., Pouwer, K., Vries, T.R., Broxterman, Q.B., Grimbergen, R.F.P., Kaptein, B., La Crois, R.M., de Wever, E., Zwaagstra, K., and van der Laan, A.C. (2003) *Synthesis*, **10**, 1626–1638.

71 Huang, .J., and Yu, L. (2006) *Journal of the American Chemical Society*, **128**, 1873–1878.

72 Kellogg, R.M., Kaptein, B., and Vries, T.R. (2007) Dutch Resolution of Racemates and the Roles of Solid Solution Formation and Nucleation Inhibition, in *Novel Optical Resolution Technologies*, Vol. 269, Springer, Berlin, pp. 134–59.

73 Kellogg, R.M., Kaptein, B., and Vries, T.R. (2007) *Topics in Current Chemistry*, **269**, 159–197.

74 (a) Addadi, L., Weinstein, S., Gati, E., Wiessbuch, I., and Lahav, M. (1982) *Journal of the American Chemical Society*, **104**, 4610–4617; (b) Weissbuch, I., Lahav, M. and Leiserowitz, L. (2003) *Crystal Growth & Design*, **3**, 125–150.

75 Deij, M.A., Vissers, T., Meekes, H., and Vlieg, E. (2007) *Crystal Growth & Design*, **7**, 778–786.

76 Hisaki, I., Watabe, T., Kogami, Y., Tohnai, N., and Miyata, M. (2006) *Chemistry Letters*, **35**, 1274–1275.

77 Tanaka, A., Hisaki, I., Tohnai, N., and Miyata, M. (2007) *Chemistry, an Asian Journal*, **2**, 230–238.

78 von Zelewsky, A., Hayoz, P., Hua, X., and Haag, P. (1994) *Coordination Chemistry ACS Symposium Series*, **565**, 293–302.

79 (a) Ziegler, M., and von Zelewsky, A. (1998) *Coordination Chemistry Reviews*, **177**, 257–300; (b) Lin, W.B., Wang, Z.Y. and Ma, L. (1999) *Journal of the American Chemical Society*, **121**, 11249–11250; (c) Fox, J.M., Katz, T.J., Van Elshocht, S., Verbiest, T., Kauranen, M., Persoons, A., Thongpanchang, T., Krauss, T. and Brus, L. (1999) *Journal of the American Chemical Society*, **121**, 3453–3459; (d) Lacroix, P.G. (2001) *European Journal of Inorganic Chemistry*, 339–348; (e) Coughlin, F.J., Westrol, M.S., Oyler, K.D., Byrne, N., Kraml, C., Zysman-Colman, E., Lowry, M.S. and Bernhard, S. (2008) *Inorganic Chemistry*, **47**, 2039–2048.

80 (a) Kumagai, H., and Inoue, K. (1999) *Angewandte Chemie-International Edition*, **38**, 1601–1603; (b) Minguet, M., Luneau, D., L'Hotel, E., Villar, V., Paulsen, C., Amabilino, D.B. and Veciana, J. (2002) *Angewandte Chemie-International Edition*, **41**, 586–589; (c) Pointillart, F., Train, C., Boubekeur, K., Gruselle, M. and Verdaguer, M. (2006) *Tetrahedron Asymmetry*, **17**, 1937–1943; (d) Coronado, E., Gomez-Garcia, C.J., Nuez, A., Romero, F.M. and Waerenborgh, J.C. (2006) *Chemistry of Materials*, **18**, 2670–2681; (e) Madalan, A.M., Rethore, C. and Avarvari, N. (2007) *Inorganica Chimica Acta*, **360**, 233–240; (f) Gu, Z.G., Zhou, X.H., Jin, Y.B., Xiong, R.G., Zuo, J.L. and You, X.Z. (2007) *Inorganic Chemistry*, **46**, 5462–5464; (g) Kaneko, W., Kitagawa, S. and Ohba, M. (2007) *Journal of the American Chemical Society*, **129**, 248–249; (h) Burkhardt, A., Spielberg, E.T., Gorls, H. and Plass, W. (2008) *Inorganic Chemistry*, **47**, 2485–2493.

81 (a) Noyori, R. (1994) *Asymmetric Catalysis in Organic Synthesis*, Wiley, New York; (b) Canali, L. and Sherrington, D.C. (1999) *Chemical Society Reviews*, **28**, 85–93; (c) Coates, G.W. (2000) *Chemical Reviews*, **100**, 1223–1252; (d) Jacobsen, E.N. (2000) *Accounts of Chemical Research*, **33**, 421–431; (e) Johnson, J.S. and Evans, D.A. (2000) *Accounts of Chemical Research*, **33**, 325–335; (f) Noyori, R., Kitamura, M. and Ohkuma, T. (2004) *Proceedings of the National Academy of Sciences of the United States of America*, **101**, 5356–5362; (g) Minnaard, A.J., Feringa, B.L., Lefort, L. and De Vries, J.G. (2007) *Accounts of Chemical Research*, **40**, 1267–1277.

82 Lacour, J., and Hebbe-Viton, V. (2003) *Chemical Society Reviews*, **32**, 373–382.

83 Favarger, F., Goujon-Ginlinger, C., Monchaud, D., and Lecour, J. (2004) *The Journal of Organic Chemistry*, **69**, 8521–8524.

84 Chavarot, M., Menage, S., Hamelin, O., Charnay, F., Pecaut, J., and Fontecave, M. (2003) *Inorganic Chemistry*, **42**, 4810–4816.

85 Hamelin, O., Pecaut, J., and Fontecave, M. (2004) *Chemistry – A European Journal*, **10**, 2548–2554.

86 Rutherford, T.J., Pellegrini, P.A., Aldrich-Wright, J., Junk, P.C., and Keene, F.R. (1998) *European Journal of Inorganic Chemistry*, **11**, 1677–1688.

87 Kolp, B., Viebrock, H., von Zelewsky, A. and Abeln, D. (2001) *Inorganic Chemistry*, **40**, 1196–1198.

88 Amouri, H., Caspar, R., Gruselle, M., Guyard-Duhayon, C., Boubekeur, K., Lev, D.A., Collins, L.S.B., and Grotjahn, D.B. (2004) *Organometallics*, **23**, 4338–4341.

89 Brown, R.H. (1985) *Origins*, **12**, 8–25.

90 Scott, R.L. (1977) *Journal of the Chemical Society-Faraday Transactions II*, **3**, 356–360.

91 Coquerel, G. (2000) *Enantiomer*, **5**, 481–498.

92 (a) Wermester, N., Aubin, E., Pauchet, M., Coste, S., and Coquerel, G. (2007) *Tetrahedron: Asymmetry*, **18**, 821–831; (b) Renou, L., Morelli, T., Coste, S., Petit, M.-N., Berton, B., Malandain, J.-J. and Coquerel, G. (2007) *Crystal Growth & Design*, **7**, 1599–1607.

93 Huang, J., Chen, S., Guzei, I.A., and Yu, L. (2006) *Journal of the American Chemical Society*, **128**, 11985–11992.

94 Gallis, H.E., Van Ekeren, P.J., Van Miltenburg, J.C., and Oonk, H.A.J. (1999) *Thermochimica Acta*, **326**, 83–90.

95 Gallis, H.E., van der Miltenburg, J.C., and Oonk, H.A.J. (2000) *Physical Chemistry Chemical Physics*, **2**, 5619–5623.

96 Gervais, C., and Coquerel, G. (2002) *Acta Crystallographica. Section B, Structural Science*, **58**, 662–672.

97 Sjöberg, B. (1957) *Arkiv for Kemi; Utgivet av K. Svenska Vetenskapsakademien*, **11**, 439; Coquerel, G. (2004) *Journal de Physique IV France*, **113**, 11–15; Klussmann, M., Izumi, T., White, A.J., Armstrong, A. and Blackmond, D.G. (2007) *Journal of the American Chemical Society*, **129**, 7657–7660.

98 Klussmann, M., Izumi, T., White, A.J., Armstrong, A., and Blackmond, D.G. (2007) *Journal of the American Chemical Society*, **129**, 7657–7660; Takahashi, H., Tamura, R., Fujimoto, D., Lepp, Z., Kobayashi, K. and Ushio, T. (2002) *Chirality*, **14**, 541–547.

99 Petterson, K. (1956) *Arkiv for Kemi; Utgivet av K. Svenska Vetenskapsakademien*, **10**, 297–323; Delépine, M. (1921) *Bulletin de la Société Chimique de France*, **29**, 656; Timmermans, J. (1929) *Recueil des Travaux Chimiques des Pays Bas*, **46**, 890.

100 Petit, M.-N., and Coquerel, G. (2003) *Mendeleev Communications*, **13**, 95–96.

101 Wermester, N., Lambert, O., and Coquerel, G. (2008) *Crystal Engineering Community*, **10**, 724–733.

102 Coquerel, G. (2007) Preferential Crystallization, in *In:Novel Optical Resolution Technologies*, Vol. 269, Springer, Berlin, pp. 1–51.

103 Dufour, F., Perez, G., and Coquerel, G. (2004) *Bulletin of the Chemical Society of Japan*, **77**, 79–86.

104 Coquerel, G., Perez, G., and Hartman, P. (1988) *Journal of Crystal Growth*, **88**, 511–521.

105 Gervais, C., Beilles, S., Cardinaël, P., Petit, S., and Coquerel, G. (2002) *The Journal of Physical Chemistry. B*, **106**, 646–652.

106 Levilain, G., Tauvel, G., and Coquerel, G. (2006) 13th International BIWIC Delft (Netherlands), pp. 244–250.

107 Querniard, F., Linol, J., Cartigny, Y., and Coquerel, G. (2007) *Journal of Thermal Analysis and Calorimetry*, **90** (2), 359–365; Larsen, S. and Lopez De Diego, H. (1995) *ACH-Models in Chemistry*, **132**, 441–450.

108 Ostwald, W. (1897) *Zeitschrift fur Physikalische Chemie-International Journal of Research in Physical Chemistry & Chemical Physics*, **119**, 227.

109 Hilfiker, R.(ed.) (2006) *Polymorphism – In the Pharmaceutical and Fine Chemical Industry*, Chap. 10. Wiley VCH.

110 Sakai, K., Sakurai, R., and Nohira, H. (2007) New Resolution Technologies Controlled by Chiral Discrimination Mechanisms, in *Novel Optical Resolution Technologies*, Vol. 269, Springer, Berlin, pp. 160–199.

111 Weissbuch, I., Popovitz-Biro, R., Lahav, M., and Leiserowitz, L. (1995) *Acta Crystallographica. Section B, Structural Science*, **51**, 115–148.

112 Ndzié, E., Cardinaël, P., Petit, M.-N., and Coquerel, G. (1999) *Enantiomer*, **4**, 97–101.

113 Atwood, J.L., Davies, J.E.D., and Mac Nicol, D.D. (1984) *Inclusion Compounds*, Vols. 1–3, Academic Press, London, UK.

114 Miyata, M., Tohnai, N., and Hisaki, I. (2007) *Accounts of Chemical Research*, **40**, 699–702.

115 Kano, K., Kato, Y., and Kodera, M. (1996) *Journal of the Chemical Society-Faraday Transactions II*, **6**, 1211–1217.

116 Ferron, L., Guillen, F., Coste, S., Coquerel, G., and Plaquevent, J.-C. (2006) *Chirality*, **18**, 662–666.

117 (a) Harata, K. (1998) *Chemical Reviews*, **98**, 1803–1827; (b) Saenger, W. (1980) *Angewandte Chemie (International Edition in English)*, **19**, 344–62; (c) Szejtli, J. (1998) *Chemical Reviews*, **98**, 1743–1753; (d) Saenger, W., Jacob, J., Gessler, K., Steiner, T., Hoffmann, D., Sanbe, H., Koizumi, K., Smith, S.M. and Takaha, T. (1998) *Chemical Reviews*, **98**, 1787–1802.

118 Ariga, K., and Kunitake, T. (2006) *Supramolecular Chemistry – Fundamentals and Applications*, Springer Verlag, Berlin.

119 (a) Schurig, V. (2002) *Trends in Analytical Chemistry*, **21**, 647–661; (b) Juvancz, Z. and Szejtli, J. (2002) *Trends in Analytical Chemistry*, **21**, 379–388.

120 Kobor, F., Angermund, K., and Schomburg, G. (1993) *Journal of High Resolution Chromatography & Chromatography*, **16**, 299–311.

121 Grandeury, A., Tisse, S., Gouhier, G., Agasse, V., Petit, S., and Coquerel, G. (2003) *Chemical Engineering & Technology*, **26**, 354–358.

122 Harata, K. (1998) *Chemical Review*, **98**, 1803–1827.

123 Yannakopoulou, K., Mentzafos, D., Mavridis, I.M., and Dandika, K. (1996) *Angewandte Chemie (International Edition in English)*, **35**, 2480–2483.

124 Grandeury, A., Condamine, E., Hilfert, L., Gouhier, G., Petit, S., and Coquerel, G. (2007) *The Journal of Physical Chemistry. B*, **111**, 7017–7026.

125 Lehn, J.M. (1995) *Supramolecular Chemistry*, VCH, Weinheim.

126 Hammond, R.B., Pencheva, K., and Roberts, K.J. (2007) *Faraday Discussions*, **136**, 91–106.

11
Switching at the Nanoscale: Chiroptical Molecular Switches and Motors

Wesley R. Browne, Dirk Pijper, Michael M. Pollard, and Ben L. Feringa

11.1
Introduction

In this chapter, recent developments in the field of chiroptical switches and nanomotors will be highlighted, with less recent examples taken where appropriate. The chapter will build on our earlier contribution, which covers the field of chiroptical molecular switches up until 2000 [1]. Not all aspects of the science involved in the illustrative examples in this chapter will be discussed and the interested reader is referred of course to the original work. However, what we hope to demonstrate in this chapter are the basic principles involved and how powerful an asset chirality and chiroptics can be in the molecular toolbox for the design of functional systems. We begin with a cursory look at some notable examples of both photo- and electro-responsive chiroptical switches and then motors, which is followed by recent advances in the application of these systems to control real nanoscale functions. The focus is on chirality and the overall goal is to demonstrate how simple concepts can be built upon to yield properties with a broad range of application in nanodevices. However, before delving into the elegant systems that build on some chiroptical switches and motors it is pertinent, perhaps, to first take an aside to define some important concepts and terms.

Drawing direct analogy between macroscopic switches and machines with their nanoscale components inevitably requires an element of conceptual flexibility. Not least because the nature of the forces and conversion of light and chemical to mechanical energy is so fundamentally different between the macro- and nanoscale world. At the macroscale, switches and motors are affected by mass and inertia, volume, friction and gravity, whereas at the nanoscale viscosity, intermolecular interactions (Van der Waals forces, hydrogen bonding, etc.) and interfacial boundaries are dominant. Hence, the definitions applied to the function and operation of nanoscale switches and devices are largely subjective and direct analogy of machine-like functions omnipresent in the macroscopic world with those at the molecular level is frequently inappropriate and sometimes even misleading [2].

Chirality at the Nanoscale: Nanoparticles, Surfaces, Materials and more. Edited by David B. Amabilino
Copyright © 2009 WILEY-VCH Verlag GmbH & Co. KGaA, Weinheim
ISBN: 978-3-527-32013-4

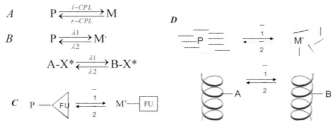

Scheme 11.1 Chiral switching between: (a) enantiomers, (b) diastereoisomers (X* = chiral auxiliary), (c) functional chiral switches (FU = functional unit), (d) macromolecular switch and switching of the organization of the matrix. P and M denote right- and left-handed helical structures, A and B denote two bistable forms of a switchable molecule. l- and r-cpl indicate left- and right-circularly polarized light, respectively.

In the present context we will hold the concept of molecular switching and machinery somewhat loosely especially since the borders between molecular switches, motors and machines can be defined only arbitrarily. Nevertheless, the ability to address functionality reversibly in a repetitive manner is essential in building molecular machines. When we consider machines, movement and the translation of mechanical motion into useful work are essential features regardless of whether it is at the nano or the macroscopic level. In molecular machines the "engine" is typically based on a switching element that responds to external stimuli that is, the input of a fuel, for example, light, protons, electrons, and so on, and, furthermore, the physical change in position, either rotational or translational, must occur with directional control. In short we must be able to operate above randomness [3]. In addition, for a (molecular) motor the repetitive nature of its mechanical movements is a distinct feature.

Optically triggered chiral switches offer fascinating opportunities as elements in molecular devices and photonic materials (Scheme 11.1). In a chiral photochromic system, where P and M represent two different chiral forms of a bistable molecule, changes in chirality, that is, from P to M and M to P, can be induced by an optical stimulus – a photon of light. The nature of the change depends on the chirality of the system. A number of chiroptical molecular switch types can be distinguished:

(A) Switching of enantiomers: Irradiation of a chiroptical switchable molecule in its enantiomerically pure form (R/S or P/M) using normal, nonpolarized light will, irrespective of the wavelength used, result in full racemization, as the two enantiomers have identical absorption characteristics. However, the enantiomers can be interconverted by irradiation at distinct wavelengths with left- or right-handed circularly polarized light (l- or r-CPL). In this way, it is possible to switch in either direction in an enantioselective manner.

(B) Switching of diastereoisomers: the compound exists in two diastereomeric forms, for instance, a P (right-handed) and M' (left-handed) helical structure, which are thermally stable but can undergo photoisomerization at two different wavelengths,

λ_1 leading to photoisomerization towards the one and λ_2 towards the other diastereoisomer. This can be accomplished also by linking a chiral auxiliary X* to a photochromic unit A, which itself can be either chiral or achiral. In such an A–X*, B–X* system, the change in chirality during the switching process is controlled by the auxiliary X*.

(C) Functional chiral switches: As these photochromic systems are often multifunctional in nature, the induced changes in chirality can simultaneously trigger a particular function to be modulated, such as fluorescence properties, molecular recognition events or motion. Often, these effects result from a change in the molecular geometry or the electronic properties of the system.

(D) Switching of macromolecules or supramolecular organization: By attaching a photobistable molecule (either chiral or achiral) covalently to, for example, a polymer or by making it part of a host–guest system. In these systems, photoisomerization of the switch unit induces property changes on the macro- or supramolecular level, for instance by controlling the helical twist sense of the backbone of the polymer or the organization of "a surrounding matrix," such as a chiral and helically organized phase of a liquid-crystalline material or a gel. With such systems, the chiral response of this macro- or supramolecular structure or organization offers the possibility to record the photochemical process indirectly also.

Finally, before discussing chiroptical systems it is perhaps useful to remind ourselves of what we mean by the terms chiroptic/al. For this we refer to the IUPAC Goldbook [4] definition: referring to optical techniques used to investigate chiral substances including measurements of optical rotation at a fixed wavelength, optical rotatory dispersion (ORD), circular dichroism (CD), and circular polarization of luminescence (CPL).

11.2 Switching of Molecular State

Molecular switching is, at its foundation, based on bistability [2, 5, 6, 8], that is, where a molecule exists in two distinct forms and in which each state can be converted to the other reversibly upon application of an external stimulus. Importantly switching between the two states must be controlled and not proceed spontaneously (within the time frame of the experiment). The external stimulus can be anything that changes the free energy of the molecule, for example, chemical, electrochemical or photochemical. In short, to be a switch a molecule should show bistability. The lack of bistability has proven to be a major hurdle in the development of photochromic materials, in particular for application in molecular-based data-storage devices [7]. Equally important is that each state is fundamentally different from the other, so that the state of the system can be "read" nondestructively (Figure 11.1a) [8]. Nature has employed an essentially perfect balance between bistability, nondestructive read-out

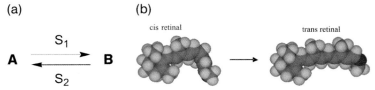

Figure 11.1 (a) The concept of a bistable switchable system. **A** and **B** represent two distinct states a molecule can exist in and S_1 and S_2 refer to the application of different stimuli. (b) cis–trans-isomerization of retinal in the initial step in the process of vision [82]. This event – a change from a bent to a more linear shape of the molecule – simple though it is, initiates a cascade of events beginning with a change in the protein and ultimately a signal transmitted to the brain.

and thermal reversion in its delicate use of retinal. The initial event in the process of vision is the absorption of a photon of light by *cis*-retinal (the wavelengths absorbed being determined by the nature of the chromophore and the local environment created by the surrounding protein) and read-out being achieved by the ion cascade that the isomerization initiates (Figure 11.1b).

In designing bistable systems it should be realized that any molecular material that has two stable identifiable states can in principle be used as a switching element to control functions at the nanoscale or as a memory element in a data storage device using binary logic. Photoreversible compounds, where the reversible switching process is based on photochemical interconversion (photochromism) play a leading role in this field. The optical and structural properties of chiral bistable molecules offer unique opportunities for nondestructive read-out (using a variety of chiroptical techniques) and control of functions, dynamics and organization. The molecules must possess a number of key qualities to be suitable for building practical nanodevices. The basic requirements for a chiral optical molecular switch are:

- photochemical switching between two chiral forms should be possible
- thermal interconversion of the stereoisomers should not occur over a large temperature range (-20 to $+100\,°C$); thermochromic processes or thermally induced isomerization would lead to loss of the information stored
- the photochemical interconversion should be fast, selective and the compounds should be fatigue resistant
- the detection technique should be sensitive, discriminative and nondestructive
- the quantum yields of photochemical interconversion should be high, allowing efficient switching to occur
- retention of all these properties when the switchable unit becomes part of a nanosized system or a macromolecular structure or device.

While light is arguably an attractive noninvasive way to address chiral bistable states, several other ways to change chirality in bistable molecular systems are known and some of these are discussed below [5].

A simple yet elegant example of a chiral bistable system is that reported recently by Breidt *et al.*, [9] which employs the axial chirality of bisphenol and the conformational chemistry of a six-membered ring (Figure 11.2a). In this case the chirality of the system is determined by the relative stabilities of the *pseudobise*quatorial **1a** and *bis*axial **1b** atropisomers. Although it is difficult to argue that this is in fact a switch in a practical sense, the perturbation, a solvent change, brings about a reversible change in the molecular state. In this example, the equilibrium and the direction of switching is controlled by changing the free energy of ground-state energy surfaces. For the rest of this chapter, however, we will focus on systems that employ external stimuli (light, redox potential, chemical conversion), not to change the relative ground-state energies of systems but instead to overcome barriers to interconversion, which are sufficiently large to prevent thermal reversal proceeding under ambient conditions, except as specific controllable steps in, for example, a cycle.

The earliest studies on photochromic organic materials [10] demonstrated that the basic condition of photochemical bistability could be fulfilled. However, application of photochemically controlled bistability has often been hampered by a lack of thermal stability (fatigue) and the destructiveness of the read-out processes involved.

An illustrative example is a chiroptical switch based on overcrowded alkenes, such as those that will be discussed below, has been reported by Chen and Chou (Figure 11.2b) [11]. The system makes use of the bistability of its two isomers in which irradiation at $\lambda = 280$ nm results in conversion of stable-**2** selectively to unstable-**2** with a diastereomeric excess of 99%. Complete reversal to stable-**2** takes place upon heating at 420 K. Photochemical conversion to the stable isomer is less efficient with irradiation at $\lambda = 254$ nm leading to a ratio of 67 : 33 of stable to unstable isomers. The development of systems suitable for applications, however, requires high efficiency in the process of switching, including both high quantum yields and fast response times [5]. A significant further challenge is to develop molecular switches that operate effectively as part of more complex multifunctional

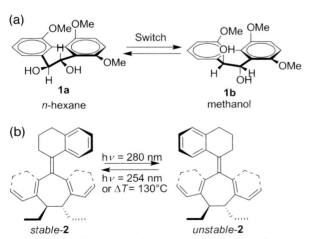

Figure 11.2 Conformational changes in a chiroptical switch induced by changes in (a) solvent [9], and (b) by light and/or heat [11].

systems and machines [2]. In the following sections we will consider several aspects of molecular switches beginning with chiral optical molecular switches designed to address some of these fundamental problems. Although adding complexity and additional functionality to these systems presented considerable scientific challenges, this approach permits access to exciting new territories in molecular design.

11.3
Azobenzene-Based Chiroptical Photoswitching

Azobenzenes have proven to be particularly useful photoresponsive units in chiroptical switching, not least due to the ability to switch between two states (*cis* and *trans*) and induce wide amplitude changes in size in doing so [12]. The switching event can be used to control a large variety of functions (Scheme 11.2), for instance binding events.

Although azobenzenes are not inherently chiral, the large structural changes, which accompany isomerization, allow for the switching process to influence chiral units [13]. Aida and coworkers have developed a series of remarkable systems in which the geometric change that accompanies *cis–trans-isomer*ization is used to drive much larger changes in intercomponent interaction over large (in a molecular sense) distances using a molecular "scissors" design (Figure 11.3) [14]. In these systems an azobenzene switching unit affects a ferrocene rotor unit. However, this system uses a ferrocene unit not only as a molecular pivot but exploits its redox chemistry also. The redox state of the ferrocene unit determines the photostationary state reached. In the reduced state the *cis-isomer* is favored upon irradiation at 350 nm, whereas in the oxidized state it is the *trans-isomer* that is the favored state. The basic design for the molecular scissors was taken a step further in coupling the movement of several molecular components in a light-powered "molecular pedal" [14b]. The change in molecular shape upon *cis–trans* photoisomerization of an azobenzene unit

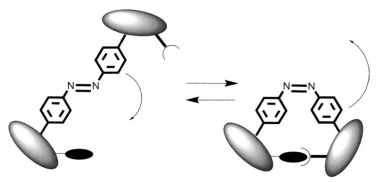

Scheme 11.2 Large-amplitude change in molecular size driven by the *cis–trans-isomer*ization of azobenzene.

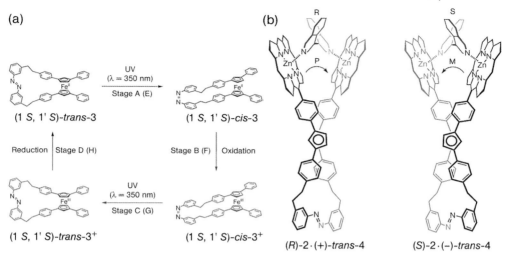

Figure 11.3 Sequential operation of a ferrocene–azobenzene scissor **3**. Sequential irradiation with UV light and redox changes of the ferrocene move the systems in a scissors-like action. Reprinted with permission from [14c]. Copyright RSC 2007. (b) Application of the molecular scissors **4**: the ferrocene pivot facilitates actuation of the movement of the Zn-porphyrins and hence the chiral guest, resulting in an inversion of the stereochemistry of the guest molecule. Reprinted with permission from [14b]. Copyright Nature 2006.

is transmitted via a pivot point (a ferrocene unit) to the orientation of two porphyrin units. The pedal-like motion of zinc porphyrin units attached to it induces a clockwise or counterclockwise rotary motion in a bound rotor guest demonstrating that small changes in molecular structure can be actuated to drive much larger mechanical changes a considerable distance from the initial point of mechanical motion (Figure 11.3b). In addition, the change in shape of the molecular switch is transmitted to the chiral guest.

A chiroptical system that can be addressed both photo- and electrochemically was developed by Wang and coworkers (Figure 11.4) [15]. In this pentahelicene-like system, photochemical control is achieved through the *cis–trans*-isomerization of an appended azobenzene unit, whereas electrochiroptical control is achieved by the reduction/oxidation of the azobenzene unit.

The first self-assembled optically active double helix based on guanidine–carboxylate binding units and incorporating photoresponsive units (azobenzenes) was reported by Yashima and coworkers [16] (Figure 11.5). The helicity of the hydrogen bonded supramolecular structure can be controlled in a reversible manner by isomerization of the azobenzene units. In the *cis* state hydrogen-bonding interactions are sufficient to maintain a supramolecular state. In the *trans* state the hydrogen bonding cannot maintain the same state and the helix inverts partially. The chirality of the system is controlled by secondary groups relatively remote from the azobenzene unit itself.

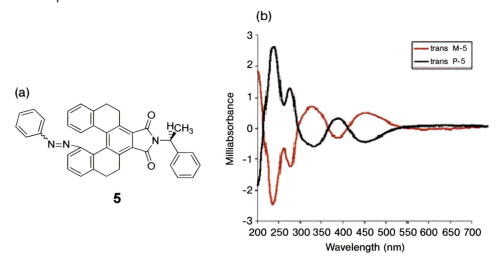

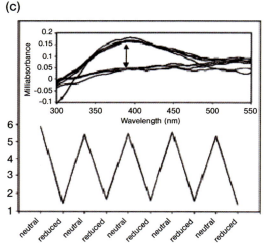

Figure 11.4 (a) The structure switchable pentahelicene **5**. (b) CD spectra of **M-5** and **P-5** in acetonitrile (10^{-5} M). (c) Electrochemically induced variation in the ellipticity of **M-5** monitored at 454 nm. (Inset) CD spectra obtained during the electrochemical modulation study. Reprinted with permission from [15]. Copyright ACS 2005.

Leigh and coworkers [17] have also taken a supramolecular approach to chiroptical switching based on the *cis–trans*-isomerization of a double bond. The large amplitude displacement of a macrocycle along a rotaxane thread moves a chromophore in and out of proximity to the chiral unit. In one state the chiroptical signal is negligible, whereas in the second state the presence of a nonchromophoric chiral group induces a chiroptical response.

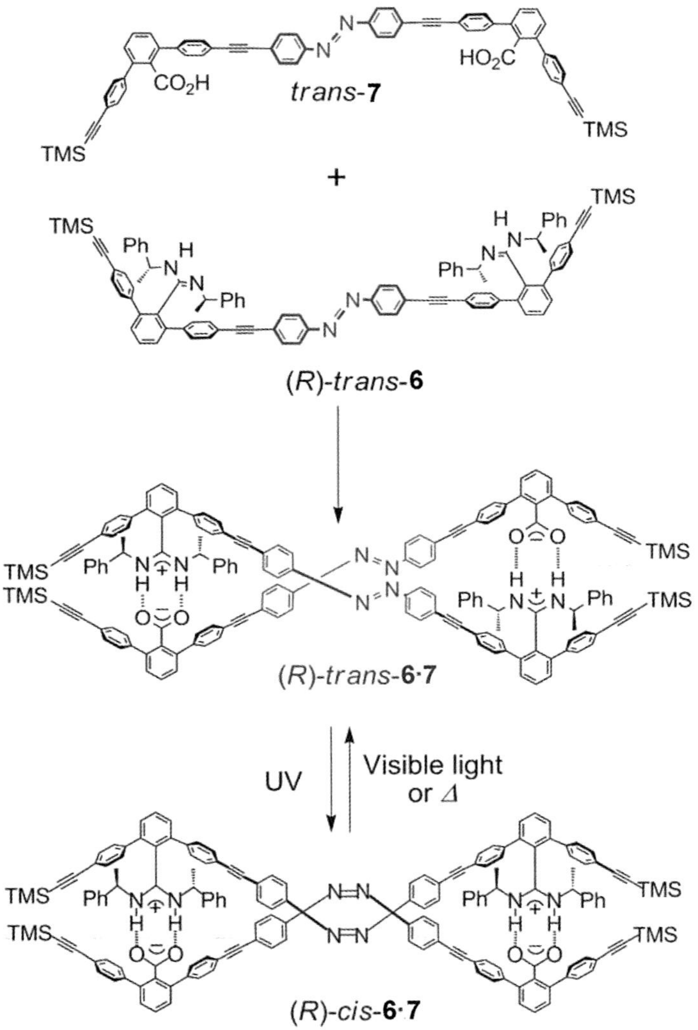

Figure 11.5 Double-helix formation and subsequent *trans–cis*-photoisomerization of (R)-*trans*-**6·7**.

Azobenzenes were also employed by Tian and coworkers in an unconventional rotaxane system built from an azobenzene, a cyclodextrin unit (the ring) and a naphthalimide unit (Figure 11.6) [18]. The azobenzene allows for multiple switching cycles to be performed as monitored by UV/vis absorption spectroscopy. The chiral environment of the cyclodextrin not only provides the torus of the rotaxane [19] but also induces a CD effect in the chromophoric units to allow the state of the system to be monitored with chiroptical techniques.

A wide range of chiral polymer-based azobenzene photochromic materials have been developed over the past decades. With azobenzene-containing polymers one

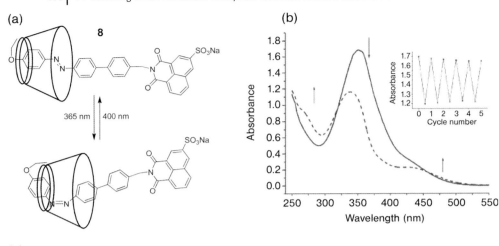

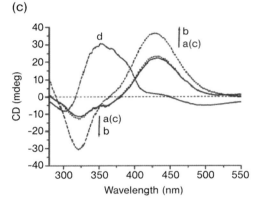

Figure 11.6 (a) Reversible configuration change of the [1]rotaxane **8** under different stimuli (b) The absorption spectra of rotaxane **8** in aqueous solution before (—) and after (---) irradiation at 365 nm for 15 min at room temperature. Insert: changes in the absorption spectra of rotaxane **8** (absorption value at around 350 nm) for several cycles. In one cycle, (c) CD spectra before: a and after: b irradiation at 365 nm and c: after irradiation at 430 nm, d: CD spectrum of *cis*-**8**. Reprinted with permission from [18]. Copyright RSC 2007.

can exploit the readily induced and reversible isomerization around the azo-bond to control the polymer's conformation and chirality and as a consequence achieve control over macroscopic properties of polymer films. For an extensive discussion on this fascinating topic the reader is referred to the relevant chapters and reviews [20].

Liquid crystals (LCs) are an important class of materials where reversible control of chirality plays a crucial role. By doping an azobenzene photochromic compound into a liquid-crystalline material, large changes in organization can be induced in the liquid-crystalline phases. This is the result of a large configurational change upon isomerization of the azobenzene optical switch. As the *trans*-form of the azobenzene-derivatives are rod-like, it usually stabilizes the LC phase. By contrast, the *cis*-form is bent and generally destabilizes the LC phase (Scheme 11.3).

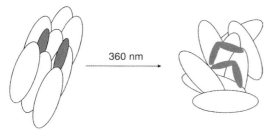

Scheme 11.3 *trans*-Azobenzene stabilizes the LC phase where *cis*-azobenzene disrupts the phase.

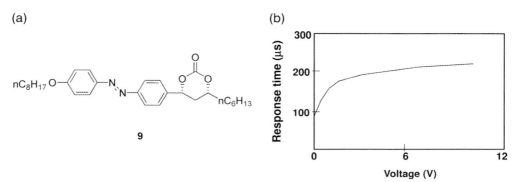

Figure 11.7 (a) Ferroelectric azobenzene-based LC compound **9**. (b) Switching response time of 90 μs.

This principle of combining azobenzene photoswitching with molecular chirality in the reversible control of LC organization was elegantly used by Ikeda and coworkers. They employed azobenzene **9** with an optically active *syn* cyclic carbonate bearing a long alkyl chain (Figure 11.7a) as dopant in ferroelectric liquid crystals (FLC) [21]. Ferroelectric liquid crystals exhibit a chiral smectic phase and respond to an electric field on a time scale of microseconds (μs) as a result of spontaneous polarization. The chiral *syn* cyclic carbonate induces large spontaneous polarization. *trans–cis* Photoisomerization ($\lambda = 355$ nm) induced a change effectively of the chiral smectic C phase (S_mC^*) into a smectic A phase (S_mA), whereas thermal *cis–trans*-isomerization restored the S_mC^* phase. The photoisomerization in the FLC phase has a fast response time in a liquid-crystalline phase, 90 μs, when voltage is not applied (Figure 11.7b) [22].

11.4
Diarylethene-Based Chiroptical Switches

A second class of photochromic compounds that can form the basis of chiroptical switches are the diarylethenes, in particular the dithienylcyclopentenes. Diarylethenes undergo a reversible ring-closing reaction of a hexatriene unit to a cyclohex-1,3-diene

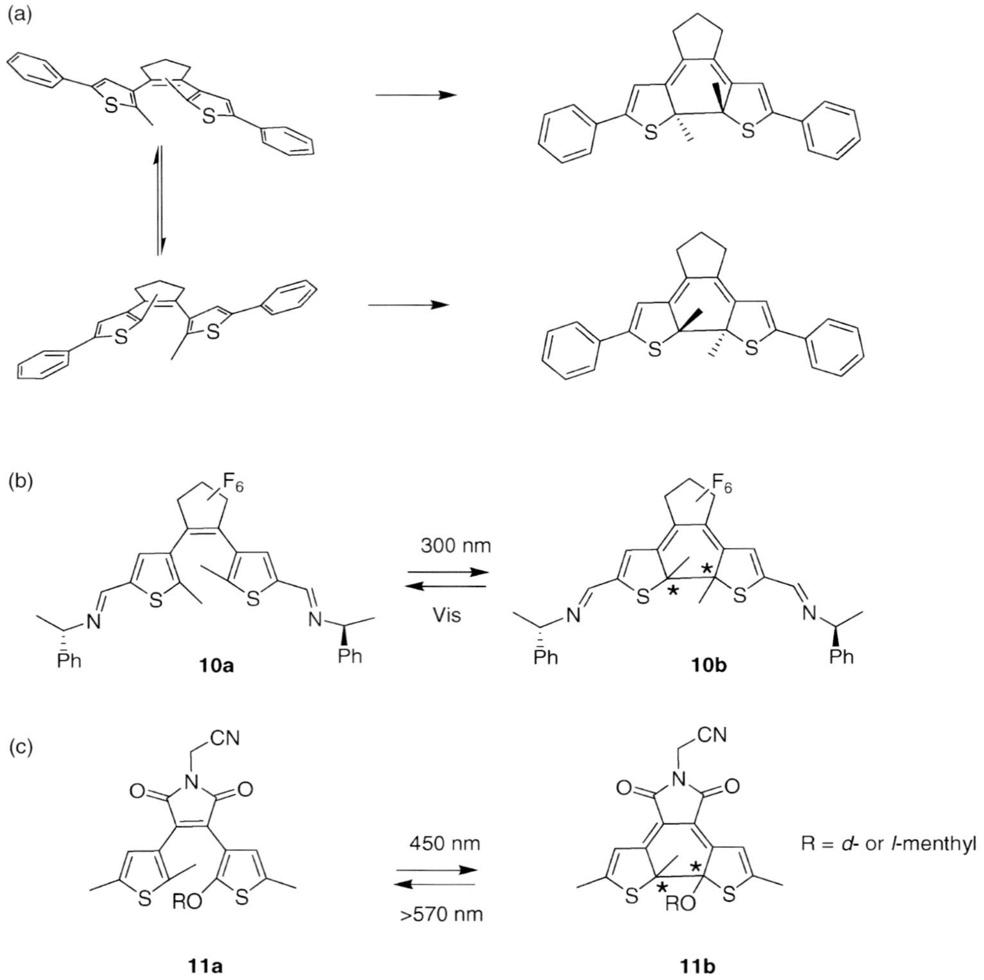

Scheme 11.4 (a) The dynamic equilibrium of the open state of a dithienylethene between a P and M helix is locked upon photochemical ring closure. When a chiral substituent is placed on the periphery (b) or at the reactive carbon centres (c) then the ring-closing reaction leads to a diastereomeric excess.

ring. The open antiparallel form exists as two rapidly interconverting dynamic helical structures (Scheme 11.4). This light-driven rearrangement is a concerted process and hence the ring closing results in the formation of two stereogenic centers, leading to a racemate of the closed form. A simple yet effective approach to modifying this class of photochromic switch is to functionalize the thienyl rings with chiral groups to bias ring closing in favor of one of the diastereomeric forms. The directing groups can either be attached at the α-carbon (of the thienyl rings) not involved directly in the

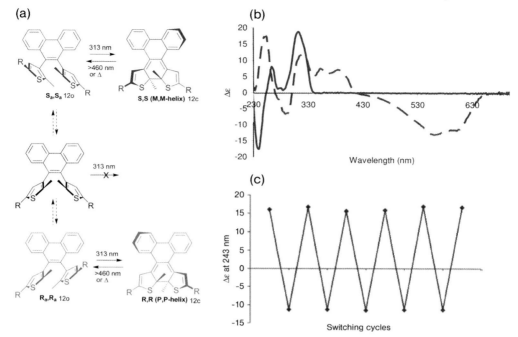

Figure 11.8 (a) Stereoisomers and thermal and photochemical processes of atropisomeric dithienylethenes (**12o**). (b) CD spectra of the M, M-helical isomer in the open (thick line) and photostationary state (dashed line) form in heptane. (c) Switching cycles observed upon alternated irradiation with UV light ($\lambda = 313$ nm) and visible light ($\lambda = 460$ nm) as detected by CD spectroscopy at 243 nm. Reprinted with permission from [25]. Copyright RSC 2007.

ring-closing reaction (Scheme 11.4b) [23] or one or both of the prochiral carbons of the thienyl ring (Scheme 11.4c) [24]. In the former case, the diastereoisomers are usually formed in low diastereomeric excess, albeit sufficient for the purposes of detection with chiroptical techniques while in the latter case the proximity of the chiral directing group to the reactive carbons results in a preference for the formation of one diastereoisomer over the other.

Recently, the isolation of individual atropisomers of photochromic diarylethenes was reported (Figure 11.8a) [25]. Due to the presence of the phenanthrene bridging moiety, the barrier for rotation about the single bond connecting the thiophene and phenanthrene units in **12o** is high (109.6–111.5 kJ mol^{-1}). The switching process between the open and closed form of each enantiomer could be followed readily by CD spectroscopy (Figure 11.8b). The system is robust, with alternating irradiation with UV ($\lambda = 313$ nm) and visible light ($\lambda > 460$ nm) allowing for several switching cycles (Figure 11.8c).

Helicenes are inherently chiral by virtue of their helical nature. Once formed the interconversion between the two helical states is difficult to control and epimerization requires heating, typically. Branda and coworkers [26] have taken the approach to

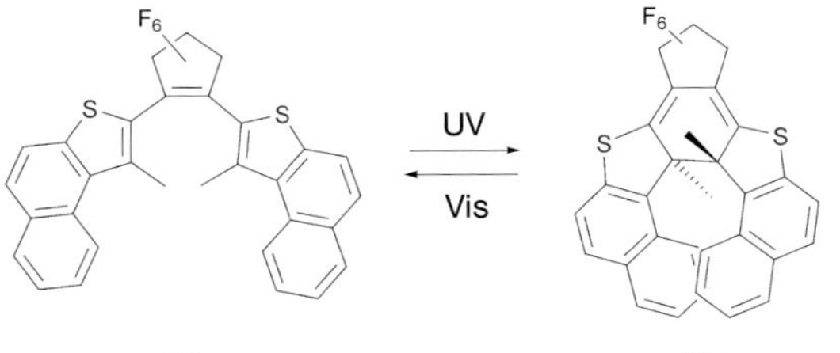

Figure 11.9 Locking and unlocking of the helicity in a photochromic helicene **13o/13c**.

make the formation of the helicene reversible by building in a photochromic diarylethene (Figure 11.9). In the ring-closed state the rate of helix inversion is, as expected, much lower than in the open state, however, in this system a racemic mixture of stereoisomers is obtained only, rendering it unsuitable in a practical sense as a chiroptical switch.

Recently, Branda and coworkers [27] have taken this approach (Figure 11.10) a step further by modifying the helicene-based dithienylethene photochromic switch with chiral pinene units. Upon ring closing one of the two possible helicenes is formed preferentially providing a large change in the CD spectrum compared with the open state. This system satisfies the requirements for a successful chiroptical photoswitch, that is, it is thermally stable in both open and closed states, it displays high stereoselectivity in its photocyclization reactions, and the changes in its CD and ORD spectral properties are sufficient to allow for application even though the photostationary states achieved are not ideal.

Low molecular weight (LMW) organogelators and hydrogelators are a remarkable class of materials with fascinating properties [28]. By applying chiral molecular switches, the properties of these gels can be controlled in a reversible manner. The LMW organogelators have a pronounced ability to self-assemble in fiber-like structures and can lead to gelation of a wide range of solvents (see chapter 4 in this book). The introduction of light-switchable functions to synthetic LMW gelators allows one to achieve reversible control over the self-assembly process.

For this purpose diarylethene photochromic moieties were functionalized with (R)-1-phenylethylamine-derived chiral amide groups (Figure 11.11a). The molecules **15o/15c** self-assemble into chiral fibers of a preferred handedness as shown by CD spectroscopy and electron microscopy (EM) due to the multiple hydrogen bonds between the amide groups (Figure 11.11b). The photochemical switching at the molecular level is attended by a major change in conformational flexibility and as a consequence a change in the aggregation behavior is observed. When a dilute solution of these switches is irradiated with UV light the closed form is obtained as a 50 : 50 mixture of stereoisomers. In contrast, when ring closure occurs in the gel state a 98 : 2 ratio of stereoisomers is observed. In this system the supramolecular chirality (fiber helicity) is essential for the observed chirality at the molecular level of the switching units. In fact there is mutual control of chirality at different hierarchical levels and the whole system is reversible due to the chiral switches present. The four-state switching cycle that was observed is shown in Figure 11.11c.

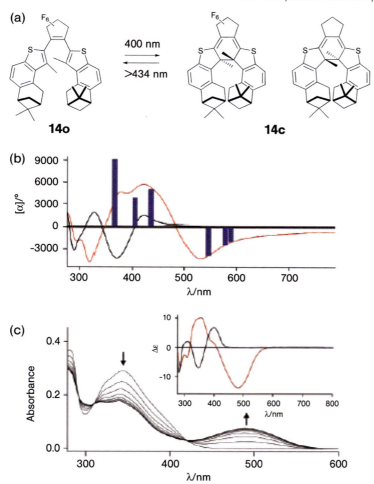

Figure 11.10 The chiral auxiliary approach. (a) A dithienylethene-based photochromic switch (14o) is modified with two chiral units. (b) The experimental and calculated ORD spectra of 14o and 14c. (c) Changes to the UV/vis absorption spectrum upon irradiation with UV light and (inset) CD spectra in the open and closed states. Upon ring closure one of the helical forms is preferred.

Of paramount importance in this system is that metastable chiral aggregates can be obtained in a fully reversible manner. Furthermore, it should be noted that the use of chiral switches allows control of self-assembly, supramolecular chirality and viscoelastic properties of these soft materials, whereas controlled mass transport allowed holographic patterning by light [29]. This system is particular illustrative in showing how the interplay of molecular communication (chiral recognition) and self-assembly operate in a dynamic way (Figure 11.11)[83].

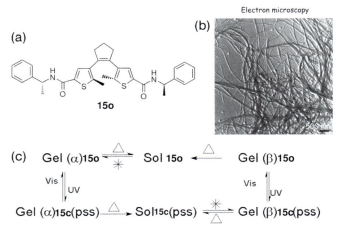

Figure 11.11 (a) A bis-amide-substituted dithienylethene low molecular weight gelator molecule (b) TEM image of the gel fibers formed upon gelation of an organic solvent. (c) The four gel state cycle. **15o** and **15c** are the ring-open and ring-closed states, respectively. Gel-α and gel-β indicate distinct gelation states, Δ indicates heating, * indicates cooling.

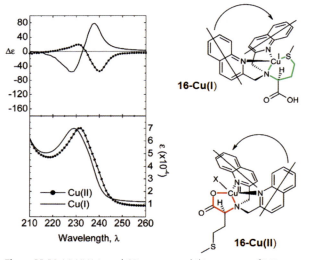

Figure 11.12 (a) UV/vis and CD spectra and (b) structure of **16** in the Cu(I) and Cu(II) oxidation states. Reprinted with permission from [30]. Copyright ACS 2003.

11.5
Electrochiroptical Switching

The effect of a change in the redox state of a molecule on the electronic structure, and hence the electronic absorption and CD spectra, has been explored recently by several groups. Canary and coworkers [30] have demonstrated that the changes in ligand environment that can accompany redox changes in inorganic complexes (in this case the Cu(I)/Cu(II) redox couple) can be used to drive chiroptical switching (Figure 11.12). In this case the dependence of the inversion in the relative stability of the oxido- and thio-copper complexes upon a change in the oxidation state is utilized. This system is an excellent example of the possibilities that transition-metal-based redox switching of the ligand coordination mode present. The change in coordination mode can be employed to drive large-amplitude structural changes in much the same way as has been achieved in the azobenzene-based systems discussed above.

A relatively simple example of redox-driven chiroptical switching is the reductive electrochemical switching of chiral *trans*-cyclohexanediol bispyrene esters (1R,2R)-**17a** and (1S,2S)-**17b** (Figure 11.13) to their dianions, as is evident from a strong change in the UV/vis and CD spectra, reported by Westermeier and coworkers [31]. The neutral compounds absorb only below 450 nm, whilst the dianions show a strong absorption band at 510 nm. This visible absorption bands shows a strong exciton split circular dichroism. Interestingly the corresponding bisamide compounds

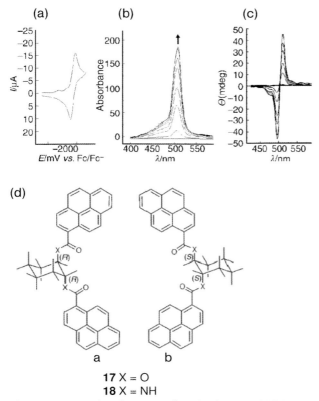

Figure 11.13 (a) Cyclic voltammetry of **17**, (b) changes in UV/vis absorption and (c) CD spectra upon reduction of **17**. (d) structure of **17a/b** and **18a/b**. Reprinted with permission from [31]. Copyright RSC 1999.

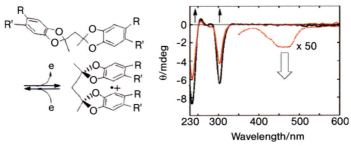

Figure 11.14 Electrochiroptical switching of **19** between a fully reduced state and a partially oxidized state followed by CD spectroscopy. Reprinted with permission from [32]. Copyright ACS 2007.

trans-cyclohexane bispyrene amides (1*R*,2*R*)-**18a** and (1*S*,2*S*)-**18b** failed to show any CD signals showing exciton coupling. Fukui *et al.* [32] have taken advantage of the interaction between identical units when in a partially oxidized state in a *bis* (catecholketal) system in which planar chirality is present (Figure 11.14). Upon one-electron oxidation of **19** a new visible absorption band appears that enables this system to be an on/off switch; this holds a distinct advantage to most chiroptical switches in which only a change in intensity of a CD signal is observed.

It is evident that in multicomponent chiroptical systems the chiral unit need not necessarily by the same as the responsive unit, for example where photoresponsive units (e.g., two anthracene units that can dimerize reversibly) are tethered to a chiral binaphthyl unit [33]. This approach offers the advantage that it allows for a much broader range of functional/responsive units to be employed. A relatively simple approach is to couple a chiral chromophoric unit with an electroactive unit or units [34]. For example Zhou *et al.* [35] have reported a simple electrochiroptical system based on a chiral binaphthyl unit connected covalently to two tetrathiafulvalene (TTF) units (Figure 11.15). Changes in the strength and nature of the interaction between the TTF units in the neutral, partially and fully oxidized states are the driving force behind the electrochemical control of the chiroptical output signal. In the partially oxidized state the two TTF^+ units are attracted to one another, reducing the dihedral angle between the naphthalene units, whereas in the fully oxidized state the TTF^{2+} units show strong electrostatic repulsion, increasing the dihedral angle with respect to the neutral state. A similar approach has been taken by Deng *et al.* using the reductive switching of the dialkyl-4,4′-bipyridyl unit in place of oxidation of the TTF unit [36]. The changes in the CD spectrum in both these systems when addressed electrochemically are at best modest. Nevertheless, these systems represent a clear example of a bicomponent functional system where the intrinsic properties of the chiroptical unit are controlled by external perturbation.

11.6
Molecular Switching with Circularly Polarized Light

Helical-shaped overcrowded alkenes have formed the mainstay of many chiroptical switches and motors over the last three decades (Figure 11.16) [37]. In these

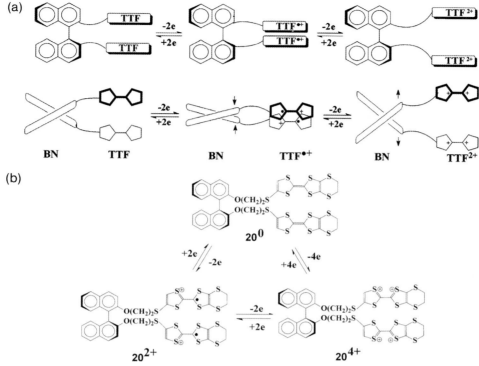

Figure 11.15 Electrochiroptical switch **20** based on interacting tetrathiofulvalene redox units. BN = binaphthyl, TTF = tetrathiofulvalene (neutral state), TTF$^+$ and TTF^{2+} = tetrathiofulvalene (partially or fully oxidized states).

molecules, switching between two states involves the (reversible) change in molecular helicity through a light-induced *cis–trans-isomeri*zation. For example, in **21**, *cis* → *trans*-isomerization results in the interconversion between right-(*P*) and left-handed (*M*) helical states (Figure 11.16) [38]. These overcrowded alkenes can function as chiroptical switches either as enantiomers (Figure 11.16; two enantiomeric

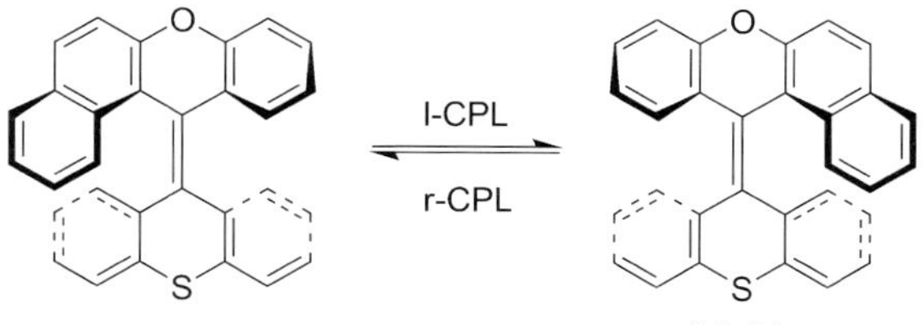

Figure 11.16 A chiroptical molecular switch based on enantiomers of **21** in which left (l-CPL) and right (r-CPL) circularly polarized light is used to achieve switching.

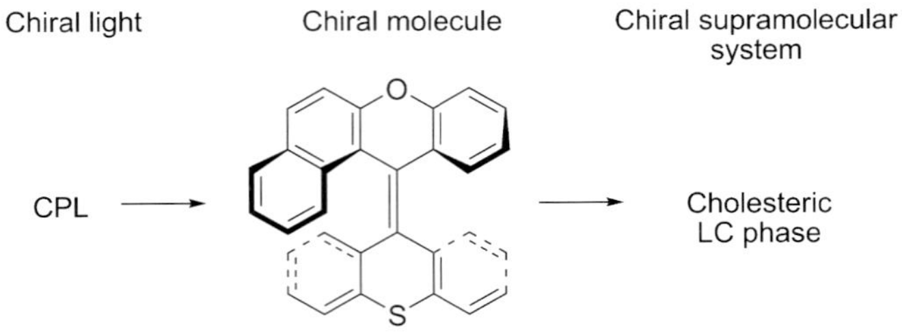

Chiral light   Chiral molecule   Chiral supramolecular system

CPL →   (P)-21   → Cholesteric LC phase

Figure 11.17 Amplification of chirality by hosting **21** in an LC phase.

states *P* and *M*) or as *pseudo*-enantiomers (Figure 11.18, two diastereoisomeric states *P* and *M'*).

When irradiated with noncircularly polarized light the photochemical isomerization between the two states will lead to a racemic mixture, whereas circularly polarized light can, potentially, achieve an excess of one enantiomer over another. Demonstrating this has presented a fundamental challenge to molecular design. Compound **21** satisfied the requirements needed to demonstrate that with right or left CPL enantioselective switching in either direction is possible (Figure 11.16b). The enantiomers of **1** are stable at ambient temperature ($\Delta G_{rac} = 25.9 \, \text{kcal mol}^{-1}$), resistant to fatigue and photoracemize rapidly upon irradiation at 300 nm with nonpolarized light ($\Phi_{rac} = 0.40$, *n*-hexane). The anisotropy factor g [39] (-6.4×10^{-3} at 314 nm) determined experimentally indicates that under optimum conditions an enantiomeric excess (ee) of 0.3% is achievable. Irradiation with right and left CPL was found to be able to switch between photostationary states with experimental ee values of 0.07% for *P*-**21** and *M*-**21**, respectively [40], allowing this system to be in three states; racemic (*P,M*)-**21**, *P*-enriched **21** and *M*-enriched **21**, with the distinction that all states can be reached at one wavelength of irradiation simply by changing the chirality of the light (Figure 11.16b)

Despite the low enantioselectivities achieved, these levels are sufficient to enable formation of (chiral) cholesteric LC films, when **21** was employed as a dopant in a nematic LC material [40]. This system represents a method of chiral amplification, as the chirality of the circular polarized light is expressed in a small excess of one enantiomer of the molecular switch (the dopant) and, subsequently, amplified in the chirality of the helical organization of the induced cholesteric LC phase (the host) (Figure 11.17).

11.7
Diastereomeric Photochromic Switches

In comparison to enantiomeric systems, higher stereoselectivities can be reached by switching between diastereomeric (or *pseudo*-enantiomeric) bistable molecules (*P* and *M'*) (Figure 11.18a) [41], as this can allow for two different wavelengths of light to

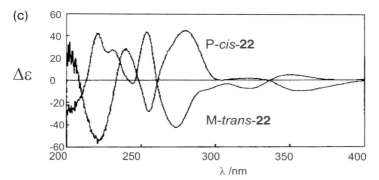

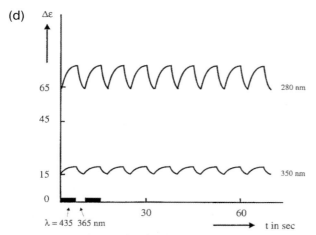

Figure 11.18 (a) Chiroptical molecular switch based on pseudoenantiomers (P and M′) as a binary storage element. (b) Irradiation of **22** at different wavelengths (λ_1 and λ_2) results in interconversion between P and M helicity. (c) CD spectra of the *pseudo*-enantiomers (P)-*cis*-**22** and (M)-*trans*-**22**. (d) Repetitive switching of **22** as detected by CD spectroscopy. Reprinted with permission from [30]. Copyright ACS 2003.

be employed, selected so that the PSS at each wavelength lies in favor of the forward and reverse reactions, respectively. This offers distinct advantages in the practical application of chiroptical switches in functional materials and devices.

The pseudo-enantiomeric nature of *P-cis*-**22** and *M-trans*-**22** is reflected in their near mirror image CD spectra (Figures 11.18b and c). The best wavelengths for the forward and reverse photoisomerization and the photostationary states (and as a consequence the ratio of *M* and *P* helices) can be tuned through the nature and position of donor and acceptor substituents [41, 42] (Figure 11.18b). The photomodulation of chirality in **22**, is achieved readily by changing the wavelength of the light used and repetitive switching cycles are monitored easily by CD spectroscopy (Figure 11.18d). Large differences in isomer composition in the photostationary state are observed with ratios of *P-cis*-**22** and *M-cis*-**22** of 30 : 70 (at 365 nm) and 90 : 10 (at 435 nm). Further insight into essential structural parameters led to the development chiroptical switches that show >99% stereoselectivity in both directions [43].

11.8
Chiroptical Switching of Luminescence

Due to its versatility as an output signal, for example, in demonstrating logic functions, fluorescence has formed the basis of a wide range of applications in particular sensing [5, 8]. Switching of fluorescence is also of value in chiroptics with many chiroptical systems showing changes in their emissive properties in addition to the changes in their electronic absorption spectra. For example, in several of the overcrowded alkenes studied in recent years the multistep sequence of changes driven alternately by light and heat (*vide infra*) provides for multistep switching of fluorescence properties and are hence promising candidates in the design multistate luminescence switches. For example *P-trans*-**22** and *M-cis*-**22** are weakly and strongly fluorescent, respectively, while in their protonated states they are nonfluorescent (Figure 11.19) [44].

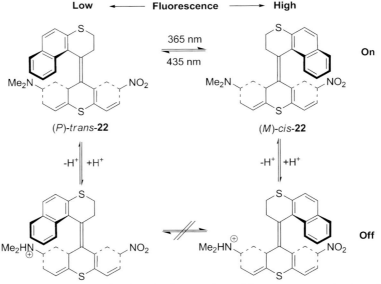

Figure 11.19 (pH)-Gated dual-mode photoswitching of donor–acceptor molecular switch **22**.

11.8 Chiroptical Switching of Luminescence | 371

In this system additional functionality is provided by the amine group that allows for reversible blocking of photochemical switching through protonation.

This system utilizes donor (Me₂N-) and acceptor (-NO₂) substituents to enable stereoselective photoisomerization to be achieved. Protonation of the amine moiety converts **22** to a nonphotoreactive acceptor–acceptor (ammonium and nitro)-substituted system [45]. Overall, **22** is a dual-mode switch where the on mode (switching) and off mode (no switching) are obtained by deprotonation or protonation, respectively. In addition, circularly polarized luminescence studies on this system have revealed the remarkable phenomenon that a single enantiomer of this chiroptical molecular switch can emit either left or right circularly polarized light [44]. It was observed that, depending on the solvent used, either r-CPL or l-CPL was emitted depending on the helicity (state) of the switch molecule, and as a consequence it is possible to switch the sense of chirality of the emitted CPL light (Figure 11.20).

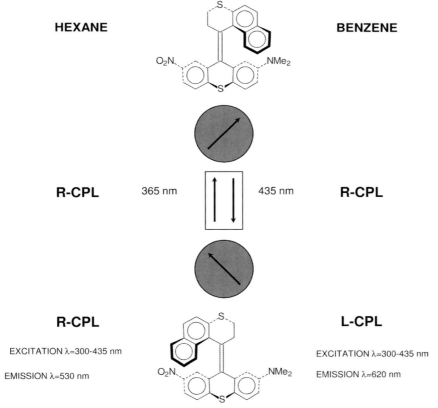

Figure 11.20 Switching of chirality and CPL emission and solvent dependence of CPL emission. In hexane, both *P* and *M* helical switch molecules emit r-CPL, whereas in benzene switching of molecular helicity results in switching of the sense of the CPL emission between r-CPL and l-CPL.

11.9
Switching of Supramolecular Organization and Assemblies

Dynamic self-assembly involving simultaneous control of organization at several hierarchical levels are central to numerous biological systems [46]. Probing these complex phenomena and mimicking them in wholly artificial systems is essential to understanding the fundamental nature of molecular interactions and can open up fascinating opportunities in the development of smart functional materials. Switchable (responsive) molecules offer considerable advantages in controlling organization of large molecular assemblies using external triggers (Figure 11.21) [40]. The key challenge in this regard is whether light-induced motion or changes in geometry, polarity, chirality or charge are sufficient to induce changes in the orientation or interaction with other molecules.

Liquid-crystalline materials are especially well suited as host materials to demonstrate how molecular switching can control supramolecular properties [47]. Doping LC materials with photoswitchable chiral guest molecules enables control of the organization and hence properties of the LC films through changes in molecular orientation of strongly anisotropic mesogenic host molecules [48]. Photoswitching between different LC states has been demonstrated already, for example, with a P-cis-23 or a M-trans-23 derivatized with a long aliphatic chain as chiral dopant (1 wt %) in nematic 4'-(pentyloxy)-4-biphenylcarbonitrile 24 (Figure 11.21) [49].

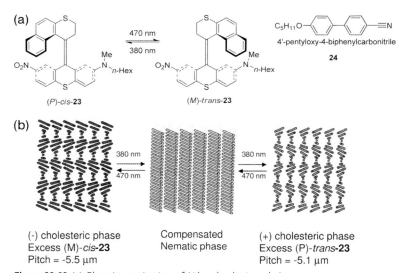

Figure 11.21 (a) Photoisomerization of N-hexyl substituted cis- and trans-isomers of chiroptical switch 23, and structure of calamitic mesogen 24. (b) Schematic representation of the switching of the handedness of a doped cholesteric liquid-crystal film.

The addition of the chiral guests (*P*)-*cis*-**23** or (*M*)-*trans*-**23** to the nematic LC phase of **24** induces the formation of chiral cholesteric (twisted nematic) phases with opposite handedness. Irradiation at 470 nm and subsequently at 380 nm results in the reversal of the helical screw sense of the cholesteric phase and photomodulation of the pitch. Furthermore, irradiation using appropriate wavelengths [48] or irradiation times leads [49] to a mixture of (*P*)-*cis*-**23** and (*M*)-*trans*-**23** of close to 1 : 1 (a pseudoracemate), which allows a compensated nematic phase to be generated. Upon prolonged irradiation at either 380 nm or 470 nm, the chirality of the mesophase can be switched on again. Photoswitching of LC phases offers an alternative to existing (electronic) methods to address display materials.

11.10
Molecular Motors

It is arguable that the most critical component of any molecular machine is its motor unit, which enables the energy input to be converted to directed mechanical motion and hence to perform a task [2, 6, 50]. In a molecular machine the motor is essentially a switching function between three or more discrete states and it should be repetitive in nature. In the previous sections, the use of photoresponsive switching units to drive changes in molecular and supramolecular behavior in a manner that can be monitored by chiroptical techniques was discussed. In the following sections the development of switch-type molecules that use intrinsic chirality to drive motor-like functions in molecular machines will be described.

In contrast to macroscopic machines and motors, at the molecular scale motors experience relentless thermal (Brownian) motion under ambient conditions [3, 51]. Operating in such a turbulent world requires that the molecular motor must either exploit or overcome Brownian motion. Nature uses Brownian motion in so-called Brownian ratchets, that is, linear and rotary protein motors, where the energy input restrains random Brownian motion [50, 52]. Artificial molecular-motor design requires that mechanical motion is directed and controlled. Furthermore, stimuli-induced motion of motor components must be detected and distinguished above random thermal motion. Achieving a biased change in motion at the molecular level is far from trivial, and it is critical to realize that, both in biological motors and synthetic molecular motors, conformational mobility is central to their function [2, 50].

Hindered bond rotation in molecular rotors and coupling of rotary motions in "gearing systems" has been explored since the early days of conformational analysis [53]. A relatively simple example is the "molecular brake" shown in Figure 11.22 [54, 55]. This system (**25**) was designed to couple a switch (an overcrowded alkene) to a rotor (xylyl group) unit. The rate of rotation around the biaryl single bond can be controlled by photochemical *cis–trans-isomer*ization of the alkene component. Surprisingly, the rate of biaryl rotation increases upon conversion from *cis*-**25** to *trans*-**25**, which was attributed to the flexibility of the naphthalene moiety. This highlights an important difference between macroscopic and molecular systems;

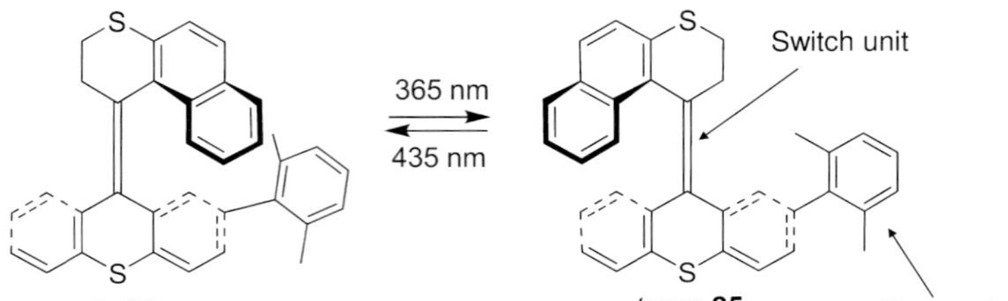

Figure 11.22 Molecular brake; the kinetic barrier to aryl–aryl single bond rotation is controlled by the switch unit in **25**.

although it is tempting to consider molecules as rigid entities it should not be forgotten that they are in fact in a constant state of movement in several directions and conformational changes play a distinctive role.

11.11
Chiral Molecular Machines

Light-fuelled unidirectional rotary motion allows for repetitive rotary motor function in wholly molecular systems, that is, overcrowded alkenes [56]. In demonstrating this motor function, we are faced with three key challenges: (i) the use of light as the source of energy, (ii) unidirectional rotary motion, (iii) repetitive 360° rotation. There are two primary principles that form the cornerstones in the realization of the light-driven molecular rotary motors. (i) The isomerization of an alkene from the *trans* to the *cis*-isomer photochemically is typically very fast and usually does not proceed thermally. (ii) The concerted action of two chiral elements (i.e. the stereogenic center and the axial helicity) in a single event, by virtue of the diastereomeric nature, leads to handedness in both structure and overall motion.

The molecular motor $(3R,3'R)$-(P,P)-*trans*-**26** has additional stereogenic centers, that is, the two methyl substituents, in comparison to the chiroptical molecular switches discussed above (Figure 11.23). The helicity can be inverted by photoisomerization. Whereas the methyl substituents can adopt an axial or equatorial orientation both in the *trans*- and the *cis*-isomers, the difference in stability between stable $(3R,3'R)$-(P,P)-*trans*-**26** with an axial methyl orientation and unstable $(3R,3'R)$-(M,M)-*trans*-**26** with an equatorial methyl orientation is 35.9 kJ mol^{-1}. Similarly, the *cis*-isomer has a stable $(3R,3'R)$-(P,P)-*cis*-**26** and unstable $(3R,3'R)$-(M,M)-*cis*-**26** form with a difference in stability of 46.0 kJ mol^{-1}. The difference in stability is assigned to increased steric interactions in the isomers with a methyl group in an equatorial orientation.

Applying light and heat drives the unidirectional rotation of the upper half (the propeller) with respect to the lower half (the stator) in a four-step cycle. Two reversible photochemical steps (steps 1 and 3), convert the stable isomers into unstable isomers. Each of the photochemical isomerization steps is followed by an irreversible thermal step (steps 2 and 4) that converts the unstable isomers into

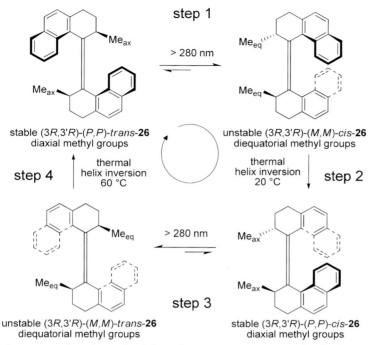

Figure 11.23 Photochemical and thermal isomerization processes of molecular motor **26**.

stable isomers. Photochemical *trans* → *cis*-isomerization of (3R,3'R)-(P,P)-*trans*-**26** generates the unstable (3R,3'R)-(M,M)-*cis*-**26**. This results in two major stereochemical events: (i) the configuration of the alkene is inverted from P to M and a clockwise rotation of the propeller part relative to the stator occurs, which inverts the helicity. (ii) The methyl substituents are forced to adopt an unfavorable equatorial orientation. In step 2 the strain is released by thermal helix inversion to give (3R,3'R)-(P,P)-*cis*-**26** in which the methyl groups adopt a more favorable axial orientation, while the propeller has continued to rotate in the same direction. In step 3 photoisomerization of (3R,3'R)-(P,P)-*cis*-**26** to (3R,3'R)-(M,M)-*trans*-**26** once again results in clockwise helix inversion and the methyl group is forced to adopt an unfavorable orientation. Finally, in step 4 the original stable isomer (3R,3'R)-(P,P)-*trans*-**26** with axial methyl groups is obtained through a thermal helix inversion releasing the strain in the molecule. The final isomerization step needs, however, heating to 60 °C. The thermal steps in the cycle shown in Figure 11.23 are depicted as single steps. However, it has been demonstrated that by changing the size of the alkyl substituents at the stereogenic centers in **26**, the thermal step comprises of two sequential helix inversions (*vide infra*).

The directionality of the rotary motion can be monitored readily by CD spectroscopy following the change in helicity that takes place in each step (Figure 11.24). While the forward steps are due to the excitation by light and thermal relaxation of unstable forms, the reverse thermal steps are effectively blocked. Irradiation

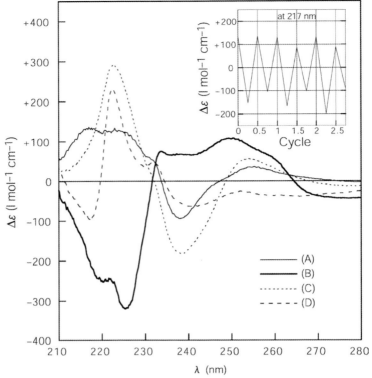

Figure 11.24 CD spectra in each of four stages of switching (see text). Trace A, (P,P)-*trans*-**26**; trace B (M,M)-*cis*-**26**; trace C, (P,P)-*cis*-**26**; trace D, (M,M)-*trans*-**26**. Inset, change in CD signal during full rotation cycle of (P,P)-*trans*-**26** monitored at 217 nm. Adapted with permission from [56]. Copyright Nature 1999.

at >280 nm and temperatures >60 °C therefore results in continuous unidirectional 360° rotation. Even at steady state a net clockwise rotation takes place as long as external energy (heat and light) is supplied. The choice of enantiomer of the motor dictates whether a clockwise or counterclockwise propeller rotation takes place.

Two aspects particularly important in the design of the next generation of motor molecules should be mentioned (Figure 11.25). In several applications of light-powered motors rotation at an appreciable speed will be required. As the photochemical steps are extremely fast (<300 ps) changes in the design were focused on the rate-limiting thermal isomerization steps. For applications in more complex systems, molecular motors with distinct upper and lower halves (Figure 11.25) are considered particularly useful as the symmetric lower half can be used for connection to other molecules or surfaces, while the upper half can still function as a propeller [57].

A major challenge is to reduce the barrier to thermal helix inversion without compromising unidirectionality to accelerate the rotary motion of the motor [58].

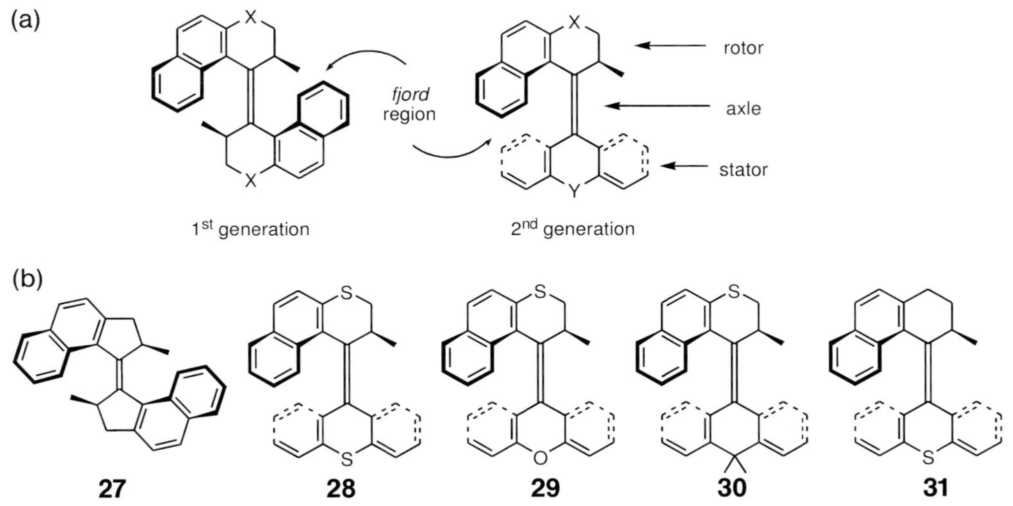

Figure 11.25 (a) General structure for first- and second-generation molecular rotary motors. (b) Structures of first- (**27**) and second- (**28–31**) generation molecular motors cited in the text.

In molecular switches and motors based on overcrowded alkenes the geometry and conformation around the central double bond (the axis of rotation) and the steric hindrance around the *fjord region* are key parameters. When we compared six- and five-membered ring systems **26** and **27** it is likely that due to reduced steric hindrance, the helix inversion will be facilitated in **27** [59]. Indeed it turned out that the thermal barriers for helix inversion in the new motor **27** were decreased considerably compared with the six-membered analog **27**, with a half-life of 78 min for the slowest and 18 s for the fastest step at room temperature [59].

More dramatic enhancements of rotary speed were achieved with second-generation motors **28–31** (Figure 11.25) [60]. In addition, in these molecules the methyl substituent adopts a more favorable axial orientation in the stable forms and a more crowded equatorial orientation in the unstable forms as is demonstrated by X-ray analysis and NMR spectroscopy. A study of the dynamic processes revealed that also in these systems two photochemical steps and two thermal steps add up to a full 360° rotary cycle. The repetitive and unidirectional nature of the rotation process can be followed easily by the distinct changes in helicity after each isomerization step as monitored by CD spectroscopy. Figure 11.26 illustrates a typical case for motor **32** comprising a (2′R)-methyl-2,3-dihydronaphthiopyran propeller and a 2-methoxy-thioxanthene stator.

Comparing the photochemical and thermal isomerization processes of these second-generation motors with the first-generation ones it became clear that the presence of single stereogenic center is sufficient to achieve unidirectional rotation. Both steric and electronic effects have been studied, using a large collection of these new motor molecules, in order to achieve a further acceleration in the rotary motion [61].

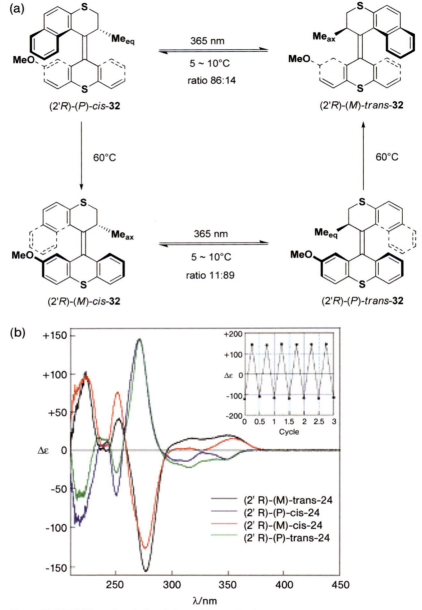

Figure 11.26 (a) Photochemical and thermal isomerization processes of motor **32**. (b) CD spectra of each of the four stages of rotation. Black line, (2R)-(M)-trans-**32**; blue line, (2R)-(P)-cis-**32**; red line, (2R)-(M)-cis-**32**; green line, (2R)-(P)-trans-**32**. Inset, changes in $\Delta\varepsilon$ value over a full rotational cycle, monitoring at 272 nm. Reprinted with permission from [16a]. Copyright ACS 2002.

The thermal barriers for helix inversion could be reduced further through systematic modification of the bridging atoms X and Y (Figure 11.25) [61]. In general, smaller Y and in particular smaller X atoms reduce the steric hindrance in the fjord region facilitating the movement of the propeller and stator moiety along each other. Further improvements were made by changing to fluorene stator halves and by fine tuning the size of the substituent at the stereogenic center [62]. Using this approach a 1.2 million-fold increase in the rotation speed was achieved with the second-generation motors. This allows the propeller to rotate unidirectionally on irradiation at up to 80 revolutions s^{-1} at 20 °C, which approaches the rotary speed of ATPase (135 rotations s^{-1}).

Structural modification leads sometimes to counterintuitive results with respect to the motor behavior. For instance in motor **33** the thermal isomerization is slower than in motor **34**, although the naphthalene moiety in the rotor part is replaced by the smaller benzene group (Figure 11.27) [63]. It appears that the decrease in ground-state energy of the unstable isomer is larger than the decrease in the transition-state energy for the thermal step. Apparently, ground-state stabilization also plays a role in the fastest motor **35**, which has a large *tert*-butyl substituent at the stereogenic center but shows an extremely fast thermal helix inversion ($t_{1/2} = 5.74$ ms) [64].

Electronic effects could be taken advantage of to achieve a dramatic increase in the rates of thermal isomerization in motor **36** [65]. In this example, an amine in the propeller unit and ketone functional group in the stator unit are conjugated and as a result of this, the central olefinic bond attains some single-bond character and hence faster rotation rates are observed. Recently, using the new motor **39**, the thermal

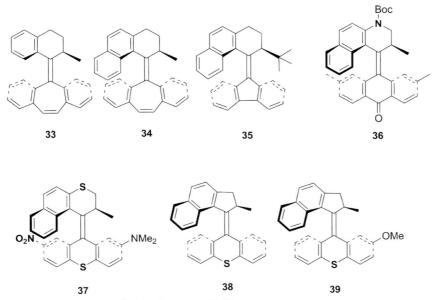

Figure 11.27 Structures of molecular motors **33–39**.

isomerization could be further reduced leading to MHz unidirectional rotation at room temperature [66].

It should be noted, however, that changes in the structure, nature and position of substituents in these motor molecules can affect all of the isomerization processes involved. Whereas in some examples thermal backward rotation has been observed, in many examples the photostationary state of the photochemical steps showing selectivities as high as 99 : 1. Finally, through the presence of donor and acceptor substituents in the stator part of the motor **37** a bathochromic shift in the optimal wavelength of irradiation similar to switch **22** was achieved allowing this motor to be operated using visible light [67]. This sets the stage for further increases in rotation speed, the introduction of additional functionality to these motors and their integration into multicomponent systems.

11.12
Making Nanoscale Machines Work

Immobilizing molecular machines on surfaces, without compromising their mechanical function, is essential to their interfacing with the macroscopic world. The functionality of the overcrowded alkene-based molecular motors discussed above were all demonstrated in solution and hence are subject to Brownian motion. Indeed, to build nanodevices using these molecules as functional elements, immobilization of our light-driven motors on surfaces is a critical next step. To this end, the second-generation motor **40** was immobilized on gold nanoparticles (Figure 11.28a) [68]. In this molecular rotary motor, the rotor axis (central alkene) is normal to the surface and the stator unit is functionalized with two legs terminated with thiol groups. The connection of the motor via two points of attachment to the stator unit precludes the noncontrolled thermal rotation of the entire motor with respect to the surface, while retaining the propeller function.

CD spectroscopy, together with ^{13}C NMR spectroscopy, was employed to characterize and verify the repetitive unidirectional rotation of the molecule with respect to the surface (Figure 11.28b). ^{13}C NMR spectroscopy was used to differentiate the two legs of the motor molecule [68]. More recently, an analogous second-generation motor with different legs and anchoring positions was assembled as monolayers on silicon and quartz (Figure 11.29) [69]. CD spectroscopy was again used to follow the isomerization steps, while XPS analysis confirmed that the anchoring to the surface was indeed via through two points of attachment. These studies addressed key issues associated with the immobilization of motor and switch molecules to surfaces.

Two systems in which the dynamic properties of the motor are used to effect a second dynamic function have been reported to demonstrate a concept of molecular transmission. The first example (Figure 11.30) builds on previous systems in which chiroptical switches were coupled to rotors [54]. In **43** a xylyl-rotor is connected to a second-generation molecular motor [70]. The motor unit isomerizes in a stepwise manner through a four-step 360° cycle (see Figure 11.26) and changes the position of

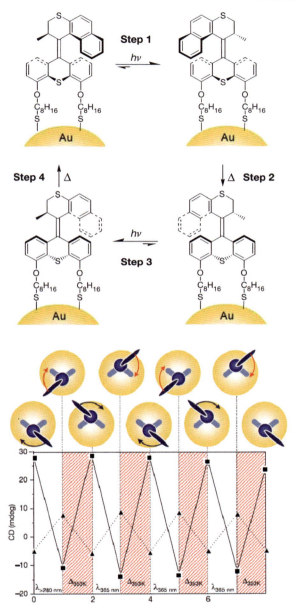

Figure 11.28 (a) Four-step rotary cycle of **40** immobilized with two alkylthiol legs to gold nanoparticles (b) Top: Unidirectional rotary motion of the propeller part of **40** viewed along the rotation axis and two four-stage 360° rotary cycles. Bottom: The change in CD absorption at 290 nm (solid) and 320 nm during the sequential photochemical and thermal steps.

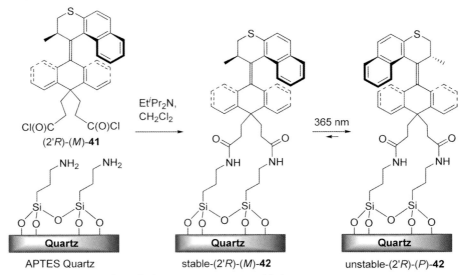

Figure 11.29 Grafting of (2'R)-(M)-**41** to APTES modified quartz and photochemical isomerization of surface-bound motor **42**. Adapted from [69]. Copyright Wiley 2007.

the propeller unit with respect to the xylyl rotor thereby affecting the latter's thermal rotary behavior. The order *cis*-stable > *cis*-unstable > *trans*-stable > *trans*-unstable was established using 2D EXSY NMR spectroscopy for the rotation around the biaryl single bond in the different isomers. Interestingly, these data point to a larger steric effect of the methyl substituent, in particular in an equatorial orientation, than that of the naphthyl part of the upper propeller unit on the xylyl rotor moiety.

In another system, the transmission of the change in chirality of the motor molecule during rotation to a change in folding of a macromolecule was studied

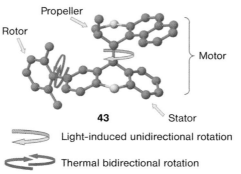

Figure 11.30 Structure of second-generation motor connected to xylyl rotor **43** and its rotary processes. Reproduced from [70]. Copyright Royal Society of Chemistry 2005.

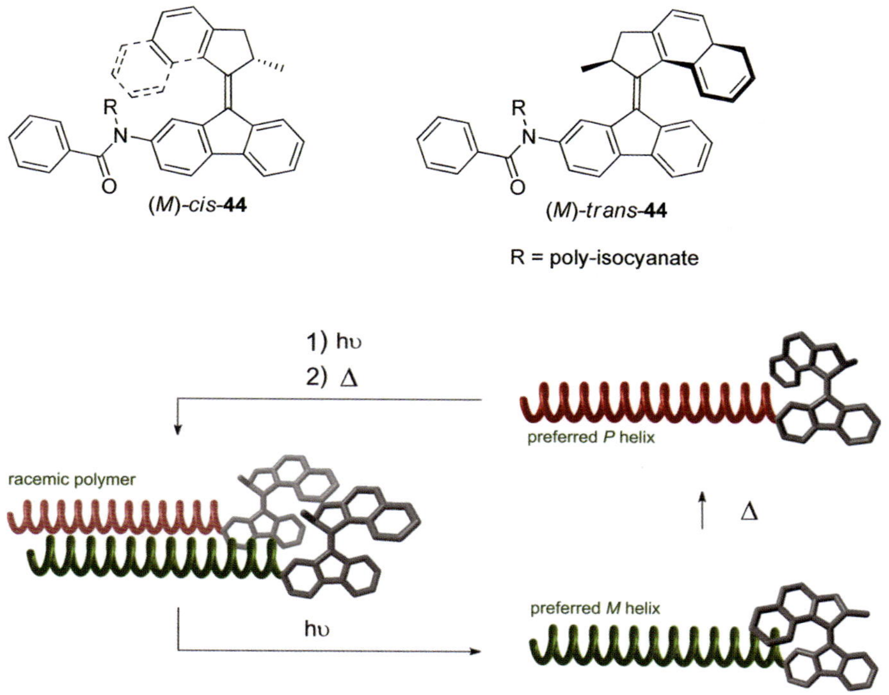

(M)-cis-**44**   (M)-trans-**44**

R = poly-isocyanate

Figure 11.31 Structure of polymer-modified second-generation rotary motor **44** and schematic illustration of the reversible induction and inversion of the helicity of a polymer backbone by a single light-driven molecular motor positioned at the terminus. Irradiation of the photochromic unit leads to a preferred helical sense of the polymer backbone. A thermal isomerization of the rotor unit inverts the preferred helicity of the polymer chain. Subsequent photochemical and thermal isomerization steps regenerate the original situation with a random helicity of the polymer backbone. Adapted from [71]. Copyright Wiley 2007.

using a dynamic helical polymer [71]. An optically active amide-substituted *trans*-motor **44** (Figure 11.31) was employed as the initiator in the polymerization of *n*-hexyl isocyanate providing a polymer showing random helicity. Photoisomerization of the motor unit to the unstable *cis*-form resulted in predominant conversion to a *M*-helical polymer, while subsequent thermal isomerization of the motor to the stable *cis*-form triggered an inversion predominantly to the *P*-helical polymer form. Subsequent photochemical isomerization followed by a second thermal isomerization resets the whole system. The motor function, in this case, is that of a multistage molecular switch, however, the triggering of polymer folding and transmission of chiral information from a single molecule to a macromolecule opens many new opportunities in addressable materials and thin films.

The ultimate goal in constructing molecular motors and machines is to use them to do work and perform useful functions [2, 72]. Controlling the macroscopic properties of materials using molecular motors has been demonstrated using liquid-crystalline

45

Figure 11.32 Molecular structure, photochemical and thermal isomerization processes of **45** and changes in the color of LC compound E7 doped with **45**. Adapted from [75]. Copyright Wiley 2006.

materials [73]. As discussed above for chiroptical molecular switches, several of the optically active overcrowded alkenes can serve as excellent chiral dopants to induce cholesteric phases in LC films [74]. The change in helicity that accompanies interconversion between the different isomers of the second-generation motors **45** is, for example, reflected in large changes in helical twisting power and concomitant changes in the helical organization of the LC phases [75]. Indeed, the color of LC films can be tuned over the visible spectrum (Figure 11.32). The excellent reversibility of this process opens up new opportunities in, for example, dynamic LC pixel generation.

Recently, microscopic movement induced by changes in the helicity of a molecular motor has been demonstrated in the reversible rotation of the texture of the surface of a LC film. Indeed using molecular motor **45** as the dopant, the physical changes are sufficiently pronounced to move microscopic objects [76].

This motor is especially effective in inducing a dynamic helical (twisted nematic) organization when applied as a chiral dopant in certain LC films (Figure 11.33). The isomerization steps induced either photochemically or thermally result in reorganization of the polygonal LC texture. AFM and optical profilometry were used to characterize the surface relief (20 nm) of these films. Remarkably, the orientation and nature of the surface relief responded to changes in the topology of molecular motors [76]. The change in surface relief (a rotation of the lines) generates sufficient torque to rotate a microscopic object, for example, a glass rod placed on the film, unidirectionally (Figure 11.33c).

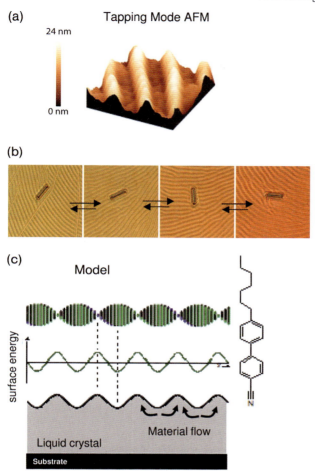

Figure 11.33 (a) Tapping mode AFM image of the surface of an LC film doped with molecular motor **45**. (b) Model showing the orientation of the LC component in the twisted nematic phase, surface energy profile and material flow accompanying the change in helical twisting power (HTP) of dopant **45** upon irradiation. (c) Physical movement of micrometer-size glass rod objects resting on the LC surface. Adapted from [76a,b]. Copyright Nature 2006 and American Chemical Society 2006.

Irradiating the LC matrix containing the motor results in a clockwise rotation of the glass rod. When the photostationary state is reached, the rotation of the rod stops. The subsequent thermal isomerization of the motor molecule results in a reverse (counterclockwise) rotation of the rod. The direction of the rotation of the rod reflects, directly, the (change in) helical chirality of the motor molecule dopants. Here, light energy and the collective action of a number of motor molecules rotates microscopic objects that are 10 000 times the size of the motors themselves. The challenge, however, is to achieve continuous unidirectional rotary motion of microscopic objects powered by these molecular motors.

11.13
Challenges and Prospects

Response time is a critical issue in the pursuit of applying molecular switches to practical applications. As for the ultrafast *cis–trans* photoisomerization of retinal in the primary step of vision, femtosecond spectroscopic techniques have observed fast isomerization in stilbenes, overcrowded alkenes [77] and diarylethenes. However, in a switchable system such as the unidirectional light-driven molecular motors the quantum yield of the process, which determines the efficiency taken to reach a photostationary state is of equally importance. For the systems discussed above quantum yields are typically between 0.07 and 0.55 [78].

Applying chiroptical switches to real devices and multifunctional systems remains a challenging goal. For example, the photochemical switching of chirality of overcrowded alkenes in thin polymer films is relevant in the context of high-density data storage [79, 80]. The polymer systems that have already been developed were applied through either doping polymers with switches or by covalent attachment of chiroptical switches to the polymer backbone.

Theoretically a compact disk with a 1 125 000 000 MB storage capacity (240 years of continuous music) is in principle possible. However, this would require high-density molecular packing as well as the possibility to address or read each individual molecule at high speed without interference, a feat far beyond current technology. A key lesson learned from these studies is that, in translating molecular systems from homogenous solutions to more complex media, matrix effects play a prominent role with respect to stereoselectivity, switching times, and so on [81].

An equally challenging goal is to use molecular switches and motors to achieve directed motion in nanoscale systems. Controlled transport and delivery of "cargo," dynamic catalytic systems and switchable and adaptive self-assembled materials and surfaces are among the challenges ahead. It goes without saying that these dynamic molecular systems will form the heart of the nanomachinery of the future.

Acknowledgments

The authors thank the Netherlands Foundation for Scientific Research (NWO-CW), the NanoNed program, and the Zernike Institute for Advanced Materials (MSC+) program and the NWO-Vidi program (WRB) for their financial support.

Glossary

AFM	atomic force microscopy
CD	circular dichroism
CPL	circularly polarized light
ee	enantiomeric excess

EM electron microscopy
HTP helical twisting power
LC liquid crystal
LMW low molecular weight
LPL linearly polarized light

References

1 Feringa, B.L., van Delden, R.A., Koumura, N., and Geertsema, E.M. (2000) *Chemical Reviews*, **100**, 1789–1816.

2 (a) Browne, W.R., and Feringa, B.L. (2006) *Nature Nanotechnology*, **1**, 25–35; (b) Kay, E.R., Leigh, D.A. and Zerbetto, F. (2007) *Angewandte Chemie-International Edition*, **46**, 72–191; (c) Stoddart, J.F. (2001) *Accounts of Chemical Research*, **34**, 410–411;(d) Balzani, V., Credi, A. and Venturi, M. (2003) *Molecular Devices and Machines – A Journey into the Nanoworld*, Wiley-VCH, Weinheim; (e) Sauvage, J.P. (2001) *Molecular Machines and Motors (Structure and Bonding)*, Springer Verlag, Berlin (f) Harada, A. (2001) *Accounts of Chemical Research*, **34**, 456–464.

3 (a) Brown, R. (1828) *Philosophical Magazine*, **4**, 171–173; (b) Rozenbaum, V.M., Yang, D.-Y., Lin, S.H. and Tsong, T.Y. (2004) *The Journal of Physical Chemistry. B*, **108**, 15880–15889; (c) Whitesides, G.M. (2001) *Scientific American*, **285**, 78–84.

4 http://goldbook.iupac.org/.

5 Feringa, B.L. (2001) *Molecular Switches*, Wiley-VCH, Weinheim.

6 Saha, S. and Stoddart, J.F. (2007) *Chemical Society Reviews*, **36**, 77–92.

7 (a) Dürr, H. and Bouas-Laurent, H. (2003) *Photochromism: Molecules and Systems*, Elsevier; (b) Irie, M. (ed.) (2000) *Chemical Reviews*, 100, issue 5 (thematic issue on photochromism).

8 Magri, D.C., Brown, G.J., McClean, G.D., and de Silva, A.P. (2006) *Journal of the American Chemical Society*, **128**, 4950–4951.

9 Reichert, S., and Breit, B. (2007) *Organic Letters*, **9**, 899–902.

10 Crano, J.C., and Guglielmetti, R.J. (1999) *Organic Photochromic and Thermochromic Compounds*, Vol **1–3**, Springer, New York.

11 Chen, C.-T., and Chou, Y.C. (2000) *Journal of the American Chemical Society*, **122**, 7662–7672.

12 (a) Takeuchi, M., Ikeda, M., Sugasaki, A., and Shinkai, S. (2001) *Accounts of Chemical Research*, **34**, 865–873; (b) Volgraf, M., Gorostiza, P., Numano, R., Kramer, R.H., Isacoff, E.Y. and Trauner, D. (2006) *Nature Chemical Biology*, **2**, 47–52.

13 (a) Carmen Carreno, M., Garcia, I., Ribagorda, M., Merino, E., Pieraccini, S., and Spada, G.P. (2005) *Organic Letters*, **7**, 2869–2872; (b) Carmen Carreno, M., Garcia, I., Nunez, I., Merino, E., Ribagorda, M., Pieraccini, S. and Spada, G.P. (2007) *Journal of the American Chemical Society*, **129**, 7089–7100.

14 (a) Muraoka, T., Kinbara, K., Kobayashi, Y., and Aida, T. (2003) *Journal of the American Chemical Society*, **125**, 5612–5613; (b) Muraoka, T., Kinbara, K. and Aida, T. (2006) *Nature*, **440**, 512–515; (c) Muraoka, T., Kinbara, K. and Aida, T. (2007) *Chemical Communications*, 1441–1443.

15 Wang, Z.Y., Todd, E.K., Meng, X.S., and Gao, J.P. (2005) *Journal of the American Chemical Society*, **127**, 11552–11553.

16 Furusho, Y., Tanaka, Y., Maeda, T., Ikeda, M., and Yashima, E. (2007) *Chemical Communications*, 3174–3176.

17 Bottari, G., Leigh, D.A., and Perez, E.M. (2003) *Journal of the American Chemical Society*, **125**, 13360–13361.

18 Ma, X., Qu, D., Ji, F., Wang, Q., Zhu, L., Xu, Y., and Tian, H. (2007) *Chemical Communications*, 1409–1411.

19 Harada, A. (2006) *Journal of Polymer Science Part A-Polymer Chemistry*, **17**, 5113–5119.

20 (a) Kumar, G.S., and Neckers, D.C. (1989) *Chemical Reviews*, **89**, 1915–1925; (b) Natansohn, A. and Rochon, P. (2002) *Chemical Reviews*, **102**, 4139–4175; (c) Shibaev, V., Bobrovsky, A. and Boiko, N. (2003) *Progress in Polymer Science*, **28**, 729–836; (d) Yesodha, S.K., Sadashiva Pillai, C.K. and Tsutsumi, N. (2004) *Progress in Polymer Science*, **29**, 45–74.

21 Kusumoto, T., Sato, K., Ogino, K., Hiyama, T., Takehara, S., Osawa, M., and Nakamura, K. (1993) *Liquid Crystals*, **14**, 727–732.

22 Negishi, M., Tsutsumi, O., Ikeda, T., Hiyama, T., Kawamura, J., Aizawa, M., and Takehara, S. (1996) *Chemistry Letters*, 319–320.

23 Denekamp, C., and Feringa, B.L. (1998) *Advanced Materials*, **10**, 1080–1082.

24 Yamaguchi, T., Uchida, K., and Irie, M. (1997) *Journal of the American Chemical Society*, **119**, 6066–6071.

25 Walko, M., and Feringa, B.L. (2007) *Chemical Communications*, 1745–1747.

26 Norsten, T.B., Peters, A., McDonald, R., Wang, M., and Branda, N.R. (2001) *Journal of the American Chemical Society*, **123**, 7447–7448.

27 Wigglesworth, T.J., Sud, D., Norsten, T.B., Lekhi, V.S., and Branda, N.R. (2005) *Journal of the American Chemical Society*, **127**, 7272–7273.

28 (a) de Loos, M., Feringa, B.L., and van Esch, J.H. (2005) *European Journal of Organic Chemistry*, 3615–3631; (b) Weiss, R.G. and Terech, P. (2006) *Molecular Gels: Materials with Self-Assembled Fibrillar Networks*, Springer-Verlag, Berlin.

29 de Jong, J.J.D., Hania, P.R., Pugžlys, A., Lucas, L.N., de Loos, M., Kellogg, R.M., Feringa, B.L., Duppen, K., and van Esch, J.H. (2005) *Angewandte Chemie-International Edition*, **44**, 2373–2376.

30 Barcena, H., Holmes, A.E., Zahn, S., and Canary, J.W. (2003) *Organic Letters*, **5**, 709–711.

31 Westermeier, C., Gallmeier, H.-C., Komma, M., and Daub, J. (1999) *Chemical Communications*, 2427–2428.

32 Fukui, M., Mori, T., Inoue, Y., and Rathore, R. (2007) *Organic Letters*, **9**, 3977–3980.

33 (a) Nishida, J., Suzuki, T., Ohkita, M., and Tsuji, T. (2001) *Angewandte Chemie-International Edition*, **40**, 3251–3254; (b) Wang, C., Zhu, L., Xiang, J., Yu, Y., Zhang, D., Shuai, Z. and Zhu, D. (2007) *The Journal of Organic Chemistry*, **72**, 4306–4312.

34 Gomar-Nadal, E., Veciana, J., Rovira, C., and Amabilino, D.B. (2005) *Advanced Materials*, **17**, 2095–2098.

35 Zhou, Y., Zhang, D., Zhu, L., Shuai, Z., and Zhu, D. (2006) *The Journal of Organic Chemistry*, **71**, 2123–2130.

36 Deng, J., Song, N., Zhou, Q., and Su, Z. (2007) *Organic Letters*, **9**, 5393–5396.

37 Feringa, B.L., and Wynberg, H. (1977) *Journal of the American Chemical Society*, **99**, 602–603.

38 Feringa, B.L., van Delden, R.A., Koumura, N., and Geertsema, E.M. (2000) *Chemical Reviews*, **100**, 1789–1816.

39 Inoue, Y. (1992) *Chemical Reviews*, **92**, 741–770.

40 Huck, N.P.M., Jager, W.F., de Lange, B., and Feringa, B.L. (1996) *Science*, **273**, 1686–1688.

41 (a) Jager, W.F., de Jong, J.C., de Lange, B., Huck, N.P.M., Meetsma, A., and Feringa, B.L. (1995) *Angewandte Chemie-International Edition*, **34**, 348–350; (b) Feringa, B.L., Jager, W.F., de Lange, B. and Meijer, E.W. (1991) *Journal of the American Chemical Society*, **113**, 5468–5470; (c) Feringa, B.L., Jager, W.F. and de Lange, B. (1993) *Journal of the Chemical Society, Chemical Communications*, 288–290.

42 Feringa, B.L., Schoevaars, A.M., Jager, W.F., de Lange, B. and Huck, N.P.M. (1996) *Enantiomer*, **1**, 325–335.

43 van Delden, R.A., ter Wiel, M.K.J., and Feringa, B.L. (2004) *Chemical Communications*, 200–201.

44 Huck, N.P.M., and Feringa, B.L. (1995) *Journal of the Chemical Society, Chemical Communications*, 1095.

45 van Delden, R.A., Huck, N.P.M., Piet, J.J., Warman, J.M., Meskers, S.C.J., Dekkers, H.P.J.M., and Feringa, B.L. (2003) *Journal of the American Chemical Society*, **125**, 15659–15665.

46 Pollack, G.H. (2001) *Cells, Gels and the Engines of Life*, Ebner and Sons, Seattle.

47 Demus, D., Goodby, J., Gray, G.W., Spiess, H-.W. and Vill, V. (eds) (1998) *Handbook of Liquid Crystals*, **Vol 1, 2A, 2B and 3**, Wiley-VCH, Weinheim.

48 Feringa, B.L., Huck, N.P.M., and van Doren, H.A. (1995) *Journal of the American Chemical Society*, **117**, 9929–9930.

49 van Delden, R.A., van Gelder, M.B., Huck, N.P.M., and Feringa, B.L. (2003) *Advanced Functional Materials*, **13**, 319–324.

50 Schliwa, M. (ed.) (2003) *Molecular Motors*, Wiley, Weinheim.

51 Astumian, R.D. (1997) *Science*, **276**, 917–922.

52 Astumian, R.D. (2001) *Scientific American*, **285**, 45–51.

53 (a) Kottas, G.S., Clarke, L.I., Horinek, D., and Michl, J. (2005) *Chemical Reviews*, **105**, 1281–1376; (b) Iwamura, H. and Mislow, K. (1988) *Accounts of Chemical Research*, **21**, 175–182.

54 Schoevaars, A.M., Kruizinga, W., Zijlstra, R.W.J., Veldman, N., Spek, A.L., and Feringa, B.L. (1997) *The Journal of Organic Chemistry*, **62**, 4943–4948.

55 See also: Kelly, T.R., Bowyer, M.C., Bhaskar, K.V., Bebbington, D., Garcia, A., Lang, F.R., Kim, M.H., and Jette, M.P. (1994) *Journal of the American Chemical Society*, **116**, 3657–3658, and references therein.

56 Koumura, N., Zijlstra, R.W.J., van Delden, R.A., Harada, N., and Feringa, B.L. (1999) *Nature*, **401**, 152–155.

57 van Delden, R.A., ter Wiel, M.K.J., Pollard, M.M., Vicario, J., Koumura, N., and Feringa, B.L. (2005) *Nature*, **437**, 1337–1340.

58 ter Wiel, M.K.J., van Delden, R.A., Meetsma, A., and Feringa, B.L. (2005) *Journal of the American Chemical Society*, **127**, 14208–14222.

59 ter Wiel, M.K.J., van Delden, R.A., Meetsma, A., and Feringa, B.L. (2003) *Journal of the American Chemical Society*, **125**, 15076–15086.

60 (a) Koumura, N., Geertsema, E.M., van Gelder, M.B., Meetsma, A., and Feringa, B.L. (2002) *Journal of the American Chemical Society*, **124**, 5037–5051; (b) van Delden, R.A., ter Wiel, M.K.J., de Jong, H., Meetsma, A. and Feringa, B.L. (2004) *Organic and Biomolecular Chemistry*, **2**, 1531–1541.

61 Pollard, M.M., Klok, M., Pijper, D., and Feringa, B.L. (2007) *Advanced Functional Materials*, 10.1002/adfm.200601025.

62 Vicario, J., Meetsma, A., and Feringa, B.L. (2005) *Chemical Communications*, 5910–5912.

63 Geertsema, E.M., Koumura, N., ter Wiel M.K.J., Meetsma, A., and Feringa, B.L. (2002) *Chemical Communications*, 2962–2963.

64 Vicario, J., Walko, M., Meetsma, A., and Feringa, B.L. (2006) *Journal of the American Chemical Society*, **128**, 5127–5135.

65 Pijper, D., van Delden, R.A., Meetsma, A., and Feringa, B.L. (2005) *Journal of the American Chemical Society*, **127**, 17612–17613.

66 Klok, M., Boyle, N., Pryce, M.T., Meetsma, A., Browne, W.R., and Feringa, B.L. (2008) *Journal of the American Chemical Society*, **130**, 10484–10485.

67 van Delden, R.A., Koumura, N., Schoevaars, A., Meetsma, A., and Feringa, B.L. (2003) *Organic and Biomolecular Chemistry*, **1**, 33–35.

68 van Delden, R.A., ter Wiel, M.K.J., Pollard, M.M., Vicario, J., Koumura, N., and Feringa, B.L. (2005) *Nature*, **437**, 1337–1340.

69 Pollard, M.M., Lubomska, M., Rudolf, P., and Feringa, B.L. (2007) *Angewandte Chemie-International Edition*, **46**, 1278–1280.

70 ter Wiel, M.K.J., van Delden, R.A., Meetsma, A., and Feringa, B.L. (2005) *Organic and Biomolecular Chemistry*, **3**, 4071–4076.

71 (a) Pijper, D., and Feringa, B.L. (2007) *Angewandte Chemie-International Edition*, **20**, 3693–3696; (b) Pijper, D., Jongejan, M.G.M., Meetsma, A. and Feringa, B.L. (2008) *Journal of the American Chemical Society*, **130**, 4541–4552.

72 (a) Oh, S.K., Nakagawa, M., and Ichimura, K. (2002) *Journal of Materials Chemistry*, **12**, 2262–2269; (b) Berna, J., Leigh, D.A., Lubomska, M., Mendoza, S.M., Perez, E.M., Rudolf, P., Teobaldi, G. and Zerbetto, F. (2005) *Nature Materials*, **4**, 704–710; (c) Huang, J., Brough, B., Ho, C.-M., Liu, Y., Flood, A.H., Bonvallet, P.A., Tseng, H.-R., Stoddart, J.F., Baller, M. and Maganov, S.A. (2003) *Applied Physics Letters*, **85**, 5391–5393; (d) Liu, Y., Flood, A.H., Bonvallet, P.A., Vignon, S.A., Northrop, B.H., Tseng, H.-R., Jeppesen, J.O., Huang, T.J., Brough, B., Baller, M., Magonov, S., Solares, S.D., Goddard, W.A., Ho, C.-M. and Stoddart, J.F. (2005) *Journal of the American Chemical Society*, **127**, 9745–9759; (e) Holland, N.B., Hugel, T., Neuert, G., Cattani-Scholz, A., Renner, C., Oesterhelt, D., Moroder, L., Seitz, M. and Gaub, H.E. (2003) *Macromolecules*, **36**, 2015–2023; (f) Hugel, T., Holland, N.B., Cattani, A., Moroder, L., Seitz, M. and Gaub, H.E. (2002) *Science*, **296**, 1103–1106; (g) Harris, K.D., Cuypers, R., Scheibe, P., van Oosten, C.L., Bastiaansen, C.W.M., Lub, J. and Broer, D.J. (2005) *Journal of Materials Chemistry*, **15**, 5043–5048.

73 Eelkema, R., and Feringa, B.L. (2006) *Organic and Biomolecular Chemistry*, **4**, 3729–3745.

74 (a) van Delden, R.A., Koumura, N., Harada, N., and Feringa, B.L. (2002) *Proceedings of the National Academy of Sciences of the United States of America*, **99**, 4945–4949; (b) Eelkema, R. and Feringa, B.L. (2006) *Organic and Biomolecular Chemistry*, **4**, 3729–3745.

75 Eelkema, R., and Feringa, B.L. (2006) *Chemistry, an Asian Journal*, **1**, 367–369.

76 (a) Eelkema, R., Pollard, M.M., Vicario, J., Katsonis, N., Ramon, B.S., Bastiaansen, C.W.M., Broer, D.J., and Feringa, B.L. (2006) *Nature*, **440**, 163–163; (b) Eelkema, R., Pollard, M.M., Katsonis, N., Vicario, J., Broer, D.J. and Feringa, B.L. (2006) *Journal of the American Chemical Society*, **128**, 14397–14407.

77 Schuddeboom, W., Jonker, S.A., Warman, J.M., de Haas, M.P., Vermeulen, M.J.W., Jager, W.F., de Lange, B., Feringa, B.L., and Fessenden, R.W. (1993) *Journal of the American Chemical Society*, **115**, 3286.

78 Jager, W. (1994) Ph.D thesis, University of Groningen, the Netherlands.

79 Green, J.E., Choi, J.W., Boukai, A., Bunimovich, Y., Johnston-Halperin, E., Delonno, E., Luo, Y., Sheriff, B.A., Xu, K., Shin, Y.S., Tseng, H.R., Stoddart, J.F., and Heath, J.R. (2007) *Nature*, **445**, 414–417.

80 Oosterling, M.L.C.M., Schoevaars, A.M., Haitjema, H.J., and Feringa, B.L. (1996) *Israel Journal of Chemistry*, **36**, 341–348.

81 (a) Schoevaars, A.M. (1998) Ph.D. thesis, University of Groningen. (b) Feringa, B.L., Huck, N.P.M. and Schoevaars, A.M. (1996) *Advanced Materials*, **8**, 681–684.

82 Mathies, R.A., and Lugtenburg, J. (2000) *Molecular Mechanisms of Visual Transduction* (eds D.G. Stavenga, W.J. DeGrip and E.N. Pugh, Jr), Elsevier, Amsterdam, the Netherlands, p. 55.

83 de Jong, J.J.D., Lucas, L.N., Kellogg, R.M., van Esch, J.H. and Feringa, B.L. (2007) *Science*, **304**, 278–281.

12
Chiral Nanoporous Materials
Wenbin Lin and Suk Joong Lee

12.1
Classes of Chiral Nanoporous Materials

Porous materials have been widely used in many fields such as ion exchange, separation, catalysis, sensing, and biological molecular isolation [1]. They are without a doubt among the most important materials in our modern world. According to the definition of the pore size by the International Union of Pure and Applied Chemistry (IUPAC) [2], porous materials can be classified into three groups based on their pore sizes: microporous (<2 nm), mesoporous (2–50 nm), and macroporous materials (>50 nm). This definition is, however, somewhat inconsistent with the convention of nanoscale objects. In this chapter, nanoporous materials refer to porous materials having large porosity with pore diameters between 1 and 100 nm [3]. With a high surface area and large porosity, nanoporous materials tend to have a high surface to volume ratio. Many nanoporous materials also present an ordered and uniform pore structures. If nanoporous materials contain chirality, they can be used for enantioselective applications such as asymmetric catalysis, chiral separation, and enantioselective sensing.

The most important parameters for nanoporous materials are the pore structures and functionalities within the pores. The former can generally be characterized by use of physical adsorption of a gas (such as nitrogen or argon gases) or by a liquid pressure method (such as mercury extrusion), whereas the latter can be inferred from chemisorption or other functional group titration studies. Amorphous and quasi-crystalline porous materials such as silica gels, aluminum oxide, and activated carbons have been widely used in industry as sorbents or catalyst supports. Because these materials possess no long-range order and have nonuniform pore structures and widely distributed pore sizes, they are not very useful for applications that require size- and shape-selectivity. On the other hand, aluminosilicate materials (i.e. zeolites) are microcrystalline materials with long-range order and uniform pore size and structure that have found widespread applications in size- and shape-selective catalysis and separations of many important commodity chemicals. For example,

Chirality at the Nanoscale: Nanoparticles, Surfaces, Materials and more. Edited by David B. Amabilino
Copyright © 2009 WILEY-VCH Verlag GmbH & Co. KGaA, Weinheim
ISBN: 978-3-527-32013-4

size-selective catalytic transformation of petrochemicals has been extensively used to improve the quality of gasoline, which significantly reduces the pollution from fossil-fuel combustions [3]. The presence of cations inside the zeolite pores also make them ideal choices for ion exchange processes. Because of the tremendous need for chiral molecules for the pharmaceutical, agrochemical, and fragrance industries, significant efforts have been devoted to the development of chiral zeolites and related porous solids for enantioselective applications. Although there have been breakthroughs in the synthesis of zeolites with larger pores, no chiral zeolites are available in enantiopure form to date. Several alternative strategies have been developed to prepare chiral nanoporous materials over the past decade.

Chiral nanoporous materials can be divided into three classes based on their compositions, namely, metal-organic frameworks, porous oxides and related composite materials, and mesoporous silica-immobilized chiral catalysts. These chiral nanoporous materials typically contain micropores and mesopores and possess diverse surface areas. They can range from totally amorphous materials to perfect single crystals. Many of them have been evaluated for enantioselective catalysis and separations. This chapter will survey the latest developments in the synthesis, characterization, properties, and enantioselective applications of these chiral nanoporous materials.

12.2
Porous Chiral Metal-Organic Frameworks

Metal-organic frameworks are crystalline hybrid materials built from metal ions or clusters (metal-connecting points) and organic bridging ligands (linkers). The metal-connecting points link the organic bridging ligands via coordination bonds to form one-, two-, or three-dimensional framework structures. Porous metal-organic frameworks (MOFs) have received significant attention because of their potential applications in many areas, including gas storage, separation, and heterogeneous catalysis [4]. MOFs are typically synthesized under mild conditions compared with microporous zeolites, which allow systematic engineering of chemical and physical properties through modifications of their components. Chiral MOFs are particularly attractive candidates as heterogeneous asymmetric catalysts for the economical production of optically active organic compounds. As a heterogeneous catalyst, chiral MOFs provide many advantages such as facile catalyst separation and recovery, high stability, and ease of manipulation. Furthermore, because of their well-defined chiral networks, they provide uniform chiral active sites, high efficiency, reproducibility, selectivity, and mild reaction conditions.

There are three general synthetic strategies for constructing chiral MOFs: (a) construction from achiral components by seeding with a chiral single crystal, (b) templating of MOFs with coordinating chiral coligands, and (c) construction from metal-connecting nodes and chiral bridging ligands. The first method utilizes totally achiral building blocks (i.e. achiral metal-connecting points and bridging ligands) to assemble chiral frameworks. Any MOF can crystallize in a chiral space group owing to a particular spatial disposition of all the building units within the MOF. Such a self-resolution process has been observed in numerous MOFs which are built from all achiral components but crystallize in chiral space groups. Bulk samples of such MOFs, however, tend to contain both enantiomorphs and are thus racemic. Aoyama *et al.* reported the synthesis of an enantiopure bulk sample of MOF built from achiral building blocks [5]. Compound $Cd(L_1)(NO_3)_2(H_2O)(EtOH)$, **1**, was obtained by treating achiral [5-(9-anthracenyl)]pyrimidine, L_1, with $Cd(NO_3)_2 \cdot 4H_2O$. Adjacent helices of **1** adopt the same handedness and are linked by interstrand H_2O-nitrate hydrogen-bonding interactions that lead to chiral crystals of **1**. Bulk samples of homochiral **1** (with the same handedness of chiarlity) were obtained by the seeding technique. Homochiral MOFs were also recently obtained via a novel chiral symmetry-breaking strategy by manipulating the statistical fluctuation inherent to the crystallization of helical coordination polymers [6].

The second method uses chiral auxiliary ligands (that do not bridge the metal-connecting points) to direct the growth of chiral frameworks that are built from achiral metal-connecting points and bridging ligands. Rosseinsky *et al.* used an optically pure 1,2-propanediol (1,2-pd) coligand to direct the formation of homochiral porous networks of the type $M_3(btc)_2$ (where btc is 1,3,5-benzenetricarboxylate) [7]. $Ni_3(btc)_2(py)_6(1,2\text{-pd})_3 \cdot [11(1,2\text{-pd}) \cdot 8(H_2O)]$, **2**, adopts a doubly interpenetrated network structure. The coordinating chiral 1,2-pd not only controls the handedness of the helices but also directs the crystal growth to afford a homochiral bulk sample.

The third method relies on the coordination of metal-connecting points to chiral bridging ligands to form homochiral frameworks. A wide variety of chiral molecules with the tendency to bridge metal-connecting points have been used to construct chiral MOFs. The use of chiral bridging ligands ensures the chirality of the resulting network structures. Most of the chiral MOFs were built from commercially available chiral bridging ligands, including naturally occurring amino acids (such as aspartic acid and glutamic acid) [8], hydroxyl carboxylic acids (such as lactic acid, malic acid, and tartaric acid) [9], nucleotide acids [10], cinchona alkaloids [11], and biotin [12].

L$_2$ **L$_3$-H$_2$**   **L$_4$**

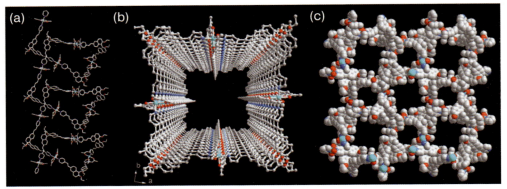

Figure 12.1 (a) Left-handed helical chain of **3** built from alternating Ni-(acac)$_2$ and **L**$_2$. (b) A schematic illustrating the interlocking of adjacent helical chains. (c) A space-filling model showing the open channels within the 3D chiral framework of **3**.

However, appropriately designed chiral multifunctional ligands are needed in order to generate chiral MOFs with the desirable functionalities for chiral separation and asymmetric catalysis. Lin *et al.* have synthesized many functional chiral MOFs using chiral bridging ligands derived from readily available chiral 1,1′-bi-2-naphthol [13]. For example, periodically ordered chiral interlocked nanotubes with the formula [Ni(acac)$_2$(**L**$_2$)]·3CH$_3$CN·6H$_2$O, **3**, were obtained by treating 1,1′-binaphthyl-6,6′-di-(4-vinylpyridine) ligands **L**$_2$ with Ni(acac)$_2$ (Figure 12.1) [13a]. The binaphthyl derivatives bridge the Ni(acac)$_2$ units by coordination of the pyridine residues to afford infinite helical chains which assemble via noncovalent interactions. The nanotubes in **3** further interlock to lead to chiral pores with 1.7 × 1.7 nm dimensions that are filled with removable H$_2$O and CH$_3$CN solvate molecules.

Interpenetration can be a problem in the preparation of MOFs even when relatively small changes in ligand structure are made. By way of example, Kesanli *et al.* have also constructed chiral MOFs using BINOL-derived dicarboxylate bridging ligands [14]. Single crystals of [Zn$_4$(μ$_4$-O)(**L**$_3$)$_3$(DMF)$_2$]·4DMF·3CH$_3$OH·2H$_2$O, **4**, were obtained by treating Zn(ClO$_4$)$_2$ with **L**$_3$-H$_2$ in DMF and MeOH at 50 °C (Figure 12.2). Compound **4** adopted a fourfold interpenetrated network structure of cubic topology built from [Zn$_4$(μ$_4$-O)(**L**$_3$)$_3$(DMF)$_2$] clusters, which critically limits the pore size of **4** and renders it unusable for asymmetric catalysis or chiral separations.

Wu *et al.* have demonstrated highly enantioselective heterogeneous asymmetric catalysis with chiral MOFs built from chiral bridging ligands containing orthogonal functional groups [15]. Metal-connecting units can link these chiral bridging ligands to form extended network structures via the primary functional groups. The orthogonal secondary chiral functionalities on the other hand cannot link the metal-connecting units and remain uncoordinated. They periodically decorate the porous MOF and can be used to introduce asymmetric catalytic sites that are uniform throughout the entire MOF sample. A crystalline chiral porous MOF [Cd$_3$Cl$_6$(**L**$_4$)$_3$]·4DMF·6MeOH·3H$_2$O, **5**, (where **L**$_4$ is (*R*)-6,6′-dichloro-2,2′-dihydroxy-1,1′-binaphthyl-4,4′-bipyridine) was recently synthesized and characterized by single-crystal X-ray

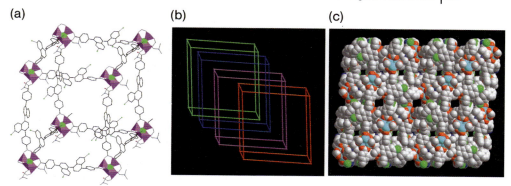

Figure 12.2 (a) A view of the cubic cavity formed by the 3D network of **4** (ethoxy groups have been omitted for clarity; the purple and green polyhedra represent the cluster-building unit). (b) A schematic presentation of the fourfold interpenetration in **4**. (c) A space-filling model of **4** as viewed down the *b*-axis.

diffraction studies [15]. The Cd(II) centers in **5** are doubly bridged by the chloride anions to form 1D zigzag $[Cd(\mu\text{-}Cl)_2]_n$ chains. Each Cd(II) center in the 1D zigzag chain further coordinates to two pyridyl groups of the **L₄** ligands and connects adjacent $[Cd(\mu\text{-}Cl)_2]_n$ chains to form a noninterpenetrating 3D network with very large chiral channels of ~1.6 × 1.8 nm cross section (Figure 12.3). X-ray powder diffraction and gas-adsorption measurements indicated that the framework structure of **5** was maintained upon the removal of all the solvent molecules and the evacuated framework of **5** retains its porosity, with a specific surface area of 601 m²/g and a pore volume of 0.26 mL/g. Treatment of **5** with $Ti(O^iPr)_4$ produced an active catalyst (designated as **5·Ti**) for the diethylzinc addition reactions with up to 93% ee. Control experiments with a series of larger aldehydes of varying sizes (from 0.8 to 2.0 nm) showed that the catalytic activity critically depended on the aldehyde size and no catalytic activity was observed when the aldehyde became bigger than the open-channel size. This result

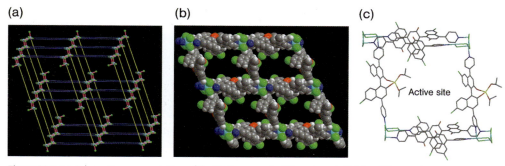

Figure 12.3 (a) Schematic representation of the 3D framework of **5** as viewed slightly off the *a*-axis. (b) Space-filling model of **5** as viewed down the *a*-axis showing the large chiral 1D channels (~1.6 × 1.8 nm). (c) Schematic representation of the active $(BINOLate)Ti(O^iPr)_2$ catalytic sites in the open channels of **5**.

demonstrates that **5·Ti** is a true heterogeneous asymmetric catalyst since both diethylzinc and the aromatic aldehyde are accessing the (BINOLate)Ti(O^{i}Pr)$_2$ sites via the open channels to generate the chiral secondary alcohol product.

Wu and Lin further demonstrated that catalytic activities of chiral MOFs are highly dependent on the framework structures. Two related MOFs [Cd$_3$(**L$_4$**)$_4$(NO$_3$)$_6$]· 7MeOH·5H$_2$O, **6**, and [Cd**L$_4$**(H$_2$O)$_2$][ClO$_4$]$_2$·DMF·4MeOH·3H$_2$O, **7**, were constructed using Cd metal-connecting points and the same ligand **L$_4$** (with different anions) [16]. Compound **6** adopts a twofold interpenetrated structure with large interconnected channels of $\sim 4.9 \times 13.1$ Å^2 along the a- and b-axes and $\sim 13.5 \times 13.5$ Å^2 along the c-axis, while **7** is formed by interlocking of two crystallographically equivalent 2D rhombic grids and possesses 1D channels of $\sim 1.2 \times 1.5$ nm^2 in dimensions (Figure 12.4). While **6** can be activated with Ti(O^{i}Pr)$_4$ to generate an

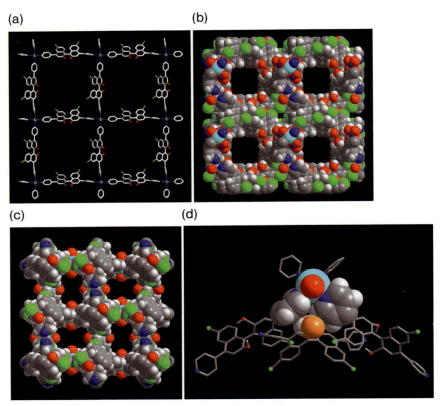

Figure 12.4 (a) The 2D square grids in **6**. (b) A space-filling model of **6** as viewed down the c-axis showing the chiral 1D channels of 13.5 × 3.5 Å^2 in dimensions. (c) A space-filling model of **7** showing the open channels running down the c-axis showing the rectangular channels with dimensions of $\sim 1.2 \times 1.5$ nm^2 along the c-axis. (d) The steric congestion around chiral dihydroxy groups of **L$_4$** ligands (orange balls) in **7** owing to the interpenetration of mutually perpendicular 2D rhombic grids via π–π stacking interactions. Cyan: Cd, green: Cl, red: O, blue: N, gray: C, and light gray: H.

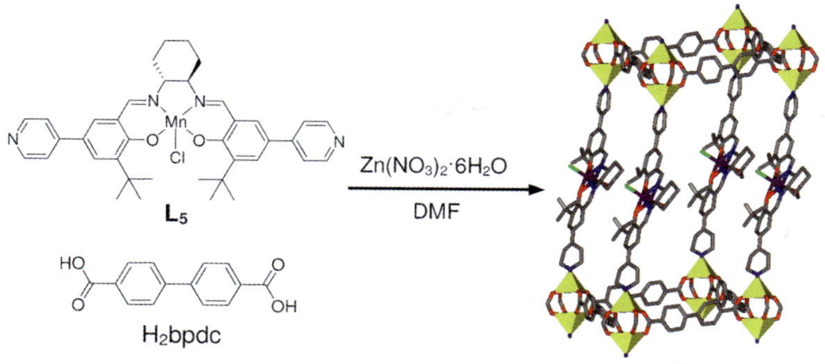

Figure 12.5 Synthesis of **8** and a POVRAY view of a single network unit for **8**.

active catalyst for the diethylzinc addition to aromatic aldehydes (with up to 90% ee), a mixture of **7** and $Ti(O^iPr)_4$ was not active in catalyzing the addition of diethylzinc to aromatic aldehydes under identical conditions. The pyridyl and naphthyl rings in **7** from mutually perpendicular, interpenetrating 2D rhombic grids in which the aromatic groups form strong π–π interactions. As a result, the chiral dihydroxy groups of **L4** ligands are held in very close proximity to the $Cd(py)_2(H_2O)_2$ hinges. The steric congestion around these chiral dihydroxy groups prevents the substitution of two isopropoxide functions by the BINOLate functionality and therefore no catalytic activity was observed. The observation of drastically different catalytic activities for these systems is remarkable since they were built from exactly the same building blocks. It is clear that the MOF framework structure is a determining factor in its catalytic performance.

Hupp and coworkers reported an interesting MOF (**8**) built from chiral [(R,R)-(-)-1,2-cyclohexanediamino-N,N'-bis(3-*tert*-butyl-5-(4-pyridyl)-salicylidene)]Mn (**L5**) struts for catalytic olefin epoxidation reactions with an achiral biphenyl derivative acting as a spacer in the framework (Figure 12.5) [17]. Although slightly less enantioselective, this catalyst was shown to be more active for olefin epoxidation reactions than its homogeneous counterparts because of framework confinement of the manganese salen entity.

Chiral porous MOFs can also contain desired chemical functionalities and pore shapes to be used for enantioselective sorption or separation processes. There is not as much research progress in this area. This is probably because the separation process is inherently disadvantageous over asymmetric catalysis owing to the loss of one of the two enantiomers during the separation process. A number of reports gave modest ee values of 10–20% in chiral separations using chiral MOFs [18].

12.3
Porous Oxide Materials

Since the introduction of aluminosilicate zeolites, porous oxide materials have become one of the most active research fields. The scope of such materials has been expanded to a large number of main-block phosphates, such as those of aluminum, gallium, indium, and tin, as well as several transition-metal phosphates, including systems based upon vanadium, molybdenum, cobalt, and iron [19]. However, the aluminosilicate zeolites are still the most popular materials for applications in catalysis, separations, and ion-exchange, because of their remarkable stability. Attempts are being made to exploit some of the exciting properties of the newly discovered systems. Particular interest is given to chiral porous oxides due to their potential applications in enantioselective sorption and catalysis [20].

Zhou *et al.* [21] have recently described a mesoporous germanium oxide, SU-M, with gyroidal channels separated by crystalline walls that lie about the G (gyroid) minimal surface as in the mesoporous MCM-48. It has the largest primitive cell and lowest framework density of any inorganic material and channels that are defined by 30 rings. One of the two gyroidal channel systems of SU-M can be filled with additional oxide, resulting in a mesoporous crystal (SU-MB) with chiral channels. SU-M is built from a unique $Ge_{10}O_{24}(OH)_3$ cluster (Figure 12.6a) with O atoms singly coordinated to Ge corresponding to OH. The cluster consists of a central core of four octahedrally coordinated Ge atoms and six tetrahedrally coordinated Ge atoms. Each

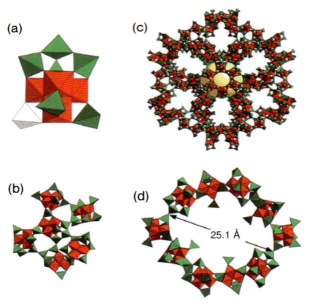

Figure 12.6 Linkage of $Ge_{10}O_{24}(OH)_3$ clusters in SU-M. (a) The $Ge_{10}O_{24}(OH)_3$ cluster built from six GeO_4 tetrahedra (green) and four GeO_6 octahedra (red). (b) Linking of a $Ge_{10}O_{24}(OH)_3$ cluster to five neighboring clusters. (c) A 30-Å-wide slab with a big cavity at the center. (d) A 30-ring window formed by ten $Ge_{10}O_{24}(OH)_3$ clusters. The diameter of the 30-ring is 10.0×22.4 Å.

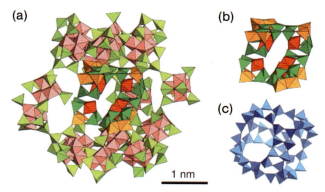

Figure 12.7 Cavities in SU-MB and faujasite. (a) The filled big tile in SU-MB. (b) The six $Ge_7O_{16}F_3$ clusters in the interior of a; orange polyhedra are trigonal bipyramids. (c) A faujasite supercage on the same scale.

cluster is linked to five other clusters (Figure 12.6b) via Ge–O–Ge bonds to form a three-dimensional framework with an overall stoichiometry of $Ge_{10}O_{20.5}(OH)_3$. SU-M contains two channels of opposite chirality. SU-MB is a chiral derivative of SU-M, prepared in the presence of hydrofluoric acid (Figure 12.7). In the structure of SU-MB, one half of the big tiles (big cavities) are filled with additional (Ge, O, F) clusters. These clusters, formulated as $Ge_7O_{16}F_3$, resemble other germanium oxide frameworks [22], including ASU-16 and SU-12. Six of these clusters (Figure 12.7b) fit inside one big tile (big cavity), with three of them connected to each 12-ring window of the main framework through the terminal atoms of the $Ge_{10}O_{24}(OH)_3$ clusters (Figure 12.7a). The occupied big tile and its contents (that is the unit shown in Figure 12.7a) now has the composition of $Ge_{222}X_{516}$ (where X is ¼ O, OH or F). The most remarkable aspect of SU-MB is the fact that only half of the cavities are filled, specifically all those of one handedness, and the symmetry is reduced to $I4_132$.

Xu and coworkers have reported a multidirectional intersecting helical channel using Zn-phosphate network [23]. This work employed chiral metal complexes as templating agents to induce a chiral environment in the host framework [24]. By using a racemic mixture of a chiral $[Co(dien)_2]Cl_3$ (where dien is diethylenetriamine) complex as the template, a new open-framework zinc phosphate, $[\{Zn_2(HPO_4)_4\}\{Co(dien)_2\}]\cdot H_3O$, **9**, with multidirectional helical channels was prepared (Figure 12.8). Of particular interest is the existence of chiral intertwined double helices in its structure. The framework of **9** consists solely of 12-membered rings, with each tetrahedrally coordinated Zn atom associated with one of them. Each 12-membered ring consists of six Zn and six P atoms, and has C_2 symmetry. Such structural units are connected together through vertex oxygen atoms to form a very open framework with a multidirectional helical pore system. It contains 12-membered ring channels that run along this direction. Each 12-ring accommodates one $[Co(dien)_2]^{3+}$ ion. Interestingly, each 12-membered ring channel is enclosed by two intertwined

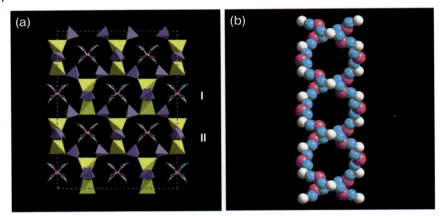

Figure 12.8 (a) A polyhedral view of the framework of **9** along the [100] direction. The $[Co(dien)_2]^{3+}$ ions reside in the 12-membered ring channels. A pair of enantiomers of $Co(dien)_2^{3+}$ cations alternate in rows I and II; (b) A space-filling diagram of two intertwined helices that enclose the 12-membered ring channel.

helices of the same handedness, which are connected through Zn-O-P linkages (Figure 12.8b).

Che and coworkers have reported surfactant-templated synthesis of ordered chiral mesoporous silica with a twisted hexagonal rod-like morphology (rod diameter 130–180 nm and length 1–6 μm) [25]. The presence of C_{14}-L-AlaA induces changes in the conformation of the amphiphile and reshapes the micelle, resulting chiral micelles (Figure 12.9) [26]. The positively charged ammonium site of N-trimethoxysilylpropyl-N,N,N-trimethylammonium chloride (TMAPS) or 3-aminopropyltrimethoxysilane (APS) interacts electrostatically with the negatively charged head group of the C_{14}-L-AlaS or C_{14}-L-AlaA. The alkoxysilane sites of TMAPS and APS are cocondensed with tetraethoxylsilane (TEOS), to lead to the silica framework. The trimethylene groups of TMAPS and APS covalently tether the silicon atoms incorporated into the framework to the cationic ammonium groups. By adjusting the HCl/C_{14}-L-AlaS and CSDA/C_{14}-L-AlaS molar ratios, a pH and composition favorable for the formation of a chiral mesostructure were obtained (Figure 12.9).

12.4
Chiral Immobilization of Porous Silica Materials

Postsynthetic modification of mesoporous materials is another powerful strategy for generating chiral porous materials for applications in heterogeneous asymmetric catalysis. MCM-41 and MCM-48 are two of the most popular mesoporous materials that comprise of amorphous metal oxide (such as silica) wall and contain uniform mesopores with long-range order [27]. Mesoporous silica materials can be easily modified since their inorganic walls have abundant hydroxyl groups that can react

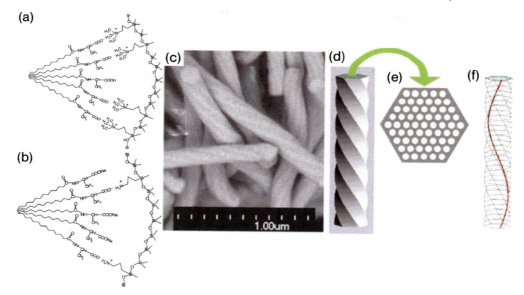

Figure 12.9 Schematic illustration of the two types of interaction between the head group of C_{14}-L-AlaS, its free amino acid C_{14}-L-AlaA and amino groups. The two types of interactions are achieved through double decomposition of negatively charged C_{14}-L-AlaS with positively charged quaternized aminosilane TMAPS (a), and neutralization of acid C_{14}-L-AlaA with primary aminosilane APS (b). (c) SEM image and schematic drawings of a structural model of chiral mesoporous silica. (d) SEM image, showing the microscopic features. (e) Schematic drawing of a structural model of the chiral mesoporous material for TEM image simulation. (f) One of the chiral channels in the material.

with active organic groups such as R-Si(OEt)$_3$, R-SiX$_3$ (X = halide Cl, Br, etc.) [28], resulting facile functionalization on the surface. For example, heterogeneous chiral catalysts for asymmetric hydrogenation, epoxidation, Aldol reaction, and Diels–Alder reaction have been prepared through immobilization of the homogeneous catalysts on organic or inorganic supports [29]. Homogeneous chiral catalysts can be immobilized on various supports through covalent grafting of chiral catalysts on mesoporous materials [30] or carbon materials [31], and ion-exchange of catalysts into charged supports [32]. Due to well-ordered pore arrays, large surface area, and uniform pore-size distributions, the mesoporous materials, such as MCM-41 and SBA-15, have attracted much attention for the immobilization of homogeneous chiral catalysts (Figure 12.10).

In order to introduce chiral catalysts into nanopores only, the Si–OH groups on the external surfaces of supports have to be deactivated and then the chiral catalysts could be immobilized in the nanopores of supports via the reaction of the catalyst with the Si–OH only in the nanopores [33]. The amount of the grafted chiral catalysts depends on the concentration of Si–OH groups of the support. For the charged groups, Al-MCM-41 [34], clays [32a], or layered double hydroxides (LDH) [32b], are usually used for the immobilization of chiral catalysts. Noncharged chiral catalysts can also be

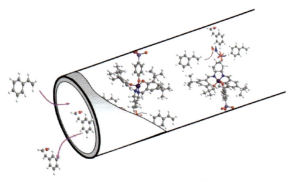

Figure 12.10 The asymmetric epoxidation of styrene catalyzed by chiral Mn(salen) catalyst immobilized in nanopores.

attached with some charged groups, which are further immobilized into nanopores of the supports.

Bigi et al. [35] reported a chiral Mn(salen) catalyst that was grafted in the nanopores of MCM-41 through a triazine-based linkage (Figure 12.11). The propyl amino group in the pore provided a flexible conformation of the Mn(salen) catalyst and the triazine group prevented the chain from folding over, thus eliminating catalyst degradation steps. This heterogeneous catalyst showed an ee of 84%, similar to that of the homogeneous counterpart (89%) for the asymmetric epoxidation of 1-phenylcyclohexene. Hutchings and coworkers [36] prepared a heterogeneous Mn(salen) catalyst by the modification of Mn-exchanged Al-MCM-41 with chiral salen ligand for the asymmetric epoxidation of (Z)-stilbene with PhIO as oxygen donor. The *cis/trans* ratios of the epoxides increased from 0.4 for the homogeneous catalyst to 4.3 for the heterogeneous catalyst. It was proposed that the nanopores of MCM-41 retarded the rotation of the radical intermediate and produced more *cis*-epoxide compared to

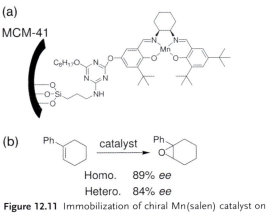

Figure 12.11 Immobilization of chiral Mn(salen) catalyst on MCM-41 via triazine-based linkage (a) and epoxidation of 1-phenylcyclohexene (b).

the homogeneous catalyst. Kureshy *et al.* [37] reported that chiral Mn(salen) catalysts immobilized in the nanopores of MCM-41 and SBA-15 showed higher chiral induction (70% ee) than its homogeneous counterpart (45% ee) for the epoxidation of styrene with aqueous NaOCl as oxidant. In addition, bulkier alkenes such as 6-cyano-2,2-dimethylchromene were also efficiently epoxidized on these supported Mn(salen) catalysts (with up to 92% ee), and the reaction results were comparable to those of the homogeneous counterparts. This heterogeneous catalyst was used four times without obvious loss of activity and enantioselectivity. A Mn(salen) catalyst was also axially immobilized in the nanopores of MCM-41 via pyridine N-oxide by Kureshy's group [38]. These immobilized catalysts showed higher ee (69%) than its homogeneous counterpart (51%) for the asymmetric epoxidation of styrene. These catalysts were also effective for the asymmetric epoxidation of bulkier substrates such as indene and 2,2-dimethylchromene (conversion~82–98%, ee~69–92%). The catalysts could be recycled for at least four times without loss of performance. The increase in ee values was attributed to the unique spatial environment constituted by the chiral salen ligand and the surface of the support.

Recently, Li and coworkers prepared the axially immobilized chiral Mn(salen) complexes in the nanopores of mesoporous materials via the phenoxyl group [39], phenyl sulfonic group [40], and propyl sulfonic group. These heterogeneous catalysts were denoted as support-PhO (PhSO$_3$ or PrSO$_3$) Mn(salen) (Figure 12.12). The phenoxyl-immobilized catalyst SBA-PhOMn(salen) exhibited 93.5% ee (*cis*-form) for the asymmetric epoxidation of *cis*-β-methylstyrene, higher than the 25.3% ee obtained for the homogeneous catalyst [39b]. These results show that the enhancement of enantioselectivity in the asymmetric epoxidation is mainly attributed to the nanopore effect of the supports. The *cis/trans* ratios of epoxides also increased from 0.46 for the homogeneous Mn(salen)Cl catalyst to 11.1 for the SBA-PhOMn(salen)

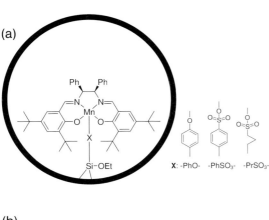

Figure 12.12 Chiral Mn(salen) heterogeneous catalyst (a) and asymmetric epoxidation (b).

catalyst. Chiral Mn(salen) complex axially immobilized in nanopores of SBA via phenyl sulfonic groups also shows higher ee value for the *cis*-epoxide and the *cis/trans* ratio than the corresponding homogeneous counterparts for the asymmetric epoxidation of *cis*-β-methylstyrene under the same conditions.

Kureshy *et al.* [36a] immobilized dicationic chiral Mn(salen) catalysts into the anionic montmorillonite clay for the preparation of the clay-supported Mn(salen) catalysts. These heterogeneous catalysts were tested for the asymmetric epoxidation of 6-nitro-2,2-dimethylchromene, indene and styrene with NaOCl as oxidant. The conversions were almost 100% and the ee values were similar to or even higher than those of the homogeneous ones (74 vs. 52% for asymmetric epoxidation of styrene). The increase in ee value was mainly attributed to the unique spatial environment in the nanopores of the support. Furthermore, the heterogeneous catalysts could be recovered and recycled four times with constant conversions and ee values.

Similarly, Anderson and coworkers [32b,c] also incorporated a chiral sulfonato-salen-Mn(III) complex in cationic Zn–Al layered double hydroxides (Figure 12.13). These catalysts were effective for the stereoselective epoxidation of (*R*)-(+)-limonene using a combination of pivalaldehyde and molecular oxygen as the oxidant. The catalyst could be recycled three times, and gave 100% conversion and 54.0% diastereoisomeric excess. This heterogeneous Mn(salen) catalyst was also found to be highly active and enantioselective for the asymmetric epoxidation of various substituted styrenes and cyclic alkenes [41]. The catalyst could be recycled without detectable loss of efficiency. These results suggest that the catalysts immobilized in the nanopores of the layered supports could be as effective as the homogeneous catalysts, and they can be separated and recycled.

Raja and coworkers [42] reported that a chiral Rh(I) complex could be confined in nanopores by ion exchange and used for the heterogeneous asymmetric hydrogenation of α-ketone. The chiral catalyst confined in nanopores of 3.8 nm showed 77% ee for the asymmetric hydrogenation; while the homogeneous chiral catalyst gave the racemic products. The increase in the nanopore size from 3.8 to 6.0 nm decreased the ee values from 77 to 61%. When the pore size was further increased to 25 nm, the

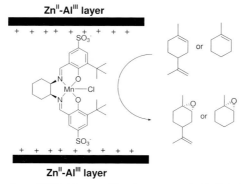

Figure 12.13 Montmorillite clay-supported chiral Mn(salen) catalyst and asymmetric epoxidation.

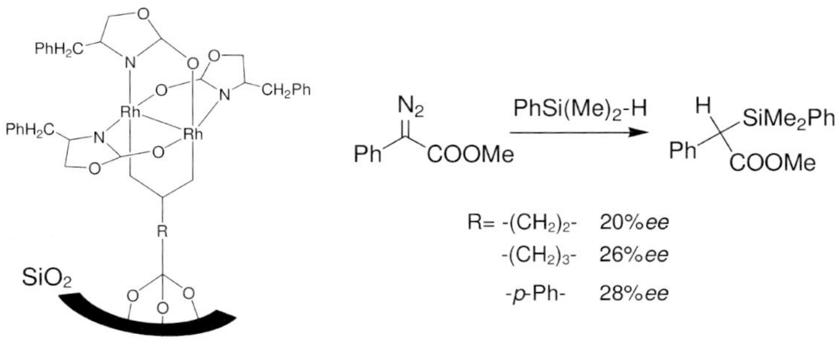

Figure 12.14 Immobilization of Rh complexes on SiO_2 and the asymmetric Si–H insertion.

ee value was decreased to zero, indicating that the nanopore effect plays an important role in the chiral induction.

Maschmeyer and coworkers immobilized chiral Rh complexes in nanopores of SiO_2 via different linkage groups (Figure 12.14) and the catalysts were tested for the asymmetric Si–H insertion reaction [43]. The homogeneous Rh catalysts produced only 2% ee for this reaction. The catalysts immobilized in nanopores of SiO_2 gave higher ee values (20–28% ee) and the asymmetric induction was affected by the linkage groups. When the linkage R was the p-C_6H_4 group, the heterogeneous catalyst gave the highest ee value of 28%. Hutchings and coworkers reported that a Cu catalyst modified with chiral bis(oxazoline) ligand could be introduced into the pores of zeolite Y via ion exchange (Scheme 12.14) [44]. This catalyst showed ee value of 77%, higher than 28% obtained for the homogenous catalyst in the asymmetric aziridination of styrene. The asymmetric catalytic reaction was performed in the zeolite cages, in which the confinement effect was considered to improve the asymmetric induction of chiral modifier.

Corma and coworkers [37b] recently reported that PMOs with a chiral vanadyl Schiff base complex in the framework (Figure 12.15) showed enantioselectivity for the asymmetric catalytic cyanosilylation of benzaldehyde with TMSCN. The hybrid nanoporous material had an ordered mesostructure with specific surface area of

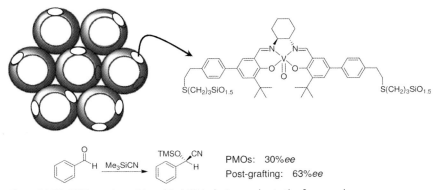

Figure 12.15 PMO catalyst with a chiral VO(salen) complex in the framework.

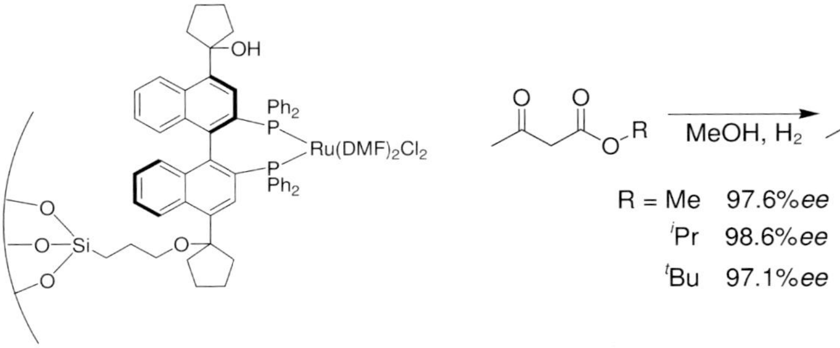

Figure 12.16 Immobilization of Ru complexes of 4,4′-substituted BINAPs on silica SBA-15 and enantioselective hydrogenation of β-ketoesters.

900 m² g⁻¹ and pore diameter of 4.2 nm. The chiral PMO catalyst showed 30% ee, which was lower than the 63% obtained for the VO(salen) catalyst covalently grafted in nanopores of mesoporous materials. Their studies showed that the decrease in ee values for the chiral PMOs may be due to the steric constraints imposed on the catalyst originated from the surrounding framework of PMOs.

Lin et al. [45] have demonstrated the immobilization of Ru complexes of 4,4′-substituted BINAPs on well-ordered mesoporous silica SBA-15 (Figure 12.16) and the catalysts were tested for the enantioselective hydrogenation of β-ketoesters. These immobilized catalysts gave very high ee values (97.1–98.6% ee) while the homogeneous Ru(BINAP)(DMF)$_2$Cl$_2$ catalyst gave much lower ee values (∼85% ee) [46].

12.5
Outlook

Nanoporous materials have had a significant impact on many applications including microelectronics, medicine, clean energy, and the environment. Chiral nanoporous materials are, on the other hand, still in their infancy but have already been used in chiral separation, enantioselective sensors, and asymmetric catalysis. There remain many challenges in the field of chiral nanoporous materials, but opportunities also abound. New chiral nanoporous materials with high catalytic activity and product selectivity are clearly needed.

Most work has so far focused on achieving high yield and selectivity, but kinetics is an equally important parameter if the chiral nanoporous material is going to find any practical use. Recent advances in controlling particle sizes and morphologies of MOFs [47], silica [48], and related materials should allow the synthesis of new generations of chiral nanoporous materials with desirable diffusion kinetics for the reactants and products. Such chiral nanoporous materials are expected to play a major role in future chirotechnology.

Acknowledgments

W.L. thanks NSF for providing funding support for nanoporous materials research efforts in the Lin lab.

References

1. (a) Bekkum, H.V., Flanigen, E.M. and Jansen, J.C. (1991) *Introduction to Zeolite Science and Practice*, Elsevier, Amsterdam; (b) Szostak, R. (1998) *Molecular Sieves: Principles of Synthesis, and Identification*, Blackie Academic and Professional, London.
2. IUPAC, Manual of Symbols, Terminology (1972) *Pure and Applied Chemistry*, **31**, 578.
3. (a) Ying, J.Y., Mehnert, C.P. and Wong, M.S. (1999) *Angewandte Chemie-International Edition*, **38**, 56; (b) Corma, A. (1997) *Chemical Reviews*, **97**, 2373.
4. (a) Batten, S.R. and Robson, R. (1998) *Angewandte Chemie-International Edition*, **37**, 1461; (b) Eddaoudi, M., Moler, D.B., Li, H., Chen, B., Reineke, T.M., O'Keeffe, M. and Yaghi, O.M. (2001) *Accounts of Chemical Research*, **34**, 319; (c) Evans, O.R. and Lin, W. (2002) *Accounts of Chemical Research*, **35**, 511; (d) Lee, S., Mallik, A.B., Xu, Z., Lobkovsky, E.B. and Tran, L. (2005) *Accounts of Chemical Research*, **38**, 251; (e) Feng, P., Bu, X. and Zheng, N. (2005) *Accounts of Chemical Research*, **38**, 293.
5. Ezuhara, T., Endo, K. and Aoyama, Y. (1999) *Journal of the American Chemical Society*, **121**, 3279.
6. Wu, S.T., Wu, Y.R., Kang, Q.Q., Zhang, H., Long, L.S., Zheng, Z., Huang, R.B. and Zheng, L.S. (2007) *Angewandte Chemie-International Edition*, **46**, 8475.
7. Kepert, C.J., Prior, T.J. and Rosseinsky, M.J. (2000) *Journal of the American Chemical Society*, **122**, 5158.
8. Ranford, J.D., Vittal, J.J., Wu, D. and Yang, X. (1999) *Angewandte Chemie-International Edition*, **38**, 3498.
9. Abrahams, B.F., Moylan, M., Orchard, S.D. and Robson, R. (2003) *Angewandte Chemie-International Edition*, **42**, 1848.
10. Sheldrick, W.S. (1981) *Acta Crystallographica*, **B37**, 1820.
11. Xiong, R.-G., You, X.-Z., Abrahams, B.F., Xue, Z. and Che, C.-M. (2001) *Angewandte Chemie-International Edition*, **40**, 4422.
12. Aoki, K. and Saenger, W. (1983) *Journal of Inorganic Biochemistry*, **19**, 269.
13. (a) Cui, Y., Lee, S.J. and Lin, W. (2003) *Journal of the American Chemical Society*, **125**, 6014; (b) Ngo, H.L. and Lin, W. (2002) *Journal of the American Chemical Society*, **124**, 14298; (c) Wu, C.-D. and Lin, W. (2005) *Angewandte Chemie-International Edition*, **44**, 1958.
14. Kesanli, B., Cui, Y., Smith, M.R., Bittner, E.W., Bockrath, B.C. and Lin, W. (2005) *Angewandte Chemie-International Edition*, **44**, 72.
15. Wu, C.-D., Hu, A., Zhang, L. and Lin, W. (2005) *Journal of the American Chemical Society*, **127**, 8940.
16. Wu, C.-D. and Lin, W. (2007) *Angewandte Chemie-International Edition*, **46**, 1075.
17. Cho, S.-H., Ma, B., Nguyen, S.T., Hupp, J.T. and Albrecht-Schmitt, T.E. (2006) *Chemical Communications*, 2563.
18. (a) Bradshaw, D., Prior, T.J., Cussen, E.J., Claridge, J.B. and Rosseinsky, M.J. (2004) *Journal of the American Chemical Society*, **126**, 6106; (b) Evans, O.R., Ngo, H.L. and Lin, W. (2001) *Journal of the American Chemical Society*, **123**, 10395.
19. (a) Davis, M.E. (2002) *Nature*, **417**, 813; (b) Cheetham, A.K., Férey, G. and Loiseau, T. (1999) *Angewandte Chemie-International Edition*, **38**, 3269;

(c) van Bekkum, H., Jacobs, P.A., Flanigen, E.M. and Jansen, J.C. (2001) *Introduction to Zeolite Science and Practice*, 2nd edn, Elsevier, New York.

20 Baiker, A. (1998) *Current Opinion in Solid State & Materials Science*, **3**, 86.

21 Zou, X., Conradsson, T., Klingstedt, M., Dadachov, M.S. and O'Keeffe, M. (2005) *Nature*, **437**, 716.

22 Plévert, J., Gentz, T.M., Groy, T.L., O'Keeffe, M. and Yaghi, O.M. (2003) *Chemistry of Materials*, **15**, 714.

23 Wang, Y., Yu, J., Guo, M. and Xu, R. (2003) *Angewandte Chemie-International Edition*, **42**, 4089.

24 Morgan, K., Gainsford, G. and Milestone, N. (1995) *Journal of the Chemical Society. Chemical Communications*, 425.

25 Che, S., Liu, Z., Ohsuma, T., Sakamoto, K., Terasaki, O. and Tatsumi, T. (2004) *Nature*, **429**, 281.

26 Tracey, A.S. and Zhang, X. (1992) *The Journal of Physical Chemistry*, **96**, 3889.

27 Kresge, C.T., Leonowicz, M.E., Roth, W.J., Vartuli, J.C. and Beck, J.S. (1992) *Nature*, **359**, 710.

28 (a) Stein, A., Melde, B.J. and Schroden, R.C. (2000) *Advanced Materials*, **12**, 1403; (b) Schroden, R.C., Blanford, C.F., Melde, B.J., Johnson, B.J.S. and Stein, A. (2001) *Chemistry of Materials*, **13**, 1074; (c) Lebeau, B., Fowler, C.E., Hall, S.R. and Mann, S. (1999) *Journal of Materials Chemistry*, **9**, 2279; (d) Margolese, D., Melero, J.A., Christiansen, S.C., Chmelka, B.F. and Stucky, G.D. (2000) *Chemistry of Materials*, **12**, 2448.

29 (a) McMorn, P. and Hutchings, G.J. (2004) *Chemical Society Reviews*, **33**, 108; (b) Song, C.E. and Lee, S.G. (2002) *Chemical Reviews*, **102**, 3495; (c) Li, C. (2004) *Catalysis Reviews: Science and Engineering*, **46**, 419; (d) Thomas, J.M., Raja, R. and Lewis, D.W. (2005) *Angewandte Chemie-International Edition*, **44**, 6456.

30 (a) Ayala, V., Corma, A., Iglesias, M., Rincón, J.A. and Sánchez, F. (2004) *Journal of Catalysis*, **224**, 170; (b) Zhang, H.D., Xiang, S., Xiao, J.L. and Li, C. (2005) *Journal of Molecular Catalysis A-Chemical*, **238**, 175; (c) Kureshy, R.I., Ahmad, I., Khan, N.H., Abdi, S.H.R., Singh, S., Pandia, P.H. and Jasram, R.V. (2005) *Journal of Catalysis*, **235**, 28.

31 Baleizão, C., Gigante, B., Garcia, H. and Corma, A. (2004) *Journal of Catalysis*, **221**, 77.

32 (a) Kureshy, R.I., Khan, N.H., Abdi, S.H.R., Ahmael, I., Singh, S. and Jasra, R.V. (2004) *Journal of Catalysis*, **221**, 234; (b) Bhattacharjee, S. and Anderson, J.A. (2004) *Chemical Communications*, 554; (c) Bhattacharjee, S., Dines, T.J. and Anderson, J.A. (2004) *Journal of Catalysis*, **225**, 398; (d) Cardoso, B., Pires, J., Carvalho, A.P., Kuzniarska-Biernacka, I., Silva, A.R., de Castro, B. and Freire, C. (2005) *Microporous Mesoporous Mater*, **86**, 295.

33 (a) Hultman, H.M., de Lang, M., Nowotny, M., Arends, I.W.C.E., Hanedeld, U., Sheldon, R.A. and Maschmeyer, T. (2003) *Journal of Catalysis*, **217**, 264; (b) Zhou, X., Yu, X., Huang, J., Li, S., Li, L. and Che, C. (1999) *Chemical Communications*, 1789.

34 Kim, G.J. and Shin, J.H. (1999) *Catalysis Letters*, **63**, 83.

35 Bigi, F., Franca, B., Moroni, L., Maggi, R. and Sartori, G. (2002) *Chemical Communications*, 716.

36 Piaggio, P., McMorn, D., Murphy, P., Bethell, D., Bullman-Page, P.C., Hancock, F.E., Sly, C., Kerton, O.J. and Hutchings, G.J. (2000) *Journal of the Chemical Society-Perkin Transactions 2*, **2**, 2008.

37 (a) Kureshy, R.I., Ahmad, I., Khan, N.H., Abdi, S.H.R., Pathak, K. and Jasra, R.V. (2006) *Journal of Catalysis*, **238**, 134; (b) Kureshy, R.I., Ahmad, I., Khan, N.H., Abdi, S.H.R., Pathak, K. and Jasra, R.V. (2005) *Tetrahedron: Asymmetry*, **16**, 3562.

38 (a) Baleião, C., Gigante, B., Das, D., Álvaro, M., Garcia, H. and Corma, A. (2003) *Chemical Communications*, 1860; (b) Baleião, C., Gigante, B., Das, D., Álvaro, M., Garcia, H. and Corma, A. (2004) *Journal of Catalysis*, **223**, 106; (c) Álvaro, M., Benitez, M., Das, D., Ferrer, B. and

Garcia, H. (2004) *Chemistry of Materials*, **16**, 2222; (d) Benitez, M., Bringmann, G., Dreyer, M., Garcia, H., Ihmels, H., Waidelich, M. and Wissel, K. (2005) *The Journal of Organic Chemistry*, **70**, 2315.

39 Xiang, S., Zhang, Y.L., Xin, Q. and Li, C. (2002) *Chemical Communications*, 2696.

40 Zhang, H.D., Xiang, S. and Li, C. (2005) *Chemical Communications*, 1209.

41 Bhattacharjee, S. and Anderson, J.A. (2006) *Advanced Synthesis and Catalysis*, **348**, 151.

42 Raja, R., Thomas, J.M., Jones, M.D., Johnson, B.F.G. and Vaughan, D.E.W. (2003) *Journal of the American Chemical Society*, **125**, 14982.

43 Hultman, H.M., de Lang, M., Nowotny, M., Arends, I.W.C.E., Hanedeld, U., Sheldon, R.A. and Maschmeyer, T. (2003) *Journal of Catalysis*, **217**, 275.

44 Caplan, N.A., Hancock, F.E., Bulman, P.P.C. and Hutchings, G.J. (2004) *Angewandte Chemie-International Edition*, **43**, 1685.

45 Kesanli, B. and Lin, W. (2004) *Chemical Communications*, 2284.

46 Hu, A., Ngo, H.L. and Lin, W. (2004) *Angewandte Chemie-International Edition*, **43**, 2501.

47 Rieter, B.J., Taylor, K.M.L., An, H. and Lin, W. (2006) *Journal of the American Chemical Society*, **128**, 9024–9025.

48 (a) Huh, S., Wiench, J.W., Trewyn, B.G., Song, S., Pruski, M. and Lin, V.S.-Y. (2003) *Chemical Communications*, 2364; (b) Nooney, R.I., Thirunavukkarasu, D., Chen, Y.M., Josephs, R. and Ostafin, A.E. (2002) *Chemistry of Materials*, **14**, 4721; (c) Cai, Q., Luo, Z.S., Pang, W.Q., Fan, Y.W., Chen, X.H. and Cui, F.Z. (2001) *Chemistry of Materials*, **13**, 258.

Index

a

absolute configuration   2, 23
absolute configuration, determination   220
absolute enantioselection   18ff
achiral components   19, 30ff, 46f, 55f, 218, 234f, 315f, 392
AFM   107, 119, 131f, 294
Ag(I)   268
aggregate-induced enhanced emission, AIEE   71
alaninate   201f
alanine   202
alignment   176
aluminosilicate zeolite   397f
amine   72
amino acids   3, 19, 264ff
– N-acyl-   253f
– Brewster angle microscopy   253
α-amino acid   129, 249, 257f
– C_{16}-Gly   249ff
– C_{29}-Lys   251
– C_n-Lys   250f
– herring-bone motif   252
– monolayer   249f
β-amino acid   129
amphiphile   248ff, 254
– 2D self-assembly   248f
– C_{18}-Glu-NCA   266f
– molecular packing   248
amplification (of chirality)   37, 116, 118ff, 138f, 147, 281, 368
angular momentum, L   10
anisotropy
– factor   76
– photo-induced circular   282ff
anticlinic tilt   278
anticonglomerate   325
antiferroelectric switching   276ff, 290

antiparticles   9, 14f
association lifetimes   29, 34
asymmetric catalysts   67, 83f
asymmetric synthesis   21, 403
atropisomers   353, 361
azobenzene   98, 158, 282, 354
azopolymer   151

b

$B4^{[*]}$ phase   292f
Bailar twist   36
band gap material   85
barbiturate   37f, 46f
benzene tricarboxamides   118
β-decay   14
β-peptides   22
bilayer aggregates   164f
binaphthyl   366, 394
BINAS   76f
BINAP   76f, 80, 85, 406
binding site   276f
BINOL   60, 395ff
bistability   351ff
block co-polymer   161, 168f
– amphihilic   165
– hydrophilic   173
blue phase   274f
bottom-up   70
Boulder model   276ff
binary system   322ff
breaking
– chiral symmetry   16
– mirror-symmetry   16

c

C_2-symmetry   276
C_3-symmetry   284ff
C_{15}-Lys/alanine   257

Chirality at the Nanoscale: Nanoparticles, Surfaces, Materials and more. Edited by David B. Amabilino
Copyright © 2009 WILEY-VCH Verlag GmbH & Co. KGaA, Weinheim
ISBN: 978-3-527-32013-4

C_{15}-Lys/serine   257
C_{15}-Lys/valine   257
calcium tartrate tetrahydrate   173
calix[4]arene   41ff, 52
camphor sulfonic acid   127, 141
capsule (molecular)   39ff
catechol   44f
cation-π interaction   33
cavity   40, 174, 204
CD, *see circular dichroism*
charge conjunction   9ff
chelate   43
chemisorption   221
chiral
– architecture   116ff
– additive   336
– adsorption   192
– aggregate   134ff, 157, 169ff
– amnesia   294
– assembly   117f
– bilayer effect   108
– bridging ligands   392ff
– capsules   37ff
– channels   399
– counterions   320f
– discrimination   87
– domains   279, 292
– external stimulus   150
– faces   7
– footprint   79
– framework   394
– memory effect   46ff, 118, 154ff
– monolayer   247ff
– solvents   124, 150ff
– superstructure   274f, 290ff
– trigger   152
chirality   1ff, 8ff, 30f, 273
– axial   274, 280
– conformational   273ff, 280ff
– control   117
– false   13ff, 18, 20
– induction   117, 136
– inversion of   294
– layers of   273, 290ff
– planar   274
– switching   117, 160
– transfer   279ff, 298
– true   13ff, 18f
chirality in two dimensions   17f
chiroclinic effect   291
chiroptics   21, 74ff, 349
chiroptical switches   161, 349ff
cholestryl-*S*-glutamate (CLG)   258ff
chromatography   81

chromophore   176ff
– perylenediimide   177
cinchonidine   84f
circular dichroism   22, 86, 98, 102ff, 110, 292ff, 356, 365
– second-harmonic generation   292
circularly polarized light   1, 151f, 160, 350ff, 366ff
– photons   19
circularly polarized luminescence   371
click chemistry   129
cluster   80ff, 196ff
– coordination   80ff
– organometallic   80f
cocrystal   316f
columnar phase (Col)   273, 282ff, 297
– helix formation   284ff
conductivity   120, 124ff
conformation chirality   273ff, 296
– hypothesis   296ff
– propeller-like   284
– twisted   296
conglomerate   2, 205, 231, 287, 313ff, 324
– dark phase   292ff, 298
conjunction   9ff
– charge, *C*   9
control
– kinetic   21
– thermodynamic   21
cooperative effects   116
coordination bonds   54
Cotton effect   52
Coulomb attraction   195
covalently connected chiral nanostructures   199ff
CP violation   15, 21
CPT theorem   10ff
o-cresol   84
crosslinking component   122, 143, 167ff
crown-ether side group   143
Cu(I)/Cu(II) redox couple   365
cubic phase   273, 276, 296ff
cyanine   169
cyanurate (CYA)   37
cyclodextrin   340, 357
cyclohexanediol   58f
cyclophane   259ff
cysteine   205

d

D-ribose   4
defect lattice   275
dense packing   225

density functional theory (DFT)
  – calculations 75, 80, 193, 201, 209f
deracemization, *see* resolution
deswelling 122
diacetylene 119
diaminocyclohexane 103
diarylethene 359
diastereomeric complexes 209f, 308, 317, 331
diastereomeric film 255
– 4-tetradecylphenethylamine 255
– 5-pentadecylmandelic acid 255
diastereoselectivity 30, 52, 54, 59, 361
dichroism 7
– electronic circular 74
– magnetochiral 7, 13
– vibrational circular 74
dimelamine 37f
dimensions 17
disc-like mesogen 271ff
dissymmetric 2ff
dopant, chiral 125, 280, 291, 385
– type I 280
– type II 280
doping 120, 134f
DNA 4f, 20, 30, 115, 169ff
double helix 4, 355, 357
double salt 332ff
drugs 29
Dutch resolution 318ff
dye 169, 171ff
dynamic equilibrium 33, 46, 52

e

ECD, *see* electronic circular dichroism
elastomers 122
electric dipole interactions 18, 23
electric field, **E** 11
electrochiroptical switching 364ff
electroclinic effect 279, 290
electron microscopy 293
electronic circular dichroism 74ff, 82
electrons
– spin-polarized 17ff
electrostatic interactions 311, 400
enantioenrichment 6
enantiomer 2ff, 14ff, 75, 80f, 266
– excess 6ff, 81
enantiomorphism 2ff, 10ff, 221, 308, 393
– time-invariant 3
– time-noninvariant 3
enantioselection 18, 403
enantioselective recognition 52f, 57
enantioselective sorption 397f
encapsulation 31, 39f, 51f, 56, 58ff

– double 58
– triple 59
entropic effects 33
entropy-driven 323
epoxidation 397, 401ff
equilibration 49, 56
equilibrium 323
eutectic 324, 332f
external stimulus 143, 150, 154f

f

Faraday rotation 2
Fe atom 197
ferroelectric switching 276ff, 290
fibers 93ff, 103ff, 305, 362, 364
fibril bundle 123ff
film 240, 247ff, 256ff
– diastereomeric 255ff
– interdigitated bilayer 261ff
– interdigitated multilayer 261ff
– Langmuir 255, 258, 268
– Langmuir-Blodgett (LB) 254, 264
– Monolayer 228, 237, 255
– Z-type 254
fluorescence 70f, 82, 168, 178, 370
foldamer 147ff, 176, 179
forces 4
– electroweak 4
fullerene 101f, 140, 143, 160

g

gas-adsorption 395
gearing systems 373
gel 93ff, 138, 143f, 160f, 174, 287
germanium oxide 398f
glucose 149
L-glutathione (GSH) 76
glycoluril 39f
gold surface 205f, 380f
graphite 219ff
grafting 146
grazing-incidence X-ray diffraction, GIXD 249ff
guest 31, 36f, 40
guest-host interaction 279

h

H-type aggregation 109
Hamiltonian 14, 16
helical 3, 6, 14, 16, 95ff, 105ff, 116ff, 295
– architecture 95, 116ff, 120, 123, 133
– assemblies 95ff, 140
– channels 399
– coordination polymers 393

– filaments   286ff
– pitch   275, 280, 294
– ribbons   287, 293
– supramolecular assemblies   117, 170, 275ff, 281ff, 297
helical twisting power (HTP)   124, 275, 384
helicene   2f, 55ff, 203, 282f
helicity   95ff, 105ff, 153ff
helix inversion   116, 154f, 375ff
heterochiral racemic assembly   58, 207, 210
heterogeneous asymmetric catalysis   392
heterogeneous equilibria   305ff, 326
hexabenzocoronene (HBC)   120
hexahelicene   2f, 8
hierarchical   120, 139, 191f, 197f, 305, 362, 372
homochiral columns   321
homochiral surface   200
homochiralty   4, 6ff
host–guest chemistry   36, 43f, 338f
hydrogels   110, 159f
hydrogen bond   31ff, 105, 108, 117, 129ff, 155, 192, 206f, 224f, 252, 262, 309ff, 355
hydrogenation   84
– aromatic   83
– asymmetric   83, 404, 406
hydrophobic effect   33

i

immobilization   380, 400ff
inclusion complex   147ff
interdigitated bi- or multilayer film   261ff
– acid-base complementarity   261
– diastereoisomeric salt   261
– crystalline trilayer   261
interdigitation   230, 262ff
– octadecanesulfonate/guandinium monolayer   262
induction (of chirality)   46f, 123ff, 149
information storage   115, 117, 154
interaction
– acid-base   137f, 141
– cooperative   143
– electron donor-acceptor   283
– electroweak   14f
– host-cation   143
– hydrogen-bonding   140, 147, 149
– hydrophobic   140, 169, 173
– ionic   140
– noncovalent   93, 116ff, 129, 137, 146, 305f
– π-   116, 125, 129, 135
– solvophobic   93, 147ff
– weak neutral current   16, 20
intermolecular forces   31f, 197, 207

interpenetration   394f
ion exchange   391f, 404
N-isobutyryl-cysteine (NIC)   75
isocyanodipeptide   129ff, 165

j

J-type aggregation   82, 100
Jaeger's dictum   19

k

kinetic control   21, 37, 46, 129, 170f, 235, 305, 333ff
kinetic resolution   85, 336

l

Langmuir monolayer   249, 254
lanthanide   314
LEED, see low-energy electron diffraction
ligand   34ff, 44, 72, 76ff, 310ff, 365, 392ff
light-induced switching   107
limonene   29, 53f, 124
linear momentum, **P**   10
liquid crystal, LC   68ff, 85f, 121ff, 159, 161, 163f, 271, 287, 298, 351, 358, 372, 384
– cholesteric   20, 125f, 138, 372
– lytropic   271, 287
– nematic   20, 124, 126
– thermotropic   271, 287
lithography   71f
"lock and key"   4, 207
low molecular weight gelators (LMWG)   93ff
low-energy electron diffraction   192, 202
luminescence   351, 370f

m

machine, molecular   350, 373f, 380
macromolecules   115f, 129f, 351
magnetic field, **B**   2f, 11ff, 20, 240
magnetochiral phenomena   7, 13f
majority rules   6, 17, 234
major symmetry directions   219ff
MALDI-MS see matrix assisted laser desorption-ionization time-of-flight mass spectrometry
mandelic acid   54, 58, 255f
matrix assisted laser desorption-ionization time-of-flight mass spectrometry (MALDI-MS)   266
Mauguin effect   281
mechanical motion   355, 373
mechanical properties   94
memory effect   46ff, 118, 154ff
menthol   123
p-mercaptobenzoic acid (p-MBA)   79f

mesogen 121ff, 271ff
– amphiphilic 273
– banana-shaped 287
– bent-core 271, 280ff, 288ff, 295ff
– block 163f
– dendritic 282ff
– disc-like (discotic) 271, 282ff
– rod-like (calamitic) 271, 295
– star-shaped 284ff
mesophase 121ff, 129, 134, 162ff, 273, 275ff, 290ff
– chiral 151, 275
– cholesteric 121ff, 133, 139, 151, 153, 155, 160ff, 275
– columnar 133, 284
– cubic 273
– helical 160
– hydrogen-bonding 177
– isotropic 276, 295
– lamellar 287
– lyotropic 129, 133, 160
– nematic 124, 139, 283ff, 295
– SmC* 163
– smectic 131, 289ff
– TGB* 151
metal–ligand coordination 33, 43, 49f, 53, 60, 197
– M_4L_6 tetrahedral assemblies 43, 50
metal-organic frameworks (MOFs) 392ff
metastable equilibria 325f, 331f, 363
micelles 165, 168, 176, 400
microscopic reversibility 21
Moiré pattern 222, 283
molecular imprinting 174
molecular modelling 203
molecular recognition 30, 51, 205, 210, 256, 339, 351
– at the air/water interface 255f
– cholesteryl-S-glutamate 258
– syn and anti serine-Cu complexes 258
molecular wire 149
monoterpene 147
montmorillonite clay 404
motors, molecular 152, 349, 373ff
N-methylnicotinium 50
multiple chiral centers 224
multicomponent assembly 237
mutarotation 149

n

nanocluster 76, 78f
nanoparticles, NP 67ff, 380
– gold 68, 380f
– inorganic 74ff
– organic 68, 71f, 81ff
nanoporous materials 391ff
nanorod 69
nanoscale segregation 161f
nanotechnology 6, 9, 12, 14, 111
nanotubes 394
nanowire 192
– silver 176
nematic phase 86, 124ff, 273ff, 281ff, 293ff, 373
noncovalent domino effect 143
noncovalent synthesis 31, 94, 308
– diastereoselective 30f
– enantioselective 30f
nucleic acids 4f
nucleation 72, 218, 248, 264, 308, 317ff, 328
nucleation, inhibition of 336

o

odd-even effect 123, 237f
olefin epoxidation 397
oligoglycine 140
oligopeptide 120, 133, 266
oligothiophene 97, 175
optical activity 1ff, 14, 16, 22f, 75ff, 127ff, 133, 150ff, 294ff
optical rotation 1ff
– natural 10ff
optical rotatory dispersion 22
OPV 105ff, 179f, 225
ORD see optical rotatory dispersion
organogel 94ff, 362
origin of life 6, 265

p

packing coefficient 51ff
packing constraints 230, 284
parity, P 4, 7, 9ff, 14f, 17ff
– violation 7, 14f
parity-violating energy difference (PVED) 15, 20
particles 9
– monodisperse 74
Pasteur 1ff, 7, 308, 318, 332
Pd(II) complex 43ff, 54ff
PEDOT 126
penicillamine 76, 78
PEO 173f
peptide 147f, 176, 256, 265ff
– homochiral 265
phenylalanine 174, 206
phase diagram 318f, 323, 331ff
phosphine 53, 56, 72

photobistable   351
photochromic   151f, 351f, 357ff, 384
photoisomerization   160, 350f, 357ff, 368, 371ff, 375
photon   11
– circularly polarized   11, 19
photonics   81, 83
photoresponsive group   153, 157, 354f, 366
physisorption   221
π-stacking   118, 120f, 284
π- π interaction   103, 105, 110, 132, 142f, 310, 321, 396
pinanediol   47, 49
pitch compensation   278
plane groups   217f, 248f
planochirality   18
polarization power   280
poly glutamic acid   169f
poly-(o-toluidine)   143
polyacetylene   124ff, 128f, 131f, 137ff, 158f, 176, 178
polyaniline   127f, 137, 141
polybithiophene   127
polycarbazole   135f
polycarbosilane   168
polyester   151
poly(meth)acrylamide   123
– gel electrophoresis (PAGE)   70, 74, 76
polyisocyanate   6, 22, 116, 143, 150f, 158
polyisocyanides   123, 129, 137, 176ff
polyisocyanodipeptide   165
polylactic acid (PLLA)   162
polymeric network   71, 121f
polymerization   116f
– electrochemical   126f
– living   131
– photo-   121, 128
– ring-opening metathesis   120, 164
– topochemical   119
polymers   6, 152, 351, 383
– chiral   115ff
– dendronized   134, 138
– imprinted   174
– nonchiral   137f
– polyvinyl   141
– surfactant soluble   72
polymethacrylate   134, 173
polymorphism   228, 312, 332, 334
polypeptides   4f, 142, 176, 178
polyphenylisocyanide   130, 141
polyphenylacetylene   131ff, 138ff, 143, 156, 173
polypropargylamide   133
polypropiolic ester   155

polypyrrole   126
polysaccharide   169, 175
polysilane   123, 136, 146f, 157
polysorbate   118
polystyrene   134, 162ff
polythiophene   134ff, 142, 147, 149
polythiophene-phenylene   125
(poly)ureidophthalimide   179
polyvinylpyridine   137
porous oxides   397f
porphyrin   43, 53f, 97, 100ff, 169ff, 199f, 226f, 355
position vector r 10
preferential crystallization   327, 329f
preorganization   33
prochiral   195, 196, 234f, 361
protein   4f, 20, 115
pseudoracemate   232, 306, 373
PVED, see parity-violating energy difference
pyrene   103, 365f
pyridyl ligands   44f, 395
pyridine-Pd (II) assemblies   54f

q
quantity
– scalar   10
– pseudoscalar   10, 14, 20ff
– vector   10
quantum size effects (QSE)   68ff

r
racemic compound   248, 306, 310ff, 325ff
racemic hosts   49, 52, 55
radiation, circularly polarized   128, 158
Raman optical activity   23
refracting media   1
relativity   16
resolution   294, 317
– CPL-induced   282ff
– electric-field driven   279, 297
– self-   393
resolution by entrainment   327
resolving agent   317, 331
ROA, see Raman optical activity
rosette   37, 46, 225
rotaxane   356
Ru(II) complex   55, 310, 320
rubrene   198
rule of reversal   337

s
saccharide   149
scanning tunnelling microscopy (STM)   192ff, 196ff, 205ff, 215f, 220ff

schizophyllan 149
second-harmonic generation CD 254, 293
sergeants-and-soldiers 6, 17, 116, 121, 284, 294
self-assembled monolayers 215, 224
self-assembly 37f, 44, 67, 93, 206ff, 263, 363, 372
SEM 122, 124ff, 167
sensing, chiral 56ff, 138, 143
separation 391ff
silica 400ff
smectic phase 273, 276, 287ff
– blue 274ff
sodium chlorate (NaClO$_3$) 16f
softball 39ff, 47ff
solid solution 324ff
solvent (effect) 41, 230, 312, 328, 333ff
space-filler concept 318
space groups 217, 306ff
spectroscopy 21
spin coating 146f
spinning cone 11f
spontaneous polarization, P_s 276ff, 288ff
spontaneous resolution 1f, 7, 16, 231, 313ff, 393
spontaneous segregation 248
stereogenic center 80
stereorecognition 40
stereoselective epoxidation 404
steric factors 100, 133f, 339
steroid derivatives 93, 96
steroid–steroid stacking 96f
stilbene 97, 102f
STM, see scanning tunnelling microscopy
stochastic asymmetric transformation 254ff
styrene 403ff
sugars 109f
supramolecular stereochemistry 22, 29f, 33, 61, 203
supramolecular host 46, 50ff
surface chirality 192, 217f
surface energy 69
surface plasmon 68ff
surface-substrate interaction 191, 235
surfactant-templated synthesis 400ff
switching 208, 349ff
– antiferroelectric 276ff, 289ff
– antipolar, see antiferroelectric switching
– bistable 278
– ferroelectric 276ff, 289ff, 295
– of superstructural chirality 291
– synpolar, see ferroelectric switching
– tristable 278, 289
symmetry 8f, 16, 191f, 217, 305

– chiral 8f, 16
– crystal 305ff
– inversion 9f
– monolayer 217ff
– operations 9ff
– spatial 8
symmetry breaking 16f, 234, 295, 312f
symmetry violation 13, 16f
– nonconservation 16

t

T violation 15
tailor-made additives 319
tartaric acid 48, 193, 200ff
Tellegen media 13f
TEM 96, 105, 108f, 119, 121, 128, 167
tennis ball 39
template 43, 169ff
– chiral 126f, 169
tetrathiafulvalene 178, 366f
time reversal, *T* 9ff, 15
thermodynamic
– considerations 46, 235, 305, 322f
– factors 21, 307
thiol 72
thiophene 166f
tilt chirality 338f
– anticlinic 289f
– synclinic 289f
transfer of chirality 194, 198
triazine triamides 119
TRISPHAT 320ff
twist grain-boundary phase (TGB) 276
2D chirality 17, 196
2D film 247ff
2D monolayers to 3D crystal 263ff
– α-glycine 264ff
– self-assembly 263
– orientated 3D crystallization 263
two-dimensional systems 17f, 196ff
true and false chirality 3, 7, 10, 12f, 18ff

u

unidirectional rotation 374f
unit-cell vector 219f

v

van der Waals 33, 96, 220, 225, 237, 318
VCD, see vibrational circular dichroism
vector
– axial 10
– polar 10
– propogation, *k* 13
– pseudo 10

– true 10
vesicles 165ff
vibrational circular dichroism 23
virus 23
– capsid 8ff
– filamentous bacterial 20
vortex motion 19

w

water
– chiral monolayer 247ff
waveguide 85
weak interactions 4, 29, 31
– weak natural current 14
Weigert effect 281ff

x

X-ray diffraction 105, 131f, 134
– grazing-incidence 249

y

z

zeolites 392f, 397f, 405
Ziegler-Natta catalyst 116
Zn-porphyrin 53f